T5-CQA-328

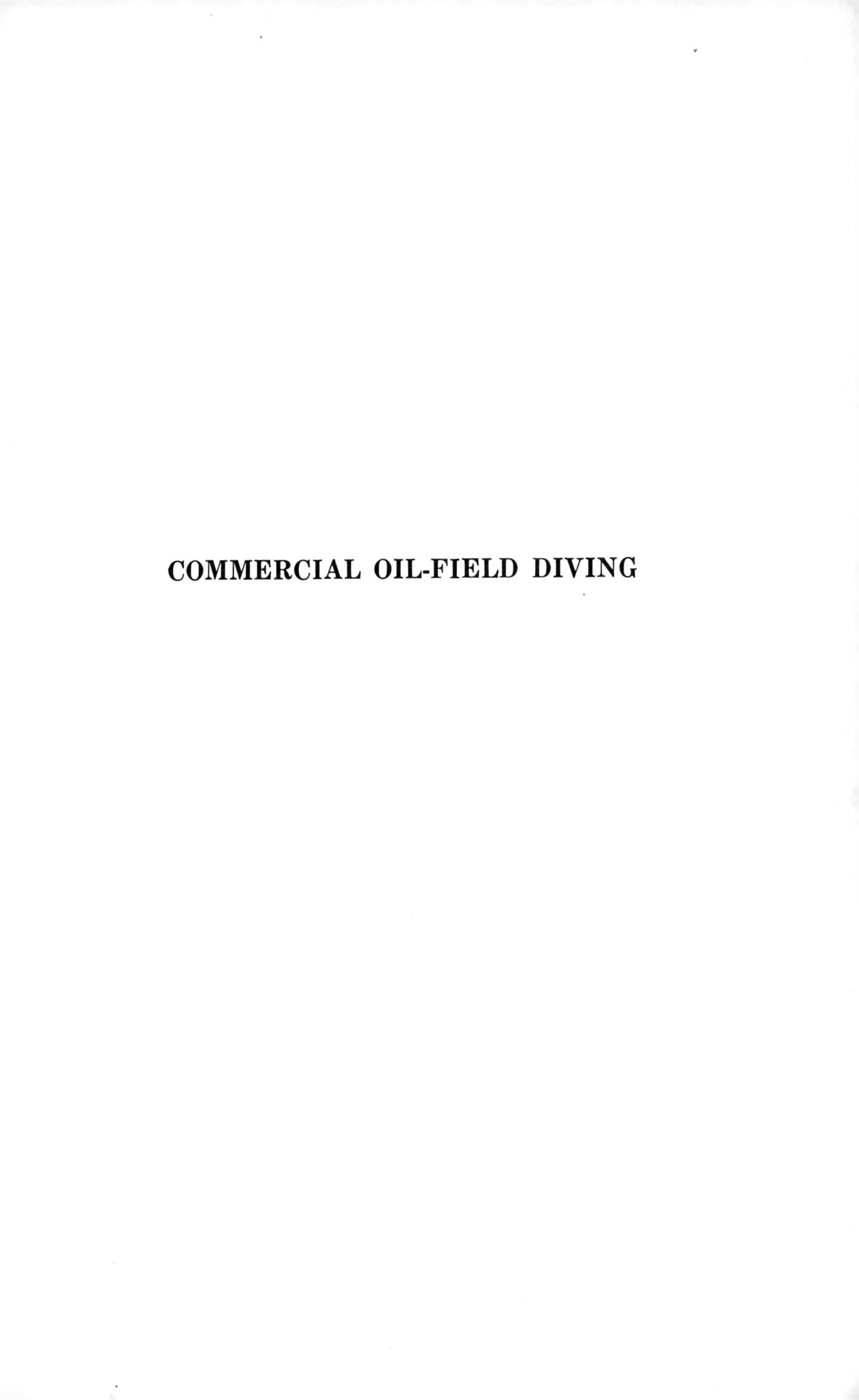

COMMERCIAL OIL-FIELD DIVING

The L. B. Meaders, one of the world's largest combination pipe-lay derrick barges, owned by Brown & Root, Inc. Photo: Bethlehem Steel Corp.

COMMERCIAL

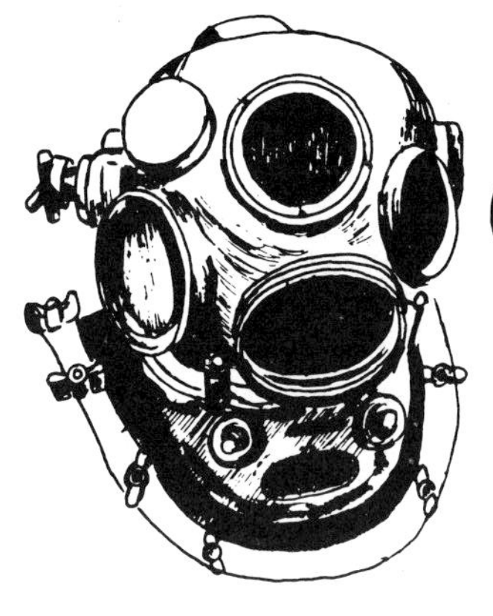

OIL-FIELD

DIVING

By

Nicholas B. Zinkowski

CORNELL MARITIME PRESS, INC.

Cambridge 1971 *Maryland*

ISBN 0-87033-157-4

Library of Congress Catalog Card Number: 70-153145

Printed in the United States of America

This book is dedicated to Julia Zinkowski

CONTENTS

Page

Frontispiece: *The L. B. Meaders* ii
Dedication v
Introduction by W. J. O'Neill xi

1. Diving As a Career 1
2. Physics and Physiology 7
3. Tending and Breaking Out 41
4. Diving Equipment 55
5. Decompression and Treatment Tables for Compressed-Air Diving 81
6. Rigging 132
7. Burning and Welding Under Water 151
8. Use of Explosives 173
9. Diving from a Pipe-Lay Barge 198
10. Diving from a Pipeline Barge, or Jet Barge 241
11. Miscellaneous Diving Applications in the Oil Field 258
12. Diving from a Drilling Rig 281
13. Mixed-Gas Diving 300
14. Diving from a Bell; Saturation Diving; Dry-Atmosphere Welding Huts 320
15. Diver's Pay; Unions; Summary of Safety Precautions . . 349

Appendix
Bibliography 359
Diving Equipment Manufacturers and Distributors . . . 363
Index 367

ACKNOWLEDGMENTS

I would like to extend thanks to the many people who helped me with this book. I am indebted to Sue Ann Walde Zinkowski for continuing encouragement and much of the typing; to Barbara Lyons Herman, for a major portion of the typing; to Arturo Herman for allowing Barbara to type and for his many excellent drawings; to George Morrissey, general manager of Taylor Diving and Salvage Company, Inc., for reading and commenting on many of the chapters; to Jerry O'Neill of Westinghouse Underseas Division, for reading and commenting on several of the chapters, for technical information and for various photographs; to Kenneth Wallace, vice-president of Taylor Diving and Salvage Company, Inc., for technical information and for making various company photographs available to me; to Gerry Todd, formerly of Associated Divers and Ocean Systems, Inc., and Peter Edel of J & J Divers, for historical information; to Torrance Parker of Parker Diving Services, for reading and commenting on several chapters and for various photographs; to H. L. Schaaf of DuPont for reading and commenting on the chapter on explosives; to Dan Wilson, president of Sub Sea International, for historical data and photographs; to Bonnie Boynton, for reading the manuscript and making the many necessary grammatical and literary corrections; to John Violette, for his many excellent photographs; to the many people and companies who supplied photographs and who are individually credited where the photographs appear; to the many companies and clients who supplied me with work through the years and afforded me the necessary experience from which to write this book; and to Jay Jones, Onil Landry, John Becksted, Vic Becker, Gene Hempstead, Lee Gates, Harry Rude, Kim Zinkowski, and all the other professional divers and tenders who provided trade tips, information and encouragement.

Finally, I am indebted to Larry Kell and Mary Kell for innumerable points of assistance; without the vehicle of their magazine for professional divers, *Undercurrents*, in which a major part of this book has been serialized, I never would have summoned up the necessary drive or sense of continuity to complete it.

N. Z.

INTRODUCTION

Diving is, by nature, an intimate personal experience; it is a field where practiced participation and observation are indispensable requisites. Written and spoken words can never adequately convey the subjective sensations, the feelings, the fears, and the joys of a solitary diver deep down on the ocean's floor. Experience is a great teacher; however, to use experience effectively toward one's own survival, a diver must also possess the ability to anticipate the problem areas. When a highly paid diver elects to forfeit that high pay, rather than to expose himself to a set of conditions which he knows have previously resulted in an accident, he makes his decision through anticipation based on experience. And so these attributes—participation and experience—are, in my opinion, indispensable to safe, effective diving.

The increasing demand for oil patch diver services occasionally presses an ill-prepared novice into situations dangerously beyond his capacity. If he survives, he is the wiser; if he does not, . . .! The bold diver who rushes into situations without sufficient forethought tends to have a short career. In other words: "There are old divers and there are bold divers, but there are no old bold divers!"

I feel honored in writing a few words of introduction to Nick Zinkowski's manual for working divers. For a number of years I have encouraged him to record, for the benefit of others, his extensive, hard-won knowledge of diving. During his working years, a true professional—and that's exactly what Nick is—can train a limited number of apprentices; however, when that professional takes the time to put his experiences on paper, thousands can benefit.

The lack of standardization of field training throughout the diving industry has tended to leave occasional gaps in the trainee's knowledge. Nick's manual contributes immeasurably to this much-needed standardization, by enabling the informed reader to seek experience in the areas which might otherwise have been omitted. For the trainee and the working diver alike, this manual fills the gap between the technical-heavy and the fun-and-games diving "know-how" books. For those with the capacity to learn from the written word as well as from personal experiences, the course for the neophite tender to the seasoned professional diver will surely be accelerated and made safer through the use of this book.

I doubt that Nick would claim that he has advanced the state-of-the-art of diving or man's exploitation of the seas. Nevertheless, this book is instrumental toward these ends.

W. J. O'Neill
Westinghouse Underseas Division,
Westinghouse, Inc.

Fig. 1. Photograph of a line drawing appearing in *History of the Sea*, by Frank B. Goodrich, LLD, to which is added: *An Account of Adventures Beneath the Sea: Diving, Dredging, Deep Sea Sounding, Latest Submarine Explorations, Etc., Prepared with Great Care by Edward Howland, Esq.;* Union Publishing House, New York, N.Y. (Publication date unknown.)

Chapter 1

DIVING AS A CAREER

Diving has attracted the individualist and the adventurer from the earliest days of its development as a specific trade. As a small and highly specialized industry, it has always been a seedbed for the growth of strong, if somewhat eccentric, personalities. The reasons are clear and remain applicable today, although to a lesser extent than in the past. Diving requires physical courage, physical strength and a type of perseverance and endurance seldom required in other lines of work.

Each time he jumps into the water, the diver knows that his very life is directly dependent upon his equipment and upon the efficiency of his topside support crew, i.e., the diver-tender, the rack operator, the crane operator, the barge shift foreman, and others. He knows that the water is likely to be brutally cold and the current is apt to be running so swiftly that most of his strength will be utilized just getting to his work and holding on. The job he is expected to accomplish will probably be something very simple like inserting a steel O-ring between two wildly oscillating flange faces and then bolting up the flanges. He knows that after he hits the water he might as well be blind, because his work will be done in a tar-black mud ditch with crumbling sides where his only reliable perceptions will be those received through his fingertips. And he knows also, possibly from bitter experience, about the insatiable appetite of pipe flanges for divers' fingers.

After his job has been completed, or he has run out of bottom time, he will have to endure possibly hours of abject teeth-chattering misery while hanging off in the cold water for his decompression stops. After his last water stop, physically sapped, he will have to make a 10- or 15-foot climb, dragging his heavy equipment with him, up the side of a rolling, heaving, wave-swept barge, on a rickety ladder with missing rungs.

On deck, his tenders will strip him of his gear and rush him into the deck chamber where he can comfortably relax for a few more hours of decompression in the dank, 120-degree heat of that confining steel prison. When the chamber door finally swings open at the end of his decompression, the diver knows (if he doesn't come down with the bends and if there are a sufficient number of divers on the job) that he will have at least twelve hours off before he will be required to do it all over again.

Even the most prejudiced among us will have to admit that any man who would willingly go through all this—day after day and year after year—to earn his bread, has to be someone special—or some kind of nut. Admittedly, it is not always this bad; there are days when the sea is calm, the current slack and the water gin-clear over a hard sand bottom; however, days like this are the exception. By its very nature diving is difficult, demanding, uncomfortable and potentially dangerous work. The wonder is that there are so many men willing to pursue it as a career.

An additional disadvantage to selecting a career as an offshore oil-field diver is exemplified in the word "offshore." Underwater oil fields of the world are located in remote areas as dramatically contrasting in temperature and environment as Lake Maracaibo in Venezuela and Cook Inlet in Alaska, or the Persian Gulf and the British North Sea. North or south, oil seems to be found in areas of extreme heat or cold and in regions of perverse susceptibility to rotten weather.

The offshore oil-field diver goes offshore to do a job, and he remains offshore until the job is completed. In this day of longer and larger-diameter pipelines, laid in ever-deeper water, this could be for months. It is not a job for a home-loving family man. The reverse side of this particular coin is the genuine opportunity that the diving business affords for paid foreign travel.

While working offshore, the diver will make his home aboard a pipe-lay barge, a pipe-dredge barge, a mammoth derrick barge, or on one of a great variety of drilling rigs. Unlike the other workers on the same rig, who work a scheduled 12-hour shift, the diver will be on call 24 hours of the day. If things are going well, he may go for many weeks making as little as one short inspection dive a day. But if things are coming unglued, he could be required to make repetitive dives every few hours around the clock for days.

His quarters, although never commodious, will vary from reasonably comfortable ones, with four men living in an air-conditioned 8 x 12-foot cubicle, to a space of the same size, unair-conditioned, where eight men must live.

For some obscure reason, the diver is never regarded or accepted as part of the regular crew, but rather as a supernumerary, an outsider, someone who must be tolerated temporarily. Quite often he is asked to make his bed on a coil of rope in a shady corner of the deck.

The food served on offshore rigs also ranges between extremes: from excellent, appetizing and wholesome, to virtually unfit, depending upon the company or the particular barge.

Entertainment and diversion are also limited. There is almost always the opportunity for fishing, for those who enjoy it. Television is poor to non-existent because of the distances from broadcasting stations, and when available it is constantly subject to intermittent and annoying interference from the rig's VHF communications equipment which distorts the TV picture and destroys the sound every time the transmitter is keyed

There are always poker, dominoes, sleep, reading, and skin magazines. Of course, liquor is strictly prohibited aboard all offshore rigs and instant unemployment is the normal punishment for a breach of this edict.

Laundry service is usually provided, as well as room stewards to occasionally hoe out the living quarters and change the sheets. Cigarettes and toilet articles can be purchased aboard most rigs.

All in all, working in the offshore oil industry is rather similar to serving in the armed forces, except for two particulars. The work is much harder and the pay is decidedly more, especially for divers. The subject of divers' pay is fully treated in Chapter 15. Depending on the number of days spent offshore and the depth of the water worked in, a good offshore diver can earn $50,000.–$60,000. per year. Although this is above average, many divers do, in fact, earn this much money. Generally and rightfully, divers are among the highest paid physical laborers or artificers in the world. But it has been proved, by evaluating the top men in the industry, that money alone is not sufficient motivation to produce a top-notch diver. To become a highly proficient diver, one has to be somewhat money-hungry, to be sure, but this has to be buttressed by a definite tendency toward egomania, paranoia, masochism and other milder forms of insanity. In essence, to be a good diver, one has to like what he is doing.

A good commercial oil-field diver must have tremendous confidence in himself, in his judgment, and in his abilities. He knows, through some as yet unexplored mystic faculty, that he is better than anybody else; that flanges jump together and risers fall into their clamps solely as a result of his personal, individual superiority. He is objective enough, however, to realize that he must learn, understand, and apply every facet of his trade technology constantly to maintain that superiority.

Tom Angel, a diver for Sanford Marine Services of Morgan City, La., has this terse motto on the wall of his office: THE MORE YOU KNOW, THE LONGER YOU LIVE. There can be no simpler nor more accurate bylaw for a commercial oil-field diver.

The diving business has grown as swiftly and as dramatically as the offshore oil industry, which is almost solely responsible for the tremendous current surge in diving activity and technology. There is virtually no phase of the offshore oil industry—exploration, drilling, pipelining—that does not require the services of dependable professional divers. The demand for divers is already great and is increasing with the continuing development of offshore petroleum and natural gas resources.

The aspiring oil-field diver should bring at least a rudimentary understanding of several other trades to his new profession: mechanics, rigging, burning, welding, use of explosives, blueprint reading, and isometric drawing. Many comprehensive books have been written on each of these subjects and can be found in any public library. Some recommended titles are listed in the Bibliography.

Underwater rigging, burning and welding, explosives' use and their application to the offshore oil industry are treated in forthcoming chapters.

THE WORKING DIVER'S TOOLS

Of the many complex functions gathered together under the broad title of mechanics, we will consider here only the use of tools. One of the primary differences between man and his evolutionary antecedents is his ability to use tools. A diver is a man who can use tools and perform work under water. It must be understood that diving is in effect a method of transportation, a means of conveying the worker from the surface to his job site under water and sustaining him there. The act of diving does require a degree of manipulative skill, varying with the type of equipment used, and it also requires the exercise of some judgment, but it is no big deal of and by itself. A diver is not paid because he can go under water. A diver is paid for the work he performs under water.

The sport of scuba diving has produced thousands of excellent divers; that is, men who understand their equipment, and are physically and emotionally conditioned to go under water, sometimes to great depths. But sporting scuba divers go under water as a diversion, to fish, explore and perhaps to photograph. Very few are capable of performing the arduous tool-wielding work required of the professional diver.

Above everything else, a commercial offshore diver must have a strong mechanical aptitude; a feel for, an understanding of, and the ability to use, both hand and power tools. Few people who are not involved in the manual trades understand the skill required to use the simplest hand tools, a hammer or a pipe wrench, effectively. Under water, the difficulty of using tools is compounded by the lack of visibility, the resistance the water presents to movement, the necessity for constantly monitoring and adjusting the diving equipment, and dozens of other distractions.

The diver's ability to use any required tool must be a reflex condition. This can be achieved only through long practice. If you have previously worked as a plumber or a carpenter's helper, or at any related manual trade, you have a head start. If you lack experience in handling tools, it is important to take advantage of every opportunity to use them if you intend to become a good diver. The better you understand the purpose and the actual use of every conceivable type of tool, the more accomplished diver you will become. If you are unable to use a certain tool or perform a specific task on the surface, it is absurd to expect to be able to do so under water. At one time or another, a diver will be called upon to use every type of standard hand tool that is manufactured. Among those most commonly used are sledgehammers, from short-handled 3- or 4-pound striking hammers to 10- and 12-pound malls; pipe wrenches; ratchet wrenches and sockets; hammer wrenches; adjustable open-end or crescent wrenches; banding tools; hydraulic jacks; and pipe cutters.

Some of the most frequently used power tools are pneumatic and hydraulic impact wrenches, pneumatic grinders, pneumatic drills and boring machines, pneumatic chipping and demolition hammers, and hydraulic pipe saws. It is wise to become familiar with these tools and their intended uses.

Pneumatic tools are generally not very effective in depths beyond 100 feet because of the difficulty in supplying the large volume of air they require at depth. Recently several companies have begun to design and manufacture hydraulic tools for the diver for use in great depths, and as each of these tools comes on the market, it is the diver's responsibility to learn how to use it properly.

The two remaining subjects on our list of auxiliary trades that will not be covered by separate chapters are blueprint reading and isometric drawing. Neither involves the use of tools under water, but together they represent a desirable adjunct to the coarser skills which make up the accomplished diver. A substantial part of the diver's work is to install, inspect, maintain and repair a variety of complicated mechanical devices under water. They include valves, tiny and huge; hand-, gas-, or electrically operated, and sometimes acoustically actuated; ocean-bottom wellheads or subsea completions, blowout preventers, etc. In view of the probability that this work will be performed in limited or zero visibility, it is imperative for the diver to know exactly what he has to do before leaving the surface. To assure this, the diver should carefully study the blueprints relating to the actual installation on which he will be working.

Most blueprints, although fundamentally simple and self-explanatory, appear bewildering to the uninitiated. Too many divers, wishing to appear more knowledgeable than they actually are, will cursorily riffle through a sheaf of blueprints, nodding their heads wisely and hoping against hope that it will all become clear to them when they reach the bottom. This is a common and foolish mistake and an unconscionable squandering of the contractor's money. In almost all underwater work it is difficult enough to do the job when you know exactly what must be done. To hit the water without all the available information pertaining to your job is being dishonest with yourself, your trade, and your employer, and in a number of situations can be physically dangerous to you. A short course in blueprint reading will render you capable of understanding the engineering plans for most of the jobs likely to be encountered and will result in your becoming a better diver.

Engineering blueprints are a precise form of communications. They tell the diver what he can expect to find on the bottom. Unfortunately, the situation under water is seldom as it appears in the blueprints. Therefore, it is mandatory for the diver to be able to interpret what he perceives, often just through his fingertips, and to be able to report his findings clearly and accurately. In this situation, one picture is truly worth a thousand words. The ability to draw is a tremendous asset to the working diver and one that is respected and appreciated throughout the industry. The two underlying themes of this chapter are important enough to be restated:

The more you know, the longer you live.
A diver is a man who can use tools and perform work under water.

Let us proceed to Chapter 2 and try to understand what happens to the human body under water.

Chapter 2

PHYSICS AND PHYSIOLOGY

There are a number of books in print that deal splendidly and in some cases exhaustively with the subjects of diving physiology and physics.

It is considerably beyond my ability to either add to or improve upon these existing works. I strongly recommend that you become thoroughly familiar with this involved and important aspect of diving. (See the Bibliography.)

GENERAL PRINCIPLES

Assuming that you have or in the future will edify yourself with a study of one or more of the books and pamphlets listed in the Bibliography, I will proceed with an elementary discussion of physics and physiology consonant with my own limited understanding of these subjects. I hope to point out what penalties the diver is certain to incur by any violations of the very rigid laws translated for us by these two sciences.

As he stands on the deck of the barge with the compressor running, ready to strap on his mask and jump overboard, the diver must realize that he is the result of evolutionary adaptations that have taken many millions of years to achieve. The decision of our ancestors to crawl out of the ocean was irrevocable. In man's present form he is ideally designed for living at or near sea level. It was not intended for him to venture too far above this level, and a trip in the opposite direction, back into the sea, with his existing natural equipment, is unthinkable. The main reason for this is that the body long ago gave up its ability to extract oxygen from sea water.

Man can live without food for weeks, without water for days, but after scant seconds of being deprived of oxygen, the processes of his brain begin to degenerate and its tissues are subject to irreparable decay within minutes. The body needs oxygen, and in the precise amount encountered where mankind has chosen to live for the past two million years—at or near sea level. The body extracts this vital oxygen from the atmosphere, which is that miles-deep ocean of air above and around us.

The atmosphere has mass; that is, it takes up space and has weight. At the bottom of this ocean of air, sea level, the atmosphere weighs or exerts a downward force of 14.7 psi. The atmosphere exerts pressure not only downward but inward upon all the surfaces of the body. Having been long accustomed to living under this pressure, we are not aware of it. The body is composed mostly of fluids and solids that are nearly incompressible. The few cavities which do exist in it—lungs, sinuses, inner ear, etc.—are filled with air, also at a pressure of 14.7 psi, so that the pressure forcing in on these spaces is compensated by an equal pressure forcing out, leaving us in a normal state of pressure balance with our environment.

The atmosphere or air is a blend of several different gases. All the gases in air combined, other than nitrogen and oxygen, occupy less than one percent of the total volume of the atmosphere. With the exception of carbon dioxide, 0.04 percent by volume of the atmosphere, they have no significance in the present discussion. Air then, for our immediate purpose, is composed of 21 percent oxygen and 79 percent nitrogen. Oxygen is vital to life and enters into many chemical processes within the body. Nitrogen is an inert gas; that is, although it is present in the body, it takes no part in any of the body's chemical activities. Its primary function is that of a diluent or mixer for the oxygen (soda for the Scotch, so to speak).

Most of the oxygen required by the body is supplied by breathing. Inhalation causes the muscles to force the rib cage up and out and pull down on the diaphragm, increasing the internal volume of the chest cavity and allowing air to rush in through the nose and mouth. Exhaling is just the reverse; the rib cage is pulled down and in and the diaphragm is pushed up, decreasing the internal volume of the chest and forcing the air out. The frequency of the breathing cycle, inhale-exhale, varies greatly among individuals. At rest one probably breathes between ten and fifteen times per minute. About one pint of air enters with each breath. This equals about two gallons, or ¼ cubic foot per minute (cfm). During heavy work the body needs more oxygen and it is obtained by deeper and faster breathing. The consumption of air can then be increased ten or more times, to 2½ cfm or more.

The diving helmet or mask is designed to keep water away from the diver's face and allow fresh air to flow across his nose and mouth. The absolute minimum volume of air available to a diver should never be less than 1½ cfm calculated to the depth at which he is to work. The delivery pressure of a diver's air supply should always be at least 50 lb. greater than the pressure existing at the depth of his work. In this age of economical power, diesel and electric, there is no justification for requiring a man to dive with less than 4½ cfm of available air at depth and 75–100-pound pressure over bottom pressure. A man requires a far greater volume and pressure of air under water than he does on the surface and this requirement increases with every foot of depth at a standard rate. Sea water weighs a good deal more than air. It exerts a pressure of 0.445 psi, or a little less than ½ pound, for every foot of depth. At a depth of 33 feet, sea water exerts a pressure almost exactly equal to the pressure exerted by the miles-deep atmosphere at sea level. Each 33-foot increment of depth from the surface is equal to an additional atmosphere. For every 33 feet a diver descends below the surface, he is subjected to an additional pressure of 14.7 psi. For reasons that we will explore later, compressed-air diving is seldom carried on at depths greater than 200 feet. Because 198 feet of depth is exactly equal to 6 atmospheres (atm) of depth, or 6 times 33 feet, we will use this depth for an example. The pressure under 198 feet of sea water is 6 times 14.7 pounds plus an additional 14.7 pounds for the weight of the atmosphere, or almost exactly 103 pounds. Another way of calculating the pressure would be to multiply the depth times 0.445, which is the pressure of sea water per foot of depth, and add 14.7 pounds for the weight of atmosphere. With a minimal fractional difference, either method of calculation brings us to a pressure of a little under 103 psi at a depth of 198 feet. Complying with our requirement of at least 50 pounds of air pressure over bottom pressure, a compressor delivering air between 175 and 200 psi would be acceptable for a dive to this depth. In order to determine the volume of air that a compressor should produce for this depth, we will have to take a different approach.

BOYLE'S LAW

Many years ago, a man named Boyle formulated the law governing a specific behavior of gases. He stated that at a constant temperature, the volume of the gas varies inversely as the pressure, while the density varies directly as the pressure. Disregarding the effects of temperature, which are minimal, this means that pressure applied to a unit of gas squeezes it into a lesser volume, proportional to the pressure applied, and at the same time increases its density. This idea can best be expressed with a hypothetical balloon.

Because it is the recommended volume of air per minute to be supplied to a diver, our balloon will contain 4½ cu. ft. of air at the surface. Seventy-nine percent of the contents of the balloon will be nitrogen. It will be easier to visualize this nitrogen as 79 units represented by green dots. The remaining 21 percent of the volume inside the balloon is oxygen and this is represented by 21 red dots. All gases mix readily, and in actuality the red and green dots in our balloon would be equally interspersed, but for clarity we will keep the red dots and the green dots segregated. At sea level the pressure inside the balloon would have to be slightly higher than 14.7 psi to overcome the resistance of the rubber, but we will disregard this minute difference. If we pull our balloon 33 feet down into sea water, the pressure acting upon it will be double what it was on the surface, or 29.4 psi, but our balloon will now occupy only half the space that it did on the surface, or 2¼ cu. ft. There will still be 79 green and 21 red dots within the balloon, but they will be much closer together. Although our balloon is now only half its original size, the pressure within it is double what it was on the surface and equal to the pressure of the water surrounding it. As it now contains only 2¼ cu. ft. of air, it does not meet our requirement of 4½ cu. ft. of air to supply a diver for one minute. Through a hose leading to the surface we will push an additional 4½ cu. ft. of air down into the balloon. The balloon will now occupy the same amount of space as it did on the surface, or 4½ cu. ft., but it will now contain twice as much air, or 158 green dots and 42 red dots. If the balloon at this point were returned to the surface, the pressure within the balloon would return to 14.7 psi, but it would now occupy twice as much space, or 9 feet. To maintain a volume of 4½ cu. ft. within the balloon, an additional 4½ cu. ft. of air must be added to it for each atmosphere (33 feet of depth) to which it is submerged. A man requires 4½ cfm of air at the surface and an additional 4½ cfm of air for each atmosphere or 33 feet of depth to which he is submerged. One cu. ft. of air at sea-level pressure is called a standard cubic foot or scf. At a depth of 198 feet a diver is under the pressure of six atmospheres of water depth plus one atmosphere at the surface, or seven atmospheres absolute. The compressor should deliver seven times 4½ cfm or 31½ scfm (standard cubic feet per minute) of air.

Note that the pressure doubles and the volume of air required by a diver is doubled between the surface and 33 feet. It does not double again until we reach 99 feet. The first 33 feet, therefore, represents the greatest pressure differential within the shortest distance encountered in diving. This fact is very significant to the diver in a number of ways, which will be discussed shortly.

An understanding of partial pressure is essential, and our balloon will be very helpful in explaining this term. It must be kept in mind that with each atmosphere (increase of 33 feet of depth) to which the balloon is subjected, an additional 4½ scf of air must be added to the balloon to keep its volume constant at 4½ cu. ft. depth. With each 4½ scf of air added to the balloon, 79 green dots and 21 red dots are being added. The percentages of nitrogen and oxygen remain the same, 79 and 21 respectively, but the amounts of these gases, or the densities, are increased with each addition. In other words, the *partial pressure* of these gases is increased.

Each gas within a mixture exerts a pressure which is independent of the other gases in the mixture. The pressure that the individual gas exerts is the partial pressure of that gas. This can be illustrated by once again considering our balloon at the surface. Our balloon contains 4½ cu. ft. of air at a pressure of 14.7 psi. The air in the balloon is composed of 79 percent nitrogen and 21 percent oxygen. If we remove all the oxygen from the balloon, it would then occupy a little less than ⅘ of the space it did with the oxygen in it. If we allow the oxygen to remain and remove the nitrogen, the balloon would then occupy a little more than 1/5 of its previous space. Now let us substitute for our balloon a rigid sphere containing 4½ cu. ft. of air at 14.7 psi. If we remove all the oxygen, the pressure within the sphere would be reduced to about 11.6 psi. This represents the partial pressure of nitrogen. If we leave the oxygen and remove all the nitrogen, the pressure within the sphere would be reduced to about 3.1 psi. This represents the partial pressure of oxygen. Of course, if we removed all the oxygen and all the nitrogen, we would have zero pressure within the sphere, or a vacuum. Within the sphere or balloon filled with air at 14.7 psi, we have a partial pressure of oxygen of 3.1 psi and a partial pressure of nitrogen of 11.6 psi. At 33 feet, if we want our balloon to displace 4½ cu. ft., we will have to start out at the surface with 9 cu. ft. of air, or pump down an additional 4½ cu. ft. Our balloon at 33 feet of depth, although displacing only 4½ cu. ft., will contain twice as much air as it did on the surface at twice the pressure. The total pressure within the balloon is 29.4 psi. The partial pressure of oxygen is 6.2 psi and the partial pressure of nitrogen is 23.2 psi. The partial pressure of a gas is generally referred to as its percentage of one atmosphere. Thus, 21 percent oxygen at the surface (one atmosphere) would be expressed as a pO_2 of 0.21 atmospheres.

HENRY'S LAW

Henry's Law states that the amount of gas which dissolves in a liquid is proportional to the pressure of the gas above the liquid. Because the human body is composed mostly of liquids, it is subject to Henry's Law.

The increased partial pressure of the gases, oxygen, nitrogen, carbon dioxide, carbon monoxide and helium, as encountered in diving, is responsible for a variety of potentially dangerous conditions. These include carbon-monoxide poisoning, carbon-dioxide poisoning, nitrogen narcosis, the bends, and oxygen poisoning. It is imperative that the diver recognize the symptoms of these conditions and understand how to avoid them and how to treat them if it is required.

The body assimilates the gases present in the breathing medium by diffusion through the blood capillary membranes in the lungs, and the rate at which these gases enter the blood stream is proportional to their partial pressures. The higher the partial pressure of a gas, the greater will be the amount of that gas absorbed by the blood and body tissues. The hemoglobin in blood combines chemically with oxygen, carbon dioxide, and carbon monoxide, and carries these gases by way of the blood circulating system to and from the tissues of the body. The tissues are constantly using oxygen; therefore, the partial pressure of oxygen in the tissues is less than the partial pressure of oxygen in the blood circulating past them. This causes the oxygen to move from the place of greater density, the blood, to the place of lesser density, the tissues. The tissues are constantly producing carbon dioxide and, therefore, the partial pressure of this gas is higher in the tissues than in the blood flowing past them. This inequality causes the carbon dioxide to move from the tissues to the blood. This blood is eventually circulated through the capillary bed of the lungs where once again we have a differential in partial pressures. The blood contains a higher partial pressure of carbon dioxide than the air in the lungs and so this gas moves out of the blood and into the air in the lungs. The lungs contain oxygen at a higher partial pressure than the surrounding blood and so the oxygen moves from the lungs into the blood for delivery to the tissues. At all times, the gases involved are flowing from the areas of higher partial pressure or greater density to the areas of lower partial pressure or lesser density.

It will be much easier for the author to bring all of the foregoing facts into practical focus and to explain some other important aspects of diving physics and physiology by taking the reader step by step through a dive to 165 feet.

WE DIVE TO 165 FEET

It is not recommended that your early training dives be made to 165 feet. Early diving should be confined to the comparative safety of shallow water where the novice can concentrate on mastering his equipment without concern over decompression.

Almost every hazard to diving can be encountered in a dive to 165 feet and that is why this depth has been selected; 165 feet is equal to six atmospheres absolute, so we will need a compressor that delivers $6 \times 4\frac{1}{2}$ or 27 cu. ft. of air per minute at $6 \times 14.7 + 100$, or about 189 psi. Although diving air is not exactly free, the author believes that a diver should have as much as he wants. Since we are setting up this dive, let us select a compressor that delivers 60 cfm of air at a delivery pressure between 175 and 200 psi. It is your ultimate responsibility to carefully check out all the equipment that you will be using. The compressor is of sufficient capacity for your intended dive; the fuel tank is full; both the drive engine and the compressor have lubricating oil up to the manufacturers' marks. The compressor has a filter to remove oil, water vapor and other impurities from the air; it has a large-volume tank to remove pulsations from the air flow, to further trap any water vapor, and to provide a reserve supply of air should the compressor fail while the dive is being made. There is a non-return or one-way valve between the compressor and the volume tank. This will prevent the air in the volume tank from escaping back through the compressor in case of failure. A safety relief or pressure blowoff valve is on the volume tank to prevent the tank from exploding if the automatic pressure regulator malfunctions.

Most importantly, the compressor intake filter is placed a sufficient distance from the drive engine so that there is no possibility of sucking carbon monoxide gas into the diver's air supply. The compressor should be located so that contamination of the air from other engines around the deck will not be a possibility. All the pipe fittings and hose connections are tight and in good condition. The supply hose or piping is led across the deck in such a way that the deck crane will not run over it, and anchors or machinery will not be inadvertently set on top of it. The diving hose is in good condition, made up with a telephone cable and a safety line, neatly served and taped together without bights and loops. The hose is long enough to allow you to reach your work easily. It is coiled up free for running so that you can be given slack as fast as it is needed. Spliced to the end of the safety line is a quick-release hook with an easy-to-grab lanyard. On the end of the hose is a free-flow Desco mask, not because I necessarily prefer it, but because more work has been done under water with this piece of equipment, perhaps, than with all other types combined. Other diving masks and helmets are described in Chapter 4.

Between the mask and the hose there is another one-way, or non-return valve, and its purpose will be explained presently.

THE MASK

The Desco mask has a triangular-shaped plastic faceplate or window mounted on a rigid frame. Attached to the other side of the frame is a flexible rubber gasket or boot, designed to fit under the diver's chin and around his face to provide a watertight seal. The mask has five adjustable straps to hold it on the head. Mounted on the right side of the mask is an air-control valve with which you regulate your incoming air. On the left-hand side of the mask is a non-adjustable exhaust valve through which your exhaled air escapes. Fashioned into the faceplate is a receptacle to hold a telephone speaker. The diving telephone should have fresh batteries and be hooked up and checked while operating.

Because your dive will probably require decompression, and since most companies insist on one for work in over 100 feet of water, we have a recompression chamber. The chamber will be discussed at the end of this chapter when we are decompressing and again in Chapters 3 and 4. For the present, we have a chamber in good condition and it is hooked up to the air supply. Two full cylinders of pure oxygen are hooked up to the chamber and we have several others for spares. There is a ladder extending from the deck of the barge, into the water, at least two or three feet. A descending line, or down line, leads from the deck of the barge down to your work area or to a weight on the sea floor, 165 feet below the surface. There is an experienced tender standing by. Aside from any personal equipment, you should now be ready to go.

THE WET SUIT

In almost any part of the world, the water temperature at 165 feet is cold enough to require some sort of thermal protection. A foam

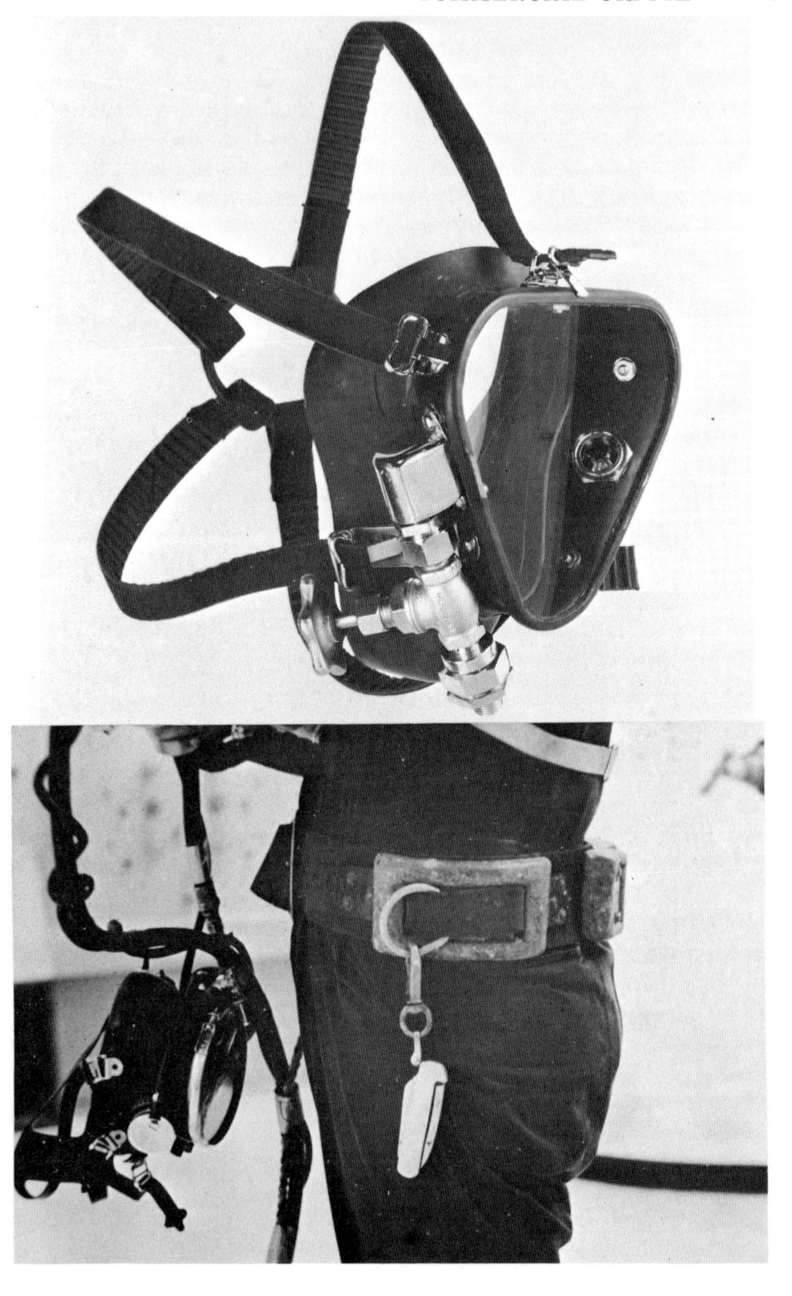

Fig. 2. (Top) A Desco mask. Photo: Diving Equipment & Supply Co., Inc. (Bottom) Rigger's jackknife attached to diver's belt with snap hook. The mask is a Scott. Photo: John Violette.

neoprene wet suit is most often used. The suits come in material thicknesses of 1/8, 3/16, 1/4, 5/8, or more. The colder the water, the thicker or heavier the wet suit needed. Other types of suits will be discussed in Chapter 4.

For this dive we will use a 1/4-inch wet suit. The wet suit material contains thousands of individual air- or gas-filled cells. These cells provide the required insulation between the diver's body and the surrounding water. The suit should fit snugly over all of your body to eliminate any large concentrations of water and to prevent the free circulation of water between your body and the inside of the suit.

Fig. 3. (Left) The author dressing out in wet suit and flippers. Photo: Paul J. Tzimoulis, *Skin Diver Magazine*. (Right) Diver Bob Grantz breathing oxygen in the recompression chamber. Photo: John Violette.

Ideally there should be a thin film of water between the diver's skin surface and the suit. The body heat will quickly warm up this film of water. The suit should fit snugly, but it should not be tight. A suit which is too tight will reduce or restrict your blood circulation and impose mechanical resistance to your breathing. Be sure the suit is comfortable and that you can inhale fully and normally with it on. A custom-tailored wet suit is worth every dollar it costs. Talcum powder or cornstarch liberally sprinkled inside a wet suit when it is dry will make it much easier to put on, but the easiest way to get into a wet suit is in the shower with the help of some soapsuds. A pair of cotton coveralls are well worth the little extra expense and provide excellent protection for an expensive custom-tailored suit. The pockets are often invaluable for carrying extra nuts, washers, small tools, etc.

When using a diving mask, it is a good idea to wear a wet suit hood. It not only keeps your head warm, but it is a big help in keeping baby crabs and other undesirable objects out of your ears. When working on the bottom, it is also advisable to wear a pair of galoshes or other foot protection. Work gloves are a necessity.

THE WEIGHT BELT

A law of physics states that an object submerged in water is buoyed upwards by a force equal to the weight of the water displaced by the object, minus the weight of the object. The human body comes in such an incredible assortment of shapes, weights, and sizes, that it is impossible to state with certainty but it is quite likely that you float in water. While wearing the wet suit you will positively need a weight belt to get under the water. How much weight is required is a matter for individual experimentation and personal preference. The work to be done under water and existing conditions will also determine how much weight should be used. If the work is to be on the bottom in a strong current, obviously more weight will be needed than when working on a riser somewhere between the surface and the bottom. Any alteration in the type of equipment you are going to use as well as in the thickness of your wet suit will require a compensating adjustment of the weight of your belt. The belt should be of a type to allow you to readily add or remove weights and you should conscientiously adjust your belt weight to the requirements of the specific dive. As you descend, your wet suit will be compressed in accordance with Boyle's Law. Therefore, the deeper you go, the heavier you will become. The variations in this condition due to the thickness of the wet suit, the compressibility of the wet suit material, the size of the wet suit, the individual's natural buoyancy or lack of it, and the depth of the dive are impossible to tabulate. It is possible to begin a dive with just enough weight at the surface to overcome the buoyancy, and then find yourself at 165 feet, dragging your belly on the bottom and moving (if at all) with great difficulty. This is an uncomfortable and possibly dangerous situation and it is worth a little experimentation with a weight belt to avoid it. It is best to be slightly buoyant at the surface, achieving neutral buoyancy at about 33 feet. Quite often it will be necessary to swim from the side of the barge 20 or 30 feet to a structure leg or riser before you descend, and some buoyancy is a big help when doing this. Too much buoyancy at the start of a dive will result in too much energy being expended to hold yourself down during your shallow water decompression stops, and also, if you are not coming up a line it could cause a too-rapid rate of ascent. When the weather is rough, negative buoyancy at the surface is helpful in carrying you through the turbulent water. Your belt should, obviously, be easy to remove but the use of the quick release buckle commonly used for scuba diving is not recommended. These buckles require a slight tension on the belt to keep them hooked. As you descend, your suit compresses and the belt becomes looser. It is easy for the buckle to come unhooked resulting in the loss of your weights. This could be dangerous by forcing you to float or ascend against your will, and furthermore, it is annoying and expensive to lose your weight belt.

DIVING KNIFE

Another important piece of diving equipment that is affected by the compressing of your wet suit with increasing depth is your diving knife. Most divers wear the knife scabbard strapped to the calf, thigh, or to the upper arm. In any of these places, as the suit compresses, the scabbard straps will loosen and the knife simply will not be where you expect it to be when you need it. A sloppy, flapping knife scabbard is a needless additional distraction under water. If you prefer to wear your knife strapped to an extremity, replace the buckle straps with Bungi cord or surgical tubing. Pull a little tension in the cord or tubing and secure it with a square knot. As you descend and your suit compresses, the easing tension in the Bungi cord or surgical tubing will automatically take up the extra slack and keep your knife where you want it. A piece of large-diameter copper tubing flattened so that it will accept the knife blade with just a little bit of pressure makes an acceptable sheath when it is attached to your belt. Some divers prefer to use a large jackknife tucked under the wet suit at the wrist. An ideal diving knife is a stainless-steel rigger's jackknife with one blade and a small marlinespike. Made up with a snap hook and hanging from a ring on the weight belt, this type of knife is always available. The marlinespike is very useful for starting small shackle pins, as well as for prying oysters off of jacket legs.

EMERGENCY DEVICES

I strongly recommend the use of an emergency flotation device or inflatable life jacket and a bailout bottle for all compressed-air diving in over 100 feet of water. Nemrod and Bouey-Fenzy combine both of these desirable features in one excellent piece of equipment. These items will be discussed presently. For the time being, if you are wearing a bailout bottle or a compressed-air inflatable vest, it is important to check the cylinders fully charged. If you are wearing a CO_2 cartridge type vest, the cartridges must be checked unused.

CHECKING THE EQUIPMENT

Not only because we are assuming this to be your first dive, but also because it is a good idea to do it before every dive (your life depends on it), let us check everything over one more time.

We have a compressor of sufficient volume and pressure for the dive. The tender has checked the fuel and the lubricating oil in both the drive engine and the compressor. He has started the engine and,

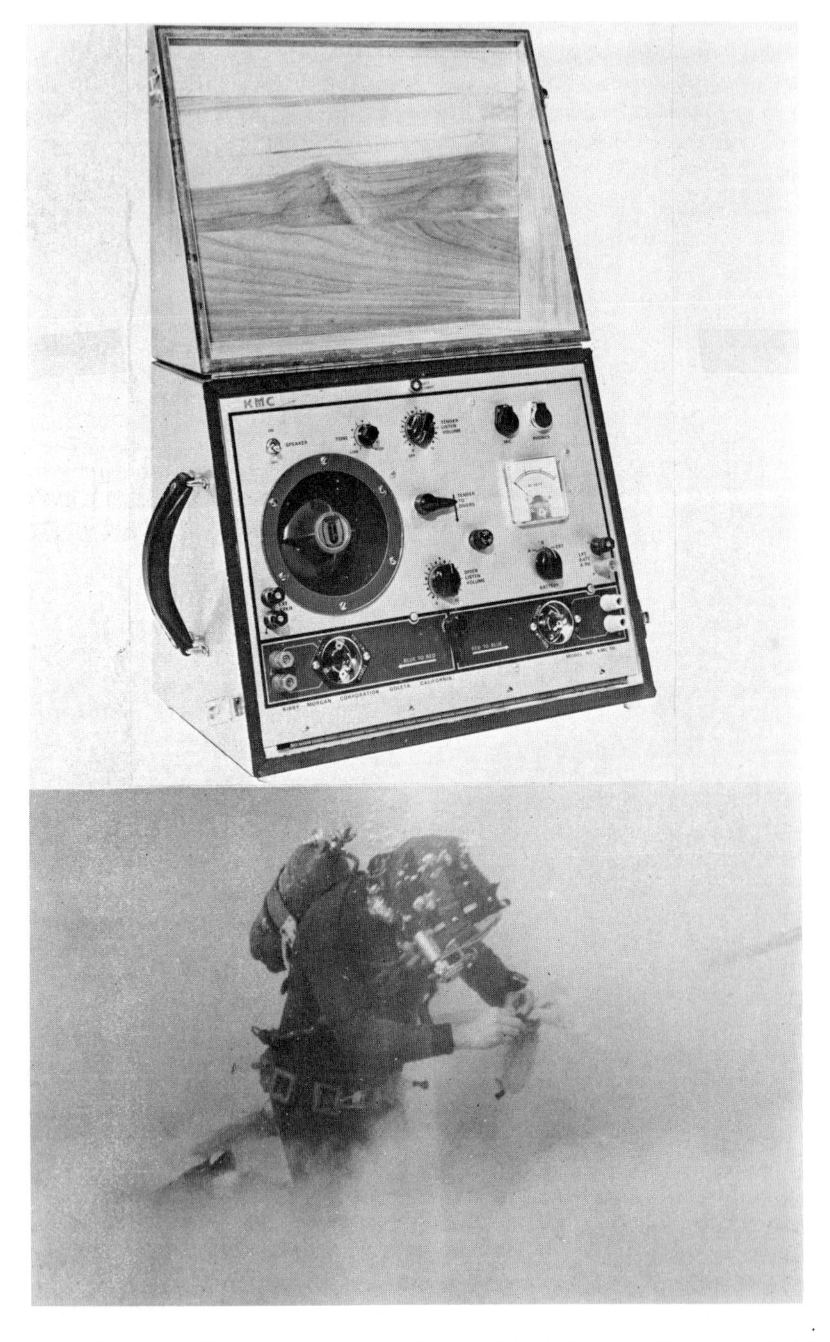

Fig. 4. (Top) A Kirby-Morgan diver's telephone. Photo: Commercial Diving Division, U.S.Divers. (Bottom) A diver working on a muddy bottom, wearing a Kirby-Morgan band mask. Note the "bailout" bottle. Photo: Divcon, Inc.

if it has one, engaged the clutch. If it is so equipped, the tender has screwed in the butterfly nut on the automatic pressure regulator so that the compressor is, in fact, pumping air. All condensed moisture has been bled off from the filters and also from the bottom of the volume tank. The gauge on the volume tank registers the required pressure and the proper valves leading to the recompression chamber and to the diving station have been opened. It is a good idea to check the air-supply valve at the recompression chamber to make sure it is not cracked open and needlessly bleeding off air.

The intake filter of the compressor is located so that it will be drawing in clean fresh air, in no way contaminated by the compressor engine or by any other engine. All other valves on the compressor and volume tank, bleed-off valves, petcocks, etc., have been checked shut. All fittings have been checked tight and not leaking. The supply hoses to the chamber and to the diving station have been led across the deck in such a way that it is reasonable to expect that nothing will be run over them or be set down on them. The diving hose is properly coiled down and the non-return valve on the mask has been checked and is functioning properly.

The phone has been hooked up, the batteries are fresh and the communications are operating. In the event of a communications failure, both you and your tender fully understand diver's line-pull signals.

The ladder is overboard and secured. The descending line is down and it is close enough to the ladder for you to reach it conveniently.

You are decked out in your perfect-fitting custom-tailored wet suit. Since your work will be on the bottom, you are wearing a pair of boots or galoshes. A good sharp knife is attached to you or your equipment in one of the optional ways already discussed. You are wearing a cartridge-type or compressed-air buoyancy device, and the cylinders or cartridges have been checked full. You are wearing a good weight belt and, from previous experimentation at the bottom of the diving ladder, it is just heavy enough to allow you to be a few pounds buoyant at the surface. If your equipment is not rigged with a pneumofathometer it would be a good idea to wear a wrist depth-gauge.

ATTACHING THE DIVING HOSE

Assured that everything is as it should be, you step grimly forward and your tender passes a loop of diving hose around you and fastens it back onto itself with a quick-release hook in the area of your stomach or lower chest. Some divers prefer to attach the diving hose to a ring on their weight belt, especially when using one of the light-weight helmets. When this method is used, it is critically important that the weight belt and its point of attachment be more than strong enough to support the full dead weight of the diver and his equipment, in case he has to be hauled bodily up the side of the barge. The disadvantage of attaching the hose to the weight belt is immediately recognized in an emergency ascent. Ditching your weight belt will make it that much easier and faster for your tender to haul you to the surface. But if your hose is secured to your belt you now have no place of firm attachment between the hose and your body, and by pulling on your hose, the tender is liable to rip the mask or helmet off of you. A number of divers have taken the trouble to have stout nylon or leather harnesses made with a rugged ring, located at about armpit level, to which the diving hose is attached. This is an excellent method and superior to passing the hose around your waist since the point of pull is higher up on your body, tending to move you up through the water in a near-vertical position, thereby considerably lessening drag or water resistance. The quick-release hook should be provided with an easily grasped lanyard.

ADJUSTING THE MASK AND HOOD

After the hose is attached by one of the foregoing methods, your tender will hand you your mask. With the Desco, it is easiest to grab the lower part of the mask with your left hand. The first thing to do is to open the air-control valve and assure yourself that you do, indeed, have air. Then, hold the mask up against your face in the position in which it is meant to be worn and with your tender's help adjust the five head straps, starting with the bottom two, until the mask is held firmly but comfortably against your face. When you think it is properly secured, shut off your air and try to inhale. If you detect air seeping in anywhere around the perimeter of the mask, tightening the strap a notch or two in that area will probably correct this condition. When inhaling creates a vacuum, pulling the mask tighter against your face, you probably have a proper watertight seal. Some divers put the mask on with the head straps on the outside of their wet suit hood, creating a seal between the mask and the rubber of the hood. This method can possibly lead to an ear squeeze which we will discuss later. Other divers put the mask on with the straps on the outside of the hood, but make the seal between the mask and the face. When using this method it is important that no part of the edges of the hood intrude between the face and the mask. A third method is to pull on the hood, then push your head through the face opening in the hood and attach the mask, making the seal directly against your face. When you have a seal and your mask straps are adjusted, the hood is pulled back over your head, over the straps, and the edge of the face opening is arranged around the edge of your mask. The last method interferes somewhat with additional strap adjustments once you are in the water.

After you have a seal and your mask and hood are comfortably adjusted, the next step is to check out your communications by saying something original like, "How do you hear me?" If the phone is on and working properly, your tender will confirm it with a wave of his hand or a nod of his head. One of the principal disadvantages of the free-flow Desco mask is that without an additional speaker rigged over the ear, the diver can hear nothing until he is in the water, and then only when he turns his air off.

BEGINNING THE DESCENT

If you are ready to go, adjust your air-control valve so that you are breathing normally, slip your hands into your gloves, which your tender is holding open for you, and back down the ladder into the water. Admittedly, it is far more dashing to leap from the deck of the barge into the water, but this is inadvisable for a number of reasons, and most certainly not until you know a little more about what is going on. For the time being, use the ladder. Grab the descending line with your left hand (your right hand is kept free for your air control valve), and pull yourself down a few feet. In rough water, to avoid being smashed against the side of the barge, it is important to drop down below the bottom of the barge as quickly as possible. Shut off your air and check your mask for leaks. If it is leaking around the seal, adjust your straps until the leak is stopped, but be careful not to pull the mask too tight. A Desco mask adjusted too tightly is almost unbearably uncomfortable after a few minutes. If you cannot seal your mask, or if it is leaking from some other cause, abort the dive. A leaking mask is annoying and can be dangerous for the beginning diver. If your mask is O.K., check out your communications again. You should be able to hear your tender reasonably well if you shut off your air and hold your breath. If the phone is not working, check the terminals on the front of your mask and tighten them if necessary. If you still do not have communications, abort the dive until the problem is located and corrected. If everything is O.K., begin your descent, looking below you occasionally and taking care not to descend in a spiral around the down line.

The first unusual sensation you will experience, probably between 6 and 10 feet, will be an increasing pressure on your ears. You have felt a similar sensation, but to a lesser degree, descending in an elevator in a tall building, or landing in an airplane. It is caused by increased external pressure on your eardrum. The inner and middle ear are natural air-filled cavities connected to the throat by a long narrow tube called the eustachian tube. At the surface, the pressure acting on the outside of your eardrum is 14.7 psi, but the pressure acting against the inside of your eardrum is also 14.7 psi, leaving your eardrum balanced between equal pressures. At a depth of 10 feet, water has seeped into your external ear canal, exerting an additional pressure of about five pounds on your eardrum. The pressure in your lungs and throat is also five pounds higher, but if your eustachian tube is closed as it normally is in most people, the pressure inside your middle and inner ear cavities is still only 14.7 psi or five pounds less than the external pressure. This results in the eardrum being forced inward, causing mild discomfort. If you continue your descent without popping your ears, the pain will increase and you will finally rupture your eardrum.

The pressure in your inner ear is equalized by periodically forcing the eustachian tubes open during your descent. Most people can do this simply by swallowing. If this does not work for you, try working your jaw from side to side and snorting sharply. A number of divers consistently have trouble clearing or "popping" their ears and can only do so by pinching their nostrils shut and snorting or blowing into the nose. In most helmets it is possible to get at least one nostril at a time pushed up against some part of the inside of the helmet to help you clear your ears. Some masks are equipped with a sliding "nose stopper," a brass rod acting through a watertight gland in the front of the mask, with a rubber pad on the end of it, which can be pushed up under the nostrils. Some diving masks are flexible enough so that a finger or thumb can push the mask rubber in far enough to close off a nostril. None of these remedies applies to the Desco mask, which is so constructed that it is impossible to get your nose to bear on any part of it. Occasionally a diver will wear a pair of rubber nose clips but this must be uncomfortable. With the Desco mask, it is possible to loosen the bottom straps, tilt your head far back and open your air-control valve so that a rush of air is escaping out of the bottom of the mask. Then insert your fingers under the mask to block off the nostrils. This is not recommended as standard procedure and certainly not until you are quite familiar with your mask. It will be wise to practice this against a possible future need while hanging on to the bottom of the diving ladder. Doing so will increase your mastery of the equipment. If it is normally difficult for you to pop your ears and you are going to use a Desco mask, it might be wise to do a little customizing. George "Dog" Taylor, a barnacle-encrusted diver with over 25 years in the trade, has overcome this particular problem in a very simple and efficient way. He has glued a short piece of rubber diving hose into the lower part of his Scott mask, located horizontally directly under the normal position of his nose. A slight pressure up and in on the front of his mask forces his nostrils against this hose and ear-popping is accomplished.

If you normally have no difficulty popping your ears, but are now having trouble, you may have a head cold. It is unwise to dive with a head cold. However, as a commercial diver, there are many times when you will be compelled to at least make an attempt. By forcefully popping your ears when you have a head cold you will be driving nasal mucous into the inner ear and sinus passages, with a serious possibility of rupturing delicate membranes and exposing these areas to dangerous infection. With a mild head cold any of the patented decongestants, such as "Contac," will usually keep the nasal passages dried up, allowing you to dive. A squirt of nasal spray just before the dive is also helpful.

EAR SQUEEZE

When discussing hoods, the possibility of ear squeeze was mentioned. This can also cause a ruptured eardrum, but for reasons directly opposite from those operating in the case of a rupture caused by failure to pop the ears. If your hood is tight-fitting and does not allow water to enter the external ear canal while you are descending, the pressure against the inside of your eardrums will be greater than that on the outside, forcing the drums outward, with the possibility of a rupture. Water or air at ambient pressure, i.e., the external pressures surrounding you, must be allowed free access to your external ear canals. Ruptured eardrums from any cause also pose the problem of a serious internal infection from bacteria in the water entering your ear canals. You should not dive until a ruptured eardrum has healed. Rely on a doctor's advice in this matter.

Now that you have popped your ears, you can continue your descent. You will notice that the deeper you go, the less frequently you will have to pop your ears. As you descend, you will have to open your air-control valve wider to compensate for the increasing pressure and the greater volume requirements. Remember the balloon. When you have descended to 33 feet (look at your wrist depth-gauge), stop for a while. Get the feel of your gear, watch the barracuda, and let me explain a few things of significance at this point.

FACE SQUEEZE

If, as you were entering the water, you left the ladder, reached for the down line and missed your grip; and if your weight belt had been excessively heavy, and your air control valve had been shut off and you were unable to open it; and if your tender did not have sense enough to stop you before you fell 33 feet, the consequences would have been disastrous. You would have been subjected to a severe, possibly fatal, face squeeze. If you think there are too many "ifs" in the above proposition, you are mistaken. The same, or similar combinations of events have occurred many times. Again, the danger and consequent severity of a face squeeze brought about by an involuntary fall is greatest within the first 33 feet, and proportionately less during any succeeding 33 feet. The air trapped in your mask would have been at a pressure of, let us say, 15 psi. When your tender stopped your fall at 33 feet, the pressure in your mask would still be 15 psi, because you were unable to open your air-control valve and add more air. The surrounding water pressure at 33 feet is approximately 30 psi (including the pressure of the atmosphere), which means that your mask is being pushed into your face and the tissues of your face are being pushed into your mask by a differential pressure of 15 psi. This would result, at the very least, in severe tissue damage and bruising of your face, but the most bizarre consequence of such an accident is that your eyeballs could be literally sucked out of their sockets into the mask. If, at the beginning of your fall, you were able to pull a part of the rubber seal of your mask away from your face, allowing the water to enter your mask, you would have avoided these ugly results. During the latter part of your fall this would be impossible because the mask would be squeezed too tightly to your face. It would then, of course, be imperative to get your air-control valve open or for your tender to pull you immediately to the surface.

A face squeeze brought about by a fall can be avoided by not wearing too much weight, training your tender and yourself to operate with just the minimum hose slack required between you, and by paying a little more attention to your control valve. Your air-control valve, regardless of what type it is, should never be jammed shut. When you are under water it should never be completely shut off. The packing nut on the valve stem should be tightened so that it requires a firm, but not hard pressure on the valve wheel to operate the valve. A valve that operates too easily is also undesirable as it is forever being opened or closed accidentally. Falling under water is always dangerous; take every possible step to avoid it.

You can also get a face squeeze without falling. If the crane operator should drop an anchor and rupture your air supply pipe, or if a fitting or splice should pull out of your diving hose, or simply if any part of the air supply system between you and the compressor should be accidentally broken or cut, allowing the air to escape, you will get a face squeeze unless your non-return valve is working properly. Gas, unless contained, rushes from a place of high pressure to a place of lower pressure.

In the above situation, all the air in your mask would rush up the hose and out the break, leaving the space within your mask at a much lower pressure than the water surrounding you. This would result in a face squeeze identical to that described above. In this situation, the deeper you are, the more severe the resulting face squeeze. Check your non-return valve frequently. Checking the non-return valve before every dive is certainly the wisest course to follow.

The non-return valve is a spring-loaded valve with a rubber or leather disk or washer acting against a machined metal seat. The valve is normally in the closed position with the spring holding it shut. Air pressure, coming through your hose into the mask, overcomes the spring tension and forces the valve open. When the air pressure is removed, the spring forces the valve closed, preventing the air in your mask from escaping up the hose. Frequent causes of a malfunctioning non-return valve are: worn, broken or cocked valve spring; debris between valve seat and washer; worn, damaged, or dried-out washer. It is part of a tender's job to frequently check the non-return valve. It is the diver's responsibility to see that he does so.

A squeeze can occur for the reasons discussed when using lightweight helmets or full deep-sea gear. With the latter equipment, a squeeze could result in your entire body being forced into the helmet cavity. *Think about that for awhile!*

DROPPING TO THE BOTTOM

The crane operator is taking a nap, your tender has a firm grip on your hose, you have a firm grip on the down line, and you personally checked the non-return valve, just before the dive. As they say in Houston, all systems are go, so let's drop on down to the bottom. Your wet suit is compressed to the point where your weights are exerting a gentle downward pull on you. You control your descent by your grip on the down line and by telling your tender to slack off or hold it, as you wish. The barracuda have made no aggressive moves toward you, but you still dart an occasional upward glance at them. Your ears are popping nicely and the air surges reassuringly into your mask as you gradually open the control valve to meet your increasing requirements. The water is clear and above you can see that your air hose is not twisted around the descending line. Below you can see that there are no immediate obstacles to entangle you, or, thankfully, no gape-jawed sharks. The light is gradually fading, pulling in the perimeter of your visibility; but what you do see, you see clearly. Looking up, you seem to be suspended in a bowl of lime jello. The water, through successive and noticeable stages, is becoming colder but not uncomfortably so. You congratulate yourself for buying a custom-tailored wet suit. You become momentarily apprehensive when you realize just how far away from the deck of the barge you are, but this feeling passes. Your equipment is the best available and has been thoroughly checked out. You have confidence in it. You have confidence in your tender, and you have confidence in yourself. Without this confidence the dive would be psychologically impossible. You are in no hurry and you are having a ball.

A diver's bottom time, for the purpose of computing decompression stops and time, is measured between the time he enters the water and the time he leaves the bottom, heading for the surface. Because of this, the diver is urged to get to the bottom and start working just as fast as he can. Some divers establish as a point of pride that they can descend at a rate of 100 feet a minute, or faster, and they tend to ridicule anyone who takes three or four minutes, or possibly longer, to reach the bottom. This is nonsense. To be sure, if you are a paid commercial diver, your responsibility to your employer requires you to get on with the job with a minimum of lolly-gagging, but the descending line is no place for a race against time. A precipitous descent, if it luckily does not incur a squeeze, a strained or ruptured eardrum or one of several other unpleasant possibilities, can result in severe dizziness and acute disorientation, which will require much more time to overcome than the few seconds saved during the descent. When descending, take your time, Be sure that you are in complete control of your equipment and your faculties at all times. The more experienced you become, the faster will you be able to descend in safety, but novice or pro, the time expended on a prudent descent can easily be added to the other end of your dive. *Take your time!*

NITROGEN NARCOSIS

As you pass 100 feet, if you are carefully monitoring your reactions, you will notice that you feel just a little bit differently than you did a few seconds ago. With marked variability among individuals, this "different" feeling will increase as you go deeper. You are experiencing the preliminary effects of nitrogen narcosis. The number of subjective explanations of the feeling of nitrogen narcosis is equal to the number of divers who have experienced it.

On the strength of my own impressions, coupled with those related to me by other divers, I reject wholeheartedly the term "rapture of the depths" as applied to nitrogen narcosis. I have yet to experience any rapture connected with it and I have never met anyone who has. The beginning symptom for me, usually at about 160 feet, is a feeling of lightheadedness. It is quite similar to the feeling experienced when coming abruptly out of a dark movie theater into the bright afternoon sunlight. At about 200 feet I can equate the sensation with that experienced shortly after rapidly drinking one or two bottles of beer on an empty stomach. At greater depths, my peripheral awareness is all but destroyed. I am able to concentrate on only one thing at a time, and that with great difficulty and conscious effort. There is also a loss of time sense and nothing seems very important.

There is a wide variation in susceptibility to nitrogen narcosis among divers. Few can work effectively beyond 200 feet using compressed air, and for this reason most diving contractors use an oxygen-helium mixture for diving beyond 170 feet. Some reduction in mental acuity can be expected during a dive in over 100 feet of water. The symptoms of carbon-dioxide poisoning or carbon-monoxide poisoning will be doubly difficult to detect, with disastrous results; therefore, their potential causes should be absolutely eliminated. Carbon-dioxide poisoning can be avoided if the control valve is kept wide open with more air than you deem necessary flowing through your mask. It is important also not to overexert yourself.

Because of the difficulty in remembering or concentrating when suffering from the effects of nitrogen narcosis, ponder in detail and step by step exactly what it is that you are supposed to do. Carefully consider your contemplated job several times before leaving the surface for a deep dive. As a matter of fact, it is wise to do this before every dive. The effects of nitrogen narcosis are much more severe when there is no visibility. Being able to visually focus on tangible objects is a great help in maintaining your orientation and resisting the narcotic effects of nitrogen. It is important that you exert yourself by a conscious effort of will to keep in mind where you are and why.

If you feel yourself beginning to lose sight of these two facts, do not hesitate to ascend.

The exact "why" of nitrogen narcosis is not understood, but it is obviously caused by the increased partial pressure of nitrogen at depth. The only remedy for it is to reduce that partial pressure. The easiest way for you to do this, if severely affected during a dive, is to come up. The effects decrease and rapidly disappear as you ascend. After a dive during which you were affected by nitrogen narcosis, you might very well have forgotten what you did during the dive, or even that you did anything at all. For this reason, tell your tender every step related to the progress of the job while you are working. "I have put the bolt in at 12 o'clock. There are now six bolts in the flange. The bolt at 3 o'clock is missing a nut. The backup wrench is in the tool bucket and the tool bucket is directly under the flange in the ditch," and so forth.

A commonly observed phenomenon relating to nitrogen narcosis is that you apparently build up a resistance to its effects during consecutive exposures. Many divers have commented that after not having dived for a week or so, they are definitely groggy during their first dive to 200 feet. On the second dive the effects are noticeably less and after the third or fourth dive they are unaware of any deleterious effects whatsoever. My own experiences agree with this, but whether the reasons are psychological or physiological, I am unable to say.

After touching bottom in 165 feet of water, you feel just a little different, but certainly in full possession of your faculties. Nitrogen narcosis does not seem to be a problem for you, at least at this depth. Your air-control valve is opened sufficiently and you are receiving adequate ventilation. Your phone is working and you are not at all cold. Your weight belt holds you firmly to the bottom and you can walk with ease. Your hose is not fouled around the down line. You are standing on a hard sand bottom and visibility is 15 to 20 feet. About ten feet away you can see a concrete-coated pipeline half buried in the sand. It is a 24-inch pipeline, and with the weight coating, its diameter is about 30 inches. Your job is to secure a buoy to it, marking the point where your barge will lay a crossing pipeline. That does not seem too difficult. All you have to do is dig a hole under the pipe, and when you are ready, the topside crew will slide the buoy cable and shackle down your descending line.

Most divers would dig the hole by loosening the sand with their knives and scooping it out with their hands, but some kind of digging tool makes the job much easier. Since this is a frequent job for an oil-field diver, it is a good idea to carry a large gardener's trowel or a folding foxhole shovel in your kit.

CARBON MONOXIDE AND MISCELLANEOUS AIR CONTAMINANTS

While you are busy digging, let us talk about carbon monoxide and air contaminants.

Unnoticed by your tender, the barge crew has placed a welding machine close to your compressor and its exhaust is blowing into the intake filter. If a lot of this exhaust is being sucked into your diving air you will probably pass out in a few minutes before realizing that something is wrong. If your tender does not immediately notice that you have stopped grunting and moving around, that will be that. You will, quite needlessly, die from carbon-monoxide poisoning.

The blood hemoglobin which picks up oxygen from your lungs and carries it to your tissues picks up carbon monoxide much more readily than it does oxygen. Although there is plenty of oxygen in your breathing air, your blood is rejecting it in favor of the deadly carbon monoxide. Your tissues, especially those of your brain, become oxygen-starved and unconsciousness and death quickly follow.

That is the extreme possibility. Let us consider the same proposition, but with conditions a little more in your favor. The situation on deck remains the same but a strong wind is blowing most of the welding machine exhaust away from your air intake; most, but not all of it. You are getting along reasonably well with your hole under the pipeline, but it is a more difficult job than you had anticipated and you are working quite hard. You are receiving plenty of air, but you notice at the edge of your awareness that it smells slightly different. Carbon monoxide is an odorless gas, but exhaust, either diesel or gasoline, carries its own distinctive odor due to incomplete combustion.

Intent on your work, you ignore the change in air smell. Presently you notice that you are beginning to get a headache and you are hot and really pooped although you haven't done that much work. You only realize that something is seriously wrong when you begin to get dizzy and perhaps nauseous. At this point, if you have any sense left, you shout to your tender, "I've got bad air, get me up out of here!" Your tender immediately begins to haul you up, but your headache becomes worse the higher you go and it is an effort to remain conscious. The crew drags you bodily up the side of the barge (remember that harness) and your tender rips off your mask. Fortunately, you do not require decompression and embolism has not occurred during the rapid ascent.

You are not quite unconscious and that fresh air really feels good. It would help you to breathe pure oxygen, and if you did require decompression, breathing oxygen according to the tables would be very beneficial. You have been lucky. The most serious outcome of this episode will be the worst headache you have ever experienced in your life.

Consider what would have happened to you if the wind had not been blowing so favorably; if your tender had not been alert; if your hose had been fouled.

Be sure that your compressor is set up so that exhaust gas from any source cannot be sucked into your breathing air. Be sure that the entire crew understands and respects your need for pure, uncontaminated air. Besides carbon monoxide, allow no other possible contaminants near your compressor, such as fuel, oil, gasoline, paint or paint thinner, or, for example, the ammonia-filled wash water that a galley hand might slosh on deck. If fuel is spilled when filling the compressor, be sure it is hosed down. Keep the fuel tank filled so that fuel will never have to be added during a dive. Monitor your feelings continuously during a dive. Analyze any feeling the least bit strange or unusual, and, if necessary, abort the dive before you get into serious trouble.

CARBON DIOXIDE

Let us get back to the hole you are digging. The deeper you dig, the harder you are working. You are breathing harder and faster. In fact, you are panting and you feel as though you are not getting enough air. You begin to get a headache and you feel hot and flushed. Frustrated by your inability to complete such a simple task, you dig harder and faster and breathe harder and faster and—if you don't stop immediately, you are probably going to pass out. You are being dangerously affected by carbon dioxide, brought about by overexertion. As I mentioned earlier, your tissues manufacture carbon dioxide. The harder you work, the more carbon dioxide you produce. If you are not allowing enough fresh air to circulate through your mask, you are probably rebreathing a portion of your exhaled, carbon-dioxide-rich air. Open your control valve wide and let that fresh air rush through your mask. Hard work is a concomitance to diving for a living, but do not overexert or exhaust yourself under water. Remember that getting back to the surface and aboard the barge will require a great deal of energy. Pace yourself and conserve your strength. If you are forced to work exceptionally hard, stop every few minutes and relax, flushing your mask with plenty of fresh air. When your breathing rate returns to normal, get back to your digging. A good warning of the presence of carbon dioxide is a fogged-up faceplate on your mask or helmet, although dangerous levels of CO_2 can exist without fogging.

Your best defense is to use plenty of air and to constantly monitor your feelings. If you feel in any way different or unusual, stop working, ventilate freely and try to analyze how you feel. If any of the foregoing symptoms continue after you have rested, or if they increase, surface until you can figure out what is wrong. There is always tomorrow—or there will be, if you pace yourself and abort the dive when conditions warrant. The contractor would rather lose a few minutes of bottom-time than pay for a helicopter to fly your corpse in to the beach.

LOSS OF AIR AND EMBOLISM

The following episode occupies just a few seconds in its entirety, but the possibility of its occurrence hovers threateningly over you constantly while under water. There is no way to guarantee absolutely that it will not happen to you as long as you are a diver.

While you are busily digging away with your head jammed under the pipeline, the reassuring rush of compressed air through your mask suddenly dwindles to a feeble whisper. Your first thought is that you accidentally banged your air-control valve against the pipe and shut off the air.

Reflexively your hand darts to your control valve, but it is open. You spin the wheel open a few more turns but there is no air. You leap out of the hole in a panic and you scream, "No air. Help!" You drop your weight belt, and your tender, who has just been told that the deck crane ran over your air-supply hose, begins to haul you up as fast as he can. If you remember to exhale (as impossible as that sounds) on the way up, you will probably survive. I say "probably" because there are a number of other factors that could be working against you. If you did not heed my cautionary instruction about not descending in a spiral around the down line, your hose would be fouled and you would be beyond surface help. You would then have to ditch your mask and either climb up your hose or swim to the surface completely unaided. If the failure of your air supply occurred just as you had exhaled, you would probably lose consciousness in about 15 seconds. If you could not climb up your hose far enough for your wet suit to regain its buoyancy before you passed out, it would be all over. You would release your grip and slowly drift back to the bottom, to be gently carried away by the current. By remaining attached to your diving rig, if it were not fouled, you would still be able to be helped from the surface, even if you did pass out. If your air-supply failure had occurred just as you had completed an inhalation, you would be in much better shape. Because of the high partial pressure of oxygen in your air at 165 feet, you should have enough oxygen in your lungs to remain conscious for about three minutes. That is, if you don't panic and exhaust yourself during the first few seconds of your attempted ascent.

MAKING AN EMERGENCY ASCENT

One unfortunate aspect of your last but certainly necessary communication with your tender is that you had to exhale some of that precious air while calling for help. If your air supply fails, it is up to you to recognize the seriousness of your position immediately and to take action at once. Inform your tender by voice, but be brief. "No air. Help!" Simultaneously give your hose four vigorous jerks, and start climbing. If your tender is as well trained as he should be, he will respond without delay. If for some reason or other he does not pull you up, or not fast enough, keep climbing as fast as you can, hand-over-hand up your hose, or down line. Now, within reasonable limits, you should have plenty of time to reach the surface, so don't panic. A sudden furious burst of activity could use up all the oxygen in your lungs in a few seconds. You have a long way to go, so pace yourself and conserve your energy and your oxygen.

As you begin to ascend, the most important requirement of the entire maneuver comes into play. *Do not hold your breath.* If you do, you may as well stay on the bottom. If you held your breath from 165 feet to the surface, you would induce several types of massive and fatal embolism. As you ascend, the surrounding water pressure decreases and the air in your lungs expands (Boyle's Law). If you hold your breath, there is no place for the expanding air to go. It will begin to rupture the tissues of your lungs, forcing air bubbles into your blood vessels, and also into the space between your lungs and chest wall. The higher you rise, the greater is the possibility of embolism and its consequent severity.

As you start your emergency ascent, throw your head back to form as straight a line as possible in your windpipe. This will provide an unimpeded route of escape for the expanding air in your lungs. Keep your mouth slightly open even if you have discarded your mask. Water will not get into your mouth because of the escaping expanding air. Consciously exhale with light force, increasing your expiration rate as you ascend. You will not experience the desire to inhale because the expanding air in your lungs will still contain sufficient oxygen, and because the partial pressure of CO_2 will be steadily decreasing as you rise. There is a danger of overexhaling and thereby depleting your oxygen supply, but this is a lesser danger than not exhaling enough. It is also impossible for you to willingly expel all the air in your lungs. There will be no warning or painful sensation in your chest to alert you. Keep your head back, your mouth open and exhale while rising. It is possible to ascend at a rate of five feet per second, or as much as 300 feet in one minute. *Don't panic. Never ditch your mask, unless your non-return valve fails or your hose is fouled.* George Morrissey, General Manager of Taylor Divers, instructs his crews in these words: "Ditching your gear should be the last thing you do, and probably will be." *Climb your hose or the down line. It is easier, requiring less energy for more distance covered, to pull yourself hand-over-hand than it is to swim. Your hose or the down line is also the most reliable indication of the direction of the surface. Don't hold your breath. Consciously exhale, expelling more air the higher you go.*

There are several ways to mitigate the chances of having to make an emergency *free* ascent. Be sure that your entire air-supply system is as safe as possible from accidental damage. Tag all valves in the air-supply system, remote from the diving station, "Diver's Air, Do Not Touch." Check all fittings and connections frequently. If the barge or vessel you are working from has its own independent air supply, regardless of how inadequate it is, pipe it into the manifold at the diving station so that it can be drawn from in an emergency simply by opening a valve. For the few vital minutes involved, bad-smelling air is preferable to no air at all. It is possible to hook up a scuba bottle to the diving manifold through the first stage of a scuba regulator for an emergency air supply. An auxiliary volume tank at the diving station is another way to provide a few minutes of emergency reserve air. Use only good, tested diving hose with no splices in it. If there is a splice in your diving hose, be sure that it is strengthened by lifeline properly seized on both sides of the splice so that no strain will be exerted on the splice when the hose is pulled.

There are a number of ways to ensure a successful free ascent if you should ever be forced to make one. A CO_2 cartridge-type inflatable life vest is a big help during a forced ascent. If you should have to ditch your mask, you are sure to reach the surface even if you pass out.

A considerable improvement over this type of vest, well worth the expense and slight inconvenience of wearing one, is a Bouey-Fenzy or Nemrod compressed-air-inflatable vest. Both of these units have a high-pressure cylinder filled with compressed air attached to the vest along with a flexible hose and mouthpiece. When the cylinder valve is opened, the vest inflates and it is possible to breathe through the tube on the way up. Both of these flotation devices are equipped with a large pressure-relief valve to allow the expanding air to escape and prevent the vest from bursting.

With any of the above flotation devices, it is important to exhale forcefully, because your speed through the water is greatly accelerated as you approach the surface. You must also look above you to be sure you are not coming up under the barge. All of these devices require maintenance. They must be rinsed with fresh water after every dive and allowed to dry. They must be checked for leaks and tears, and it goes without saying that a vest with an expended cartridge or an empty cylinder is worse than no vest at all. In the author's estimation, the best insurance in the event of an air-supply failure is a bailout bottle. This is a high-pressure cylinder containing from 6 to 18 cubic feet of air at 2200 pounds pressure. The bailout bottle can be worn with a harness on the back or hung from a belt around the waist, Fig. 4. It can be connected by a T directly into your mask, operating through the first stage of a scuba regulator or it can be used autonomously with a complete scuba regulator.

I strongly recommend that a bailout bottle or one of the compressed-air vests be used for any compressed-air diving in over 100 feet of water. It would be wise to practice your emergency ascent procedure periodically with any of the devices described.

In the competitive "get-the-diver-in-the-water" atmosphere that prevails in the industry today, some people might consider bailout bottles and inflatable vests unnecessary, time-consuming to rig, inconvenient to wear, or the manifestation of an overcautious spirit. However, no safety precaution is unreasonable if the diver's welfare is an honest concern.

THE ASCENT

With all of the crises you have had to face, it is a wonder that you have accomplished anything at all. Well, things are looking up. You have finally completed the hole. The buoy cable has been sent down the down line and you have shackled it around the pipeline. You have used up almost all your bottom time (38 minutes), so check over your work, collect your tools and tell your tender that you are ready to leave the bottom. He takes in all your hose slack and, as your feet leave the bottom, you tell him, and he records the time. He pulls you steadily toward the surface at the rate of 25 feet a minute. Aware of the possibility of embolism, you breathe normally and easily as you ascend. Your first decompression stop (we are using the Navy tables for surface decompression using oxygen) is at 60 feet and it should take you about four minutes to arrive at that point.

AVOIDING STOMACH GAS

It is possible that during your dive you were producing stomach gases as a result of having eaten something disagreeable. You might also have been unknowingly swallowing air. Chewing gum sometimes causes you to swallow air, so avoid it when diving. Stomach gas or swallowed air will manifest themselves by a bloated feeling or cramps while you are ascending. If either of these should develop, stop your ascent and allow the expanding air or stomach gas to escape by one or both of the routes provided by nature.

Stomach gas is both annoying and painful. Try to time your meals so that you will not have to dive directly after eating, and also eat only foods that you know by experience do not cause you to produce stomach gas. It is also important to space your meals so that you will not have to defecate during a dive. In a wet suit, it is usually a simple matter to drop your drawers, even under water, but watch out for the fishes. If you are in a dry suit or deep-sea gear, forget it!

Urinating in your wet suit is not as terrible as it sounds and I believe that most divers do so. No one who has experienced it can deny the exquisite, sensual, but all too brief sensation of flooding warmth that accompanies this act during a long cold dive. As a matter of fact, it was the recollection of this revivifying spurt of warmth that led Jerry O'Neill to develop the concept of the hot-water diving suit and to name it the Diurine suit. The suit should certainly be rinsed with fresh water after every dive and occasionally scrubbed with soap.

VERTIGO AND EAR DISCOMFORT

Another unpleasant sensation that you might experience while ascending, especially if you had trouble clearing your ears on the way down, will be brief, possibly severe, attacks of vertigo. This is also caused by blocked or closed eustachian tubes which do not allow the expanding air in your inner ears to escape. This excess pressure somehow affects the balance mechanism in your inner ear, and in addition to vertigo, can cause severe pain from outward pressure against your eardrums and sinuses. If you feel the onset of vertigo or pain in your ears while coming up, slow down or stop. If the pressure does not equalize after a short stop, you may have to close off your nostrils periodically during your ascent by one of the methods previously described and suck in with force against your closed nostrils. This will stretch your eustachian tubes slightly and allow the pressure to equalize.

DECOMPRESSION STOPS

During your water stops, try to keep your muscles and joints relaxed and uncramped. This is much easier to do if you have a hang-off bar or stage to sit or stand on. If there are sharks around, a 55-gallon oil drum makes a comfortable and secure decompression stage.

As you know, the decompression stops are made to eliminate the nitrogen your body has absorbed during the dive and to prevent the bends. When you are comfortably ensconced in the deck chamber, we will discuss this more fully. Your water stops are extremely important, so don't shortchange yourself.

If you have been working exceptionally hard, it is a good idea to start your water decompression at the next stop deeper than the one called for in the tables. The time spent on decompression stops in the water is undoubtedly the most monotonous and uncomfortable phase of the entire dive, but it is extremely important and if you are going to be a diver you will have to endure it. This provides a splendid opportunity to review your dive. What difficulties did you have? What did you do wrong and why? What additional tools would have made the job easier? This form of introspection is important if you aspire to be a top hand, and your water stops provide an ideal time for it.

SURFACING

You have to be very careful when coming to the surface after your last water stop, especially if the sea is rough. Well-timed coordination between you and your tender is essential to gain the ladder in rough water safely.

Of course, circumstances vary greatly but an effective way to gain the ladder is for your tender to pull you firmly to the ladder as the side of the barge is in its downward roll. If you can grab a rung and hold on firmly, the upward roll of the barge will pull you free of the water and you should be able to climb high enough before the next roll to get yourself above the turbulent water. Once again, try to make sure that your hose is not fouled around the down line before you try to climb the ladder. When you reach the deck, step well aboard, away from the edge of the barge, before you begin to remove your equipment.

THE RECOMPRESSION CHAMBER

If your tender has a cigarette burning, it is OK to take a few drags while getting out of your gear, but remember, it is important to get into the chamber and under pressure as soon as possible. (See Fig. 8.) The Navy insists on a maximum of five minutes between the time you leave your last water stop and reach 40 feet in the chamber. Don't let the barge superintendent or the next diver delay you. If they have to discuss the dive with you, let them do it on the telephone when you are in the chamber.

As soon as you are in the chamber, put your oxygen mask on and breathe normally. There will generally be two oxygen masks in the chamber, so be sure the valve on the one you are not using is closed to prevent oxygen from leaking into the chamber unnecessarily. When you are at depth, 40 feet for a normal recompression, loosen or remove your wet suit. Relax your body and avoid having any of your joints in a cramped or bent position. Avoid any position that will inhibit free blood circulation, such as legs dangling over the edge of the cot or knees crossed. Relax, monitor your feelings and stay awake. A book or magazine will help to while away chamber time.

OXYGEN POISONING

Surface decompression using oxygen has two attendant hazards—the danger of fire and oxygen poisoning.

As you know, combustion is impossible without oxygen. The higher the concentration of oxygen, the greater the danger of fire. In the chamber, when breathing oxygen from a mask, your exhaled breath, still very rich in O_2, is dumped directly into the chamber atmosphere. Even with conscientious frequent ventilating of the chamber with fresh air by your tender, the threat of fire is real and constant. All light fixtures, power and telephone connections, etc., in a chamber must be spark-proof and pressure-proof. Of course, you cannot smoke in a chamber, and if you are a smoker you must warn your tender to remove any lighters, matches or other combustibles from your clothing before he puts it in the chamber. Most barges will have oil of one type or another spilled here and there around the deck, and you must be careful not to track any of this into the chamber with you.

Under no circumstances should you try to shade the light in the chamber by wrapping clothing or a magazine around it. The light gives off a certain amount of heat and is capable of igniting materials in contact with it.

In cold weather, the air rushing into and out of the chamber when it is being ventilated creates a very chilling flow and some divers intimidate their tenders into short and infrequent ventilations. This is unwise to the point of stupidity, and can be tantamount to suicide. Have a blanket and warm clothing put into the chamber if it is cold, but be sure that the chamber is being constantly ventilated.

The other potential danger in the use of oxygen for decompression is oxygen poisoning. The human body reacts unfavorably to high partial pressures of oxygen, even when breathed for comparatively short periods of time.

As is the case with nitrogen, individual reaction to oxygen is highly variable. The adverse reaction to oxygen is manifested by involuntary trembling, or twitching of the lips and facial muscles or other parts of the body. These can lead almost immediately to severe convulsions, and for this reason a diver breathing oxygen should be watched closely by his tender. The only remedy is the immediate reduction of the oxygen partial pressure.

If you feel dizzy, or experience any of the symptoms of oxygen poisoning, or feel in any way unusual, remove your mask and breathe the chamber air.

It is for this reason, as well as to reduce the danger of fire, that the chamber must be kept well ventilated with the consequent lowering of oxygen partial pressure in the chamber atmosphere.

If you prove to have a low tolerance for oxygen, you will have to complete your decompression according to the longer, surface-decompression-using-air table. The standard procedure for surface decompression using oxygen is to breathe oxygen for periods of 20 minutes, interspersed with five-minute breaks of air breathing, until the required time for breathing oxygen as indicated in the tables is completed. The Navy tables for surface decompression using oxygen are included in Chapter 5. Consult them whenever oxygen is used for decompression or treatment. Oxygen should never be used at a pressure greater than that corresponding to a sea-water depth of 40 feet during a standard decompression. Oxygen should not be used during treatment for the bends at a pressure greater than that corresponding to a sea-water depth of 60 feet except under direct medical supervision. When breathing oxygen in the chamber, relax. Any exercise at all will greatly increase the possibility of oxygen poisoning. During your air breaks, get up and move around, stretching your limbs to stimulate circulation.

If you have never previously been exposed to high-pressure oxygen, it would be prudent to test your susceptibility to it. The standard method of doing this, called an oxygen-tolerance test, is to breathe oxygen in a recompression chamber at a depth of 60 feet for 30 minutes. Because of the possibility of a convulsive reaction, someone previously tested for oxygen tolerance should accompany you in the chamber.

The masks used for breathing oxygen during decompression are the demand type. Be sure that the regulator is adjusted so that breathing is easy, and free of effort, but that the regulator is not free-flowing.

Adjust the mask so that it fits your face firmly, but comfortably, with a tight seal so that you are not breathing any chamber air during your oxygen-breathing periods. Pure oxygen should never be used for diving. The partial pressure of oxygen encountered in compressed-air diving to a maximum depth of 200 feet does not involve the danger of oxygen poisoning. Oxygen poisoning is possible in compressed-air diving in depths beyond 300 feet.

THE BENDS

Having lived at or near sea level all your life, your body is in balance with the partial pressures of the gases existing there. Oxygen and carbon dioxide play a minor and imperfectly-understood role in the bends or decompression sickness, so they will not be discussed here. Nitrogen (or helium, in helium-oxygen diving) is the principal agent of the bends. At sea level the partial pressure of nitrogen is about 11.6 psi, and your body contains about one quart of dissolved nitrogen in the blood and tissues at a tissue tension corresponding to this partial pressure. This is all the nitrogen your body can hold at sea level.

If you increase the partial pressure, as you do when diving, your body can then hold more dissolved nitrogen. The higher the partial pressure, the more dissolved nitrogen your blood and tissues can hold (Henry's Law). At all times your body is striving to keep its nitrogen content in balance with the partial pressure of this gas in your breathing air. This is accomplished in much the same way that oxygen and carbon dioxide are carried to and from your tissues.

Causes. When you reach the bottom at 165 feet, the partial pressure of nitrogen in your lungs is about six times greater than the tension of this gas in your blood and tissues. Moving from the place of greater concentration to that of lesser concentration, the nitrogen goes from your lungs, into solution in your blood, and is then distributed to your tissues. Different tissues become saturated, that is, take up all the nitrogen they can hold for a given partial pressure, at different rates of speed. You would have to remain at a given depth between 12 and 24 hours for all the tissues in your body to become saturated with the inert gas in your breathing medium. After this period of time, no matter how long you remained at that depth, days or even weeks, your tissues would not take up any more nitrogen (or helium). This fact is the basis of saturation diving, which is discussed in Chapter 14. The deeper you dive, the faster your tissues will take up nitrogen but, deep or shallow, it still takes about the same length of time to saturate all your tissues. The reason for this is that although your tissues are taking up nitrogen faster at a greater depth, the tissue capacity for holding nitrogen is correspondingly greater.

When you reduce the partial pressure of nitrogen, as you do when ascending from a dive, the excess nitrogen picked up under a higher partial pressure comes out of solution and leaves your body, traveling from tissues to blood stream to lungs and out with your expired air. Various tissues give up their excess nitrogen at different rates of speed and this is why decompression at various depths and periods of time is necessary. If the pressure is reduced too quickly, before your blood-circulating system has a chance to eliminate the extra nitrogen through your lungs (as is the case when you come directly to the surface after a deep, long dive), the nitrogen will come out of solution and form damaging bubbles in your blood and tissues. This is the bends or decompression sickness.

Decompression is the process of gradual elimination of excess nitrogen from your body before it forms into bubbles. There are two characteristics of the tissue-nitrogen interaction that are very helpful to the decompression process. First, the pressure on a diver can be reduced by approximately one-half before bubbles will form. This is the basis of stage decompression, and also why there is no time limit for diving in 33 feet of water, or less. (Non-saturation—less than 12 hours.)

Second, after the pressure has been reduced sufficiently to form bubbles, bubbles will not begin to form, in fact, for several minutes. This allows you to be brought up from your last water stop and rushed into the chamber to complete your decompression.

Breathing oxygen during decompression greatly facilitates the elimination of nitrogen from your system. Your body is under sufficient pressure, usually equal to a depth of 40 feet, to prevent the formation of bubbles. There is no nitrogen at all in your breathing gas and so the nitrogen in your blood transfers to your lungs and is eliminated much faster.

Types. The bends, or decompression sickness, is always a serious matter and can be fatal. The mildest and probably most common form is skin bends. Nitrogen bubbles form in the subcutaneous tissues and produce a rash or blotching of the skin with severe itching and perhaps tingling. When bubbles form in the tissues, they rupture delicate membranes and press against nerves causing excruciating pain. They frequently form in the joints of the arms and legs and it is from this proclivity that the name "bends" is derived. Though the pain in the joints or extremities caused by the bends is severe, this type of bends is relatively innocuous when compared with others. Bubbles can form in the major arteries or blood vessels, obstructing the blood circulation, or in the heart. This can quickly lead to death. Bubbles can form in the spinal column, destroying the nerves that control movement and other bodily functions, and permanently cripple you. Bubbles can form in the brain and kill you instantly, or permanently destroy your mental faculties or permanently damage the centers of your brain that control vital functions, such as speech.

Symptoms. Symptoms of the bends are rash and skin itching, pain in the joints or muscles of your extremities, weakness in your arms or legs, paralysis, dizziness, shortness of breath, and unconsciousness. Do not take chances. Treat any symptoms, no matter how slight.

Precautions. The U.S. Navy decompression tables for air diving are the culmination of thousands of dives, tests and experiments. They are the best available. Dive in accordance with these tables, and don't take any short cuts. If you have been working hard, or you have been exceptionally cold during your dive, it would be wise to take one stop below the recommended stop for your particular dive. The inconvenience of a little extra decompression time is far preferable to the long hours in a chamber required for the treatment of the bends. If you feel any unusual symptoms at all following a dive, with or without decompression, get into the chamber as quickly as possible and undergo short treatment according to the table. Again, a few uncomfortable hours in a recompression chamber is preferable by far to spending the rest of your life on crutches. Treat any symptoms, no matter how slight, immediately.

After completing a dive that required decompression, do not leave the barge for several hours, though you may be anxious to get ashore. Bouncing around in a crew boat, two hours away from a chamber, is no place to be hit with the bends.

People unfamiliar with diving (and this might include the foreman, the barge captain or the project superintendent), tend to minimize the seriousness of the bends. For reasons that I will never understand, some are inclined to deprecate an afflicted diver, indicating by their attitudes that a bent diver is malingering in some way. Perhaps they feel this way because when a diver has to be treated, the job often must be shut down. It would obviously be unwise to continue a diving operation when other divers might come down with the bends while the chamber is in use.

If you are not responding to treatment quickly enough, a foreman or superintendent, ignorant of the seriousness of your situation, might magnanimously offer to send you in to a doctor on the crew boat or helicopter. If symptoms persist, don't leave the chamber under any circumstances. Stay under sufficient pressure to relieve your symptoms, and have the doctor flown out to you. Of course, the doctor should have a knowledge of submarine medicine.

If you should be put in a position where you are required to dive while a man is being treated in the chamber, try to explain as patiently and clearly as possible why this would be unwise. There will be many times in your career when people in charge will try to induce you to dive in dangerous or downright impossible situations. Only confidence in your own ability and judgment will enable you to refuse.

Remember, dive by the tables and treat any symptoms of the bends immediately.

The Navy decompression tables and treatment tables are included in Chapter 5.

When your decompression is terminated and you are coming up, remember that an embolism is just as possible in the chamber as it is in the water. *Breathe normally.*

LEAVING THE CHAMBER

When you are on the surface and the hatch is open, hang up your mask and shut off the oxygen-supply valve. When you leave the chamber, take all your garbage, soggy towels, coffee cups, etc., with you. Have your tender police up the chamber and be sure it is clean and serviceable for the next diver.

The time that you leave the chamber should be recorded. You will not be clean (that is, your body will not return to its normal sea-level tissue tension of nitrogen) for 12 hours. Any dive made before this period is over is termed a repetitive dive and must be made in accordance with the repetitive dive tables. Personally, the author is not convinced of the safety of more than one repetitive dive, even when it is made in accordance with the tables. Unless absolutely necessary, avoid making more than one repetitive dive without a full intervening 12-hour rest period.

The preceding discussion of diving physics and physiology is meant to serve primarily as an introduction to these subjects. I again refer you to the very worthwhile publications listed in the Bibliography. Additional practical aspects of diving will be covered in the following chapters.

Chapter 3

TENDING AND BREAKING OUT

The cliché will have to be tolerated, for, in truth, the best way to learn how to dive is to dive. If you are rich, this will be no problem. Finish reading this book, buy some diving gear, use the gardener for a tender and practice in your father's swimming pool. When you think you have the hang of it, open up a diving company, or even better, buy one that is already operating. If you are not exactly rich, but have about a thousand bucks invested in your GTO, you can sell it and attend one of the diving schools listed in the Appendix. They are not presented in preferential order, but just as they come to mind. Selecting one will have to be your responsibility. To my knowledge, none of these schools is oriented towards oil-field diving and this is an unfortunate oversight when we consider that perhaps 90 percent of all commercial diving in the world today is being done in the oil fields. Contrary to the propaganda claims of most of these schools, it would be very unlikely for you or anyone else to jump right into a diving job with the ink still wet on your diploma. A diving school will put you in company with other people interested in diving, and it will give you a chance to use several different kinds of diving gear. Proficiency at anything comes from practice and experience, and to become a diver, experience accumulated while actually doing things under water is what you need. If you can afford it, I strongly recommend going to a diving school.

THE DIVER TENDER

With or without diving-school training, the best way to break into diving is to begin by working for a diving company as a diver's tender. It will probably be easier for you to get such a job if you have attended diving school. Diver tending is a specialized trade and many people work at it full time, as a career. A professional tender who has no aspirations to become a diver is by far a better tender, from a diver's point of view, than some eager young fellow who is constantly importuning for a chance to get into the water. Nowadays, there are far too few professional tenders.

There are many advantages to beginning a diving career as a tender. The pay is good and you will become familiar with diving equipment and involved with diving personalities and diving work. If you are fortunate enough to have an opportunity to tend for a truly professional diver, you will have an ideal teacher. Even tending for a bum has some advantages, because you can learn from his mistakes. The greatest advantage in beginning a diving career as a tender is that you will be working on actual diving jobs, in a position to watch and learn. If you are tending for a good diver, confident in himself and in you, it won't be long before he will let you do most of his shallow-water work. Sooner or later, you will be on a job where all the divers have run out of time. If you have acquitted yourself well, and people trust you, you will be asked to make a working dive. If you perform well at each of these opportunities, you will be on the payroll as a diver before you know it.

DUTIES AND RESPONSIBILITIES

Diver tending is a specialized and seriously responsible job. Your diver's life, to a great extent, depends on how well you know and perform your work. As a tender, you will have to know as much about the trade as a diver. Considering the total job, a tender undoubtedly works harder and certainly longer hours per given day than his diver.

To begin with, when the call to go to work comes in (almost always around one o'clock in the morning), the tender will have to know which bar his diver is probably hanging out in. The tender will have to drive to the diver's house or apartment and load his personal diving gear, wet suit, mask, weight belt, etc. After all this, he will have a relaxing three- or four-hour drive to the crew boat dock with the diver mumbling, or snoring on the back seat.

Cheer up! You'll be a diver some day and have a tender of your own to abuse.

When you reach the crew boat dock, you will have to load all the gear aboard, and park the car. It is also up to you to be sure that you are on the right crew boat. After a four- or five-hour crew boat ride, if your diver is lucky, he will not be first to get into the water, and he will be down below in his bunk, while you check out the job and get his equipment ready.

After the last job, perhaps four or five days previously, your diver's gear may have been stuffed sopping wet into a sea bag and marinating all this time in the trunk of his car. Smelling like a fish market on Sunday morning, his wet suit and coveralls must be scrubbed with soap and fresh water and hung out to dry. On the last job, your diver may have lost his knife and torn his hood and broken the laces of his boots. He, of course, has been too busy while on the beach to take care of any of this and as an ambitious, conscientious tender, you always carry spare knives, hoods, wet suit booties and other expendable items. For some mysterious reason, the diver never makes any attempt to supply his own gloves, and this is another responsibility you, as a tender, must assume.

After the diver's wet suit and coveralls have been scrubbed and dried, they should be inspected for tears or other damage and patched or repaired if they require it. The zippers on the wet suit and coveralls should be periodically treated with silicone lubricant to keep them operating smoothly, and the diver's knife should be sharpened before and after every dive. If the diver's weight belt, or other harness, is leather, these items should be frequently treated with neat's-foot oil to keep them supple. If the diver has a hand light, it should be checked for fresh batteries. The diver's telephone should be hooked up to a speaker and checked operating and also supplied with fresh batteries. If your diver is using a bailout bottle, or a Fenzy or Nemrod inflatable life vest, the bottles should be checked for a full charge. If he is using a CO_2 cartridge vest, it should be checked for unused cartridges and the tripping mechanism should be lubricated. Check any vest for rips, tears or fabric deterioration.

If the diver is supplying his own mask or diving helmet, this should be checked out and cleaned up. The non-return valve should be removed and tested, and once or twice a week it should be completely disassembled for an inspection of the springs and washers. The telephone speakers should be checked in good condition and the communications hooked up and tested long before your man is ready to dive. The faceplate should be polished clean and it is a good idea to keep a suitable rag in your pocket for this purpose.

After the diver's personal gear is checked out and made ready, the tender should eyeball the entire operation. Check out the compressor and be sure the installation is satisfactory. Don't take another tender's word for it, but check the fuel yourself before your diver goes into the water. If no one is in the water or in the chamber (and be doubly sure of this), shut down the compressor and check the lubricating oils and drain the filters.

Go over the supply piping, valves, hoses, etc. Check out the recompression chamber. Make sure that there will be oxygen for your diver and that the chamber is clean.

Chambers, as is the case with all mechanical equipment, have their own individual idiosyncrasies. Contact a tender who has been operating this particular chamber and have him check you out on it. If no one familiar with the chamber is available, take it down on a dry run and assure yourself that it is operating properly. Check out the chamber lights and communications. Assure yourself that all depth gauges are in calibration and reading equally.

Next, check out the diving station, manifold, pneumofathometer, diving hose, etc. Be sure the quick-release hook on the hose is in good condition, has a lanyard and operates freely. If the taping on the hose has come undone so that there are bights of telephone cable and life line hanging about, retape the hose as necessary. If you find anything amiss during your preliminary check, set it right, if you can do so without stepping on anyone's toes. If it is something of a serious nature, beyond your ability or authority to rectify, bring it to the attention of your diver.

A very important part of your decision-making responsibility as a tender is to refuse to tend a diver if the conditions are unsafe. This is unquestionably a very difficult stand to take, but sometimes neces-

Fig. 5. Arming the diver (tenders hard at work). Photograph of a line drawing appearing in *History of the Sea*, by Frank B. Goodrich, LLD, to which is added: *An Account of Adventures Beneath the Sea: Diving, Deep Sea Sounding, Latest Submarine Explorations, Etc., Prepared with Great Care by Edward Howland, Esq.;* Union Publishing House, New York, N.Y. (Publication date unknown.)

sary, and certainly preferable to contributing to the death or disablement of a diver.

To continue with your pre-dive chores, you should determine just what your diver will be doing, so that you can round up any tools or materials he might require.

You should know the approximate depth of water your diver will be working in and you should have a complete set of legible tables on your person at all times during a dive. Know beforehand what his optimum bottom time will be and never allow a foreman or a superintendent to bully you into extending your diver's bottom time or interfere in any way with his decompression schedule. You will, of course, need a good watch and you should have a notebook and pencil in your pocket to record your diver's times and decompression stops and to otherwise keep a log of his activities.

DIVER HAND SIGNALS

A tender should know a basic set of diver hand signals and these should correspond with the ones the diver uses. The most commonly used signals are the following:

From the tender to the diver: One Pull—Are you all right?, or answer the telephone or, if the diver is moving over the bottom or descending or ascending, stop. Four Pulls—Come on up, immediately.

From the diver to the tender: One Pull—Answer the telephone or, if the diver is coming up or going down, stop. Two Pulls—Give the diver hose slack. Three Pulls—Take up the diver's hose slack. Four Pulls—Pull up the diver immediately. A series of frantic jerks on the hose numbering four or more means—Pull the diver up as fast as you can.

THE TENDER'S KIT

A good tender should have his own kit and it should include the following articles: A good watch with a second hand, or a stopwatch, a jackknife, one or two spare knives of the type his diver uses, a sharpening stone, an 8-inch and a 10-inch crescent wrench for hooking up hoses, diving masks, etc., gloves, rain gear, a good bound notebook and pen.

If the tender is an aspiring diver, he should carry his own wet suit and weight belt. A good tender will also carry wet suit patching material and cement, a sewing kit, a spare hood and booties, a tube of silicone lubricant, a spare non-return valve, snap hooks, spare batteries for the telephone and hand light, spare telephone speakers.

A tender must thoroughly understand the decompression tables, the oxygen decompression tables and the bends treatment tables. He should be able to recognize the symptoms of all the possible physiological hazards that a diver is exposed to and he should be able to administer first aid and artificial resuscitation. *Note:* An injured, unconscious or lifeless diver should always be put into a chamber and under pressure immediately before any time-consuming diagnostic process begins or before injuries are attended to or artificial resuscitation is attempted.

Fig. 6. Putting the diver overboard. Photograph of a line drawing appearing in *History of the Sea*, by Frank B. Goodrich, LLD, to which is added: *An Account of Adventures Beneath the Sea: Diving, Deep Sea Sounding, Latest Submarine Explorations, Etc., Prepared with Great Care by Edward Howland, Esq.*; Union Publishing House, New York, N.Y. (Publication date unknown.)

DRESSING THE DIVER

If the job is in 100 feet of water or more, your diver will have to be up on deck, dressed, and ready to go as a stand-by diver for the man preceding him. Awaken your diver at least an hour before he is needed on deck so that he can shake out the cobwebs and have some coffee or something to eat, if he wishes. Spread out his gear for him, with his wet suit right side out and powdered with talc or cornstarch and with the zippers unzipped. Help him into his suit if he requires it. A really good tender smokes the same brand of cigarettes as his diver and keeps him well supplied while he is on deck.

When it is time for your man to make his dive, hook up his mask or helmet to the diving hose. Be sure the connections are tight and turn the air on at the manifold. Hook up his telephone and check it out. The terminals on the mask or helmet and on the telephone should be bright and free of verdigris. Be sure the terminal lugs are sufficiently tightened. If the phone wires are corroded or splayed, they should be cut back to shiny copper.

Be sure the hose is coiled neatly and free for running. When everything is checked and ready, hold the weight belt around the diver's waist until he buckles it. Pass the hose around the diver's back and fasten it back onto itself, or to his harness if he is using one. Be sure that the pin on the quick-release hook is fully engaged. Turn on the telephone, crack the air valve open and hand the diver the mask. If he is using a Desco mask you can help him by holding the straps evenly distributed over his head while he tightens them. Be sure that the edges of the face opening of his hood are not caught under the mask seal. When he is ready, hold out his gloves, one at a time, so that he can slip his hands into them. The diver's peripheral vision is greatly impaired by his mask or helmet, so take care that he doesn't back into any cables or other tripping obstacles. Be ready to support him if he stumbles.

ASSISTING THE DIVER INTO THE WATER

As your diver gets on the ladder, hold his hose short and firmly and be ready to check him if he slips or falls. Before he starts down the ladder, be sure he has any special tools or materials he intends to take with him. As your diver backs down the ladder, don't hold his hose so tightly that it restrains him, but give him just enough slack to allow him to move freely.

If your diver is the type who leaps, jumps, or otherwise launches himself off the deck of the barge, you will have to feed over enough hose slack for him to reach the water before he goes. This is done primarily to avoid having your arms pulled out of their sockets or getting yourself pulled overboard and, secondarily, to lessen the chances of the diver hurting himself. As soon as your diver disappears

Fig. 7. Conscientious tenders, looking after their diver. Photograph of a line drawing appearing in *History of the Sea*, by Frank B. Goodrich, LLD, to which is added: *An Account of Adventures Beneath the Sea: Diving, Deep Sea Sounding, Latest Submarine Explorations, Etc., Prepared with Great Care by Edward Howland, Esq.;* Union Publishing House, New York, N.Y. (Publication date unknown.)

under the water after his jump, pull up his hose slack and hold him from sinking until you are sure he is on the down line.

DUTIES WHILE DIVER IS UNDER WATER

After your diver gets under water on the down line, he will spend a few seconds checking out his gear. When he begins his descent, mark down his time. Slack his hose at a steady rate. Never throw over more hose than your diver actually needs at the moment. Always be able to feel your diver on the end of the hose. While your diver is descending, tend him from a point 10 or more feet away from the down line with a diagonal angle to his hose to keep him from unwittingly spiraling around the down line.

At no time while the diver is dressed out in his gear, either in the water, on the ladder, or on the deck of the barge close to the edge, should a tender release his hold on the diver's hose. On deck or on the ladder, this is important to prevent the diver from falling. When he is on the bottom, the hose is the only communications link with the diver if the phone quits. *Always hold on to your diver's hose.*

When your diver is on the bottom, do not hold his hose too tight. On the other hand, too much slack is an annoyance and a potential fouling hazard. Periodically, "fish" your diver. Gently draw up his slack to where you can feel him and then pay off several feet. If your diver is not a grunter or curser, and you do not feel him moving around on the end of the hose, check him out frequently on the telephone, but avoid being a chatterbox.

Quite often the tender will have to signal a crane or a derrick in accordance with his diver's instructions. Learn the proper hand signals for directing crane operations. These can be found in Fig. 23, Chapter 6.

Precautions. While your diver is down, don't engage in frivolous conversation with other people on deck. Pay attention to your job and to your diver. Be sure that any tools or equipment to be sent down to him are properly rigged and secured. Do not allow the crane or the barge itself to be moved except at your diver's express instructions, or at least with his clear foreknowledge and assent. Do not allow heavy objects to be swung over the water where your diver is working. Keep people from absentmindedly throwing things overboard near your diver. Prevent people from fishing in the vicinity where your diver is working. Know your diver's optimum bottom time for the depth at which he is working and inform him that his time is running out three or four minutes before it is time for him to start up. Some divers are reluctant to leave the bottom before their job is completed. The penalty of extra decompression for time spent at depth beyond the optimum time is rarely worth the little additional work the diver might accomplish. Know your diver. If he habitually runs over his time, lie to him a little. Lead him to believe that he has spent more time, by four or five minutes, on the bottom than he actually has. He will never know the difference and you will succeed in getting him off the bottom somewhere close to the proper time.

ASSISTANCE DURING ASCENT

When your diver leaves the bottom, mark his time. The time interval between the beginning of his descent and the time he leaves the bottom is his total bottom time. His decompression schedule is figured for his deepest depth and his total bottom time. Some divers want to be hauled off the bottom bodily, while others prefer to climb up the down line or davit chain or pipe riser by themselves. Be prepared to accommodate the particular wishes of your diver, but keep a close watch on the pneumogauge to be sure he is not exceeding the normal rate of ascent.

When decompression is to be taken with compressed air, the rate of ascent is 60 feet per minute. In the case of surface decompression using oxygen, the rate of ascent is 25 feet per minute. If for some reason the diver exceeds the ascent rate he must make up the missed time during a stop 10 feet deeper than his first scheduled stop.

Decompression Stops. If your diver has to spend a long time decompressing in the water, it would be an act of kindness to provide him with a hang-off bar to sit on. This can be a piece of three-inch pipe, about two feet long, with the U-bolt from a cable clip welded to the middle of it. A line is attached to the bar and it is slid down the down line on a shackle to the depth of the diver's first decompression stop. If your diver should lose his grip on the down line, the hang-off bar can be slid directly down his hose with an extra-large shackle.

If there are sharks around and the divers are nervous because of them, a very simple and effective diving stage and shark protector can be made by taking the top off a 55-gallon drum. The diver can stand in it, and at the threat of shark attack, he can simply settle down inside. Punch holes in the bottom of it so that it will drain when being lifted aboard.

Be sure you read the diver's depths accurately at his decompression stops. The ends of the pneumohoses are in different locations on different diving rigs. Be sure the diver is holding the end of his pneumohose at chest level when you take the reading for his depth. If there is a sea running and your diver's depth during decompression is fluctuating due to wave action and the roll of the barge, keep him several feet deeper than the table stipulates. Have his shallowest depth during any rolling equal to the decompression stop depth.

While your diver is hanging off, you will have a good chance to coil the hose and fill in your log. It will be a big help and reduce the amount of writing you will have to do during the dive if you prepare several pages of your notebook in the following fashion.

LS	left surface
RB	reached bottom
TBT	total bottom time
RFS	reached first stop
60	
50	
40	
30	time at each or any of these stops
20	
10	
LLS	left last stop
RS	reached surface
R40C	reached 40 feet in the chamber
RSC	reached surface, chamber

PREPARING THE DIVER FOR THE CHAMBER

When your diver is at his last water stop, ask him if he would like coffee, something to eat, or anything else put into the chamber. A book or a magazine is always a big help to while away the time. Arrange to have a clean towel and whatever else he may want put into the chamber for him.

If the diver's clothes are put into the chamber, be sure there are no matches, cigarette lighters or other combustibles in the pockets. When your diver reaches the surface after his last stop, pull him over to the ladder. Some divers, especially when wearing helmets or heavy gear, like to be helped up the ladder. You can do this by keeping a heavy and steady strain on his hose while he is climbing the ladder.

Some divers, however, prefer to do all the work themselves. Check your diver's preferences. When your diver reaches the deck, lead him inboard, away from the edge of the barge before you remove his gear. Help him out of his equipment. If your diver requires surface decompression, the maximum allowable elapsed time between leaving his last stop and reaching a depth of 40 feet in the chamber is five minutes. The shorter this interval, the better. Don't let your diver lollygag on deck or engage in prolonged conversation with the foreman concerning the job. Get him into the chamber as soon as possible and get him to 40 feet. He can talk to the foreman over the telephone.

Careful Observance. Time your diver's oxygen-breathing periods and air-breathing intervals accurately. If he spends an excessive amount of time yacking on the telephone, or eating when he should be breathing oxygen, add the extra time to his schedule. Don't let your diver go to sleep while breathing oxygen. Keep the chamber light on and keep an eye on him. Ventilate the chamber frequently, especially while the diver is breathing oxygen.

The sizes of decompression chambers vary so much that it is impossible to give definite ventilation rates. To be safe, ventilate the

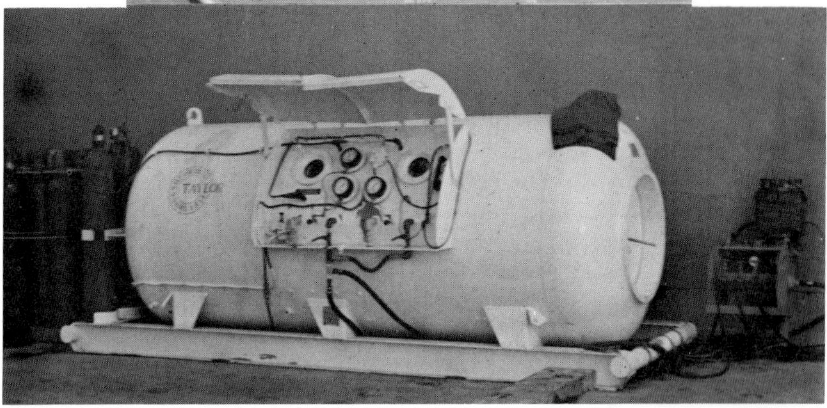

Fig. 8. (Top) A shipshape and squared-away tender. Note how carefully he is tending his diver's hose. The diver is wearing a Savoie helmet. Photo: J. Ray McDermott & Co., Inc. (Bottom) A double-lock recompression chamber. Photo: Author.

chamber as much as possible, consistent with the available air supply. The purpose of ventilating the chamber is to flush out the carbon dioxide but the main consideration is to keep the oxygen partial pressure as low as possible to reduce the fire hazard. It is for this reason that the chamber must be kept clean, free of oil and any other combustible or spark-producing materials.

The diver must never be allowed to put a towel or other object around the chamber light with the intention of reducing glare. This simple action has killed divers by fire and explosion in the past.

Chamber Layout and Operation. The actual operation of a recompression chamber is fairly simple, and does not vary much from one chamber to another. Basically, a chamber is a rigid cylinder capable of holding pressure. It is provided with an entrance hatch that opens inward, and that is forced outward against a seat and gasket when pressure is applied to the inside of the chamber. The hatch is held closed before pressure is applied, by a sort of dog. The best chambers are double-lock, or two-compartment chambers. A chamber is also provided with viewing ports, electric lighting, telephone communications, a bunk for the diver to rest on, demand-type oxygen-breathing masks, an inlet valve for applying air pressure, and an exhaust valve for venting pressure. A chamber must also have a gauge or gauges, usually reading out in feet of sea water. Recompression chambers should be connected to some sort of emergency or standby air supply.

When your diver has entered the inner lock, pull the hatch to and secure it with the dogs. Be sure the exhaust valve is shut and open up the air supply valve, watching the depth gauge all the while. When pressurizing the chamber you must keep your eye on the diver to be sure that you are not applying pressure faster than he can clear his ears. When the gauge reads 40 feet, secure the air-supply valve. The supply of oxygen should have been checked previously. Be sure the valve on the oxygen bottle is open and that the regulator is screwed in. A supply pressure of 30 or 40 pounds on the oxygen regulator should be sufficient.

Time your diver from the moment he puts on his oxygen mask, at 40 feet. The required oxygen-breathing times, as listed in the tables, should be broken up into 20-minute periods with 5-minute intervals of breathing air. There is some current speculation that the oxygen-breathing periods should be reduced to 15 minutes. Write down all the times and intervals. Don't rely on your memory.

When operating a chamber, pay attention to your job. As soon as the chamber is pressurized and your diver is on oxygen and everything is running normally, release the dogs on the hatch. The reason for doing this is that as the door tightens against the seal with the increase in pressure, the dogs may fall into a tighter position. When the pressure is released, the door will tighten against the dogs and they may be difficult or impossible to release.

While operating the chamber, constantly scrutinize the gauges to be sure that the pressure is not rising due to a leaky air-supply valve, or dropping because of a chamber leak. Ventilate the chamber frequently, manipulating the valves carefully so as not to let the depth fluctuate. Your diver will know if you are not maintaining depth because his eardrums will be popping in and out. On most jobs the chamber and the diver's air supply are hooked up to the same system. Be sure that when you pressurize the chamber, or ventilate, you do not use so much air that you are depriving a diver in the water of his supply.

Some diving companies use a system designed to prevent this from happening when initially pressurizing the chamber. The hatch on the inner lock is secured and the inner lock is pressurized to a degree that will allow the entire chamber to be brought down to 40 feet. The diver and a tender enter the outer lock and the hatch is secured. The tender then opens a valve between the locks, allowing the pressure to equalize in both locks. The diver then enters the inner lock and the tender dogs the hatch behind him. The chamber operator then brings the outer lock and the tender back to the surface. With this method the inner lock is used as a huge volume tank or air reservoir, and this also provides an excellent emergency procedure assuring that a diver needing decompression will at least be able to be put under pressure in the event of a compressor failure.

BRINGING UP THE CHAMBER

At the end of the decompression, and provided that your diver has no symptoms of the bends, bring the chamber up at the rate prescribed in the tables. Watch your diver for bends symptoms on the way up. When bringing the chamber up, the moisture in the chamber atmosphere will condense due to the dropping pressure. This will cause a heavy mist, impossible to see through in the chamber. The situation can be alleviated by adding fresh air through the air-supply valve at a rate somewhat less than you are allowing air to escape through the exhaust valve, but balanced so that you are maintaining a steady rate of ascent. In doing this you must again be careful not to take air away from a diver in the water. If your diver is an unfortunate tobacco fiend, have a cigarette ready for him when he comes out of the chamber. Be sure to mark down the time your diver leaves the chamber, because his surface interval time is calculated from this point.

When the decompression is complete, clean out the chamber. Be sure the interior valves to the oxygen masks are shut off. Be sure there is an ample oxygen supply for the next diver. Wash out your diver's gear and hang it out to dry. Put away his mask and other equipment. Sharpen his knife. Get something to eat and go to bed.

Chapter 4

DIVING EQUIPMENT

Prior to World War II, what little commercial diving equipment there was to be found in the United States bore the name Morse or Schrader, or in the case of shallow-water helmets, Miller-Dunn.

In England, the manufacturers were Seibe-Gorman and Heinke; in France, Denayrouse; in Germany, Draeger; in Italy, Galliazi and in the Orient, the Yokahama Diving Equipment Mfg. Company. At that time, except for very minor differences, all of these companies were marketing essentially the same product. The primary item in each of their catalogs was the "Hard Hat," or deep-sea diving helmet, offered in several different models depending on the intended service, i.e., harbor or wreck work, or gathering sponges or pearl oysters.

The helmet globes were made of spun copper, tinned and with bronze fittings. The helmets were provided with a front window, or faceplate, which could be opened and removed by unscrewing. All models had at least two additional windows, one on each side, and some had a window or port or light above the faceplate to facilitate looking upward. The other fittings on the helmet were a gooseneck to which the air hose was attached, an exhaust valve, and, in some rare instances, a gooseneck for a telephone jack. The incoming air was controlled either from the surface, by means of speeding up or slowing down the hand-cranked air pump, or by a control valve mounted on a short length of hose and fastened to the breastplate for use with power-driven compressors.

Almost all helmets were attached to a copper breastplate which rested on the diver's shoulders and the helmet was secured by means of an interrupted thread which required ⅛ turn to seat it on a leather gasket on the breastplate.

Most helmets were also provided with some type of safety locking device to prevent them from being accidentally screwed off under water.

The diving dress was made from thin sheet rubber sandwiched between two layers of canvas or cotton twill, and it made a waterproof seal by means of a thick rubber gasket squeezed between the breastplate and four brass straps, which were fastened down, on American helmets, with 12 wing nuts. When dressed out, the only part of the diver's body in contact with the water was his hands. Provided the dress did not leak, the only way the diver could get wet was by accidentally shipping water in through his exhaust valve or his wrist cuffs. In extremely cold water, rubber mitts would be secured to the dress, keeping the hands dry also.

Fig. 9. (Top) The author at age 19, 1948. (Bottom) Larry Kell, publisher of *Undercurrents* magazine, preparing for a dive to 90 feet to inspect a well platform in Lake Maracaibo, Venezuela. Note the small size of the compressors which will be used to supply the diver's air through a demand mask. Photos: Author.

Accessories to complete the deep-sea rig were an air pump or an engine-driven compressor, a weight belt, chafing pants to protect the dress, air hose and, depending upon an individual's preferences, weighted shoes. Until about 1943, this type of equipment was standard for diving operations throughout the world, and except for telephones, the non-return valve and the two-spring-adjustable exhaust valve, hard-hat diving equipment has remained virtually unchanged since Augustus Seibe invented it 150 years ago.

THE DEEP-SEA DIVING RIG

In this age of Jacques Cousteau, scuba diving and Man-in-the-Sea projects, many people mistakenly believe that hard-hat gear or the deep-sea rig is obsolescent and no longer used. But the deep-sea diving rig continues in wide service throughout the commercial diving industry, and justifiably so. It continues to offer many advantages over other, newer types of diving gear.

To be a master tradesman in any line of work, you have to be able to use all the tools pertaining to your trade. The deep-sea rig is effective and useful diving equipment. Learn to use it. Deep-sea diving equipment certainly looks formidable but to understand it and to be capable of using it properly requires no more intellectual effort or physical coordination than riding a motor scooter.

DRESSING OUT IN DEEP-SEA GEAR

To dress out in deep-sea gear you must have the help of a trained tender. Almost every bystander or deck hand present when a diver is dressing out will attempt to assist. Discourage this and allow only your tender and perhaps one helper to dress you. You will save yourself a skinned nose, boxed ears and chipped teeth. Dress warmly. Wear long woolen socks which you can pull up over the bottoms of your trousers to prevent them from being rolled up your legs. A loose turn or two of tape around your shirt cuffs will prevent the same thing from happening on your arms. Sit down on a low, stable bench and thrust your feet through the neck of the dress and into the legs. Pull the crotch of the dress up tight and then stand up and let the tender pull the dress up under your armpits.

Soap your hands and wrists and, with the tender holding tightly to the bib, first on one side and then on the other, slip your arms into the sleeves and push your hands through the cuffs. Rinse your hands and sit down again. If you have big wrists, the cuffs will probably make an adequate seal. If not, you can turn the edges of the cuffs under and push them up your wrists until you have a comfortable seal. Do not push the cuffs up too far or you will restrict the blood circulation to your hands. Some divers prefer to use snaps or snap tubing, very thin flexible rubber bands about three inches wide which fit over the wrist and the edge of the cuff to make a perfect seal.

The breastplate is rigid, somewhat heavy, and can be uncomfortable if worn out of the water for any length of time. You should complete all the dressing possible, before putting it on.

After your cuffs are squared away, pull on your chafing pants. Have the tender put on and tie your rubber boots or weighted shoes. Most commercial divers do not use weighted shoes, but wear oversized galoshes or overshoes to protect the feet of the diving dress and to provide some padding for the soles of the feet when climbing the ladder.

When weights on the feet are desirable, e.g., diving in a swift current or great depth, or in any situation where an involuntary blowup would be fatal, sufficient lead weights are strung or riveted to leather straps and these are buckled on over the rubber boots as ankle weights.

The breastplate can be downright painful at times. Wear sufficient sweaters and a heavy wool shirt to provide padding. If you are bony in the shoulders and neck, it would be wise for you to wear a donut of padding. Diving equipment manufacturers sell them as accessories, or you can make your own out of thick foam rubber.

The next step is for the tender to put on the breastplate. To avoid a skinned nose, be sure he is standing in front of you when he is doing this. Bend your head forward and let him slip the breastplate over it. Settle it on your shoulders and, while the tender is working to attach the rubber gasket, keep one of your hands on the breastplate, but out of your tender's way. This will keep you from getting knocked on the chin or ear by the breastplate.

Next, the tender must pull the bib up tightly through the inside of the breastplate neck ring, working completely around you, and making sure that no fold in the bib is caught between the breastplate and the dress gasket. You can help him by moving your head around, creating the widest space possible between your neck and the breastplate ring at the point where he is working.

When the bib is pulled tight, the tender, standing in front of you, will hold the breastplate under your chin and pull the rubber gasket of the dress forward, allowing the front of the breastplate to drop inside the collar. Hold the front of the gasket tightly to the rim of the neck ring while your tender works around you, pulling the rubber gasket up over the studs. It is easiest for him to start with the top stud on the left shoulder, and work around towards the back and up to the other shoulder. You can provide your tender with more slack to work with by raising your elbows out to the side. When the back half of the gasket is fitted over the studs, the front of the breastplate will come easily into place. The tender must be sure that none of the gasket holes are hanging up on the stud threads and that the gasket is resting on the breastplate sealing ring.

After all of the studs are inserted, the tender will once again firmly pull the bib up, working completely around the diver, to make certain that no part of the bib is caught between the rubber gasket and the breastplate. The tender will then put on four copper washers, one each over the studs at the front, back and each shoulder. These washers are to prevent the joints of the breastplate straps from cutting into the gasket rubber. After the washers, the straps or brails are put on over the studs. The front and back straps differ slightly from each other, and they must be fitted in the correct places. As each strap is put on, the tender should screw on at least one nut to keep the strap from falling off while he is applying the other straps. The breastplate wing nuts should be screwed up firmly by hand, taking care that the straps are properly aligned over the dress gasket. With some dresses, it may be necessary for the tender to push a part of the gasket inward with his thumbs while tightening the wing nuts to be sure that the breastplate strap is resting solidly on the gasket rubber.

When the straps are properly aligned and snugged down, they are firmly tightened with a T-wrench. The straps should be tightened down in pairs, first the front and then the back, starting at the bottom and working up to the shoulders, leaving the four wing nuts at the joints until last. When all the wing nuts are tight, they should be checked once again with the wrench, starting at the front and working completely around the diver.

The breastplate studs are only anchored in soft solder and excessive strain with the T-wrench can twist them out. Excessive tightening can also damage the dress gasket. When all the wing nuts are tight, the tender should inspect the gasket to be sure it is properly under the straps and squeezed down onto the breastplate to create a watertight seal.

The most common causes of leakage around the breastplate are the bib caught between the breastplate and gasket, and the gasket or breastplate strap holes hung up on the threads of the studs. When the breastplate is completely secured (a four- or five-minute job for an experienced tender), the diver should stand up and adjust the straps of the chafing pants over his shoulders.

Most commercial divers use their chafing pants straps instead of a jock strap to keep the helmet anchored over the shoulders when the suit is inflated. These straps must never be too tight, as the breastplate will ride hard on your shoulders very painfully, and it will be difficult to walk, bend down, or move around. They should be adjusted so that the breastplate just barely lifts off your shoulders when you are in the water.

The Navy insists that divers be completely dressed in on deck, while most commercial divers put on the weight belt and helmet after they are on the ladder. It is imperative that you have a life line attached to you before getting close to the edge of the barge or on the ladder, when your helmet is not on.

All of the previously discussed rules, precautions, and hazards of diving apply to diving with deep-sea gear. The possibility of a fall, either on deck or in the water, is much greater with deep-sea gear, because it is more cumbersome and much heavier than other types. Deep-sea gear, including the weight belt and shoes, can weigh from 120 to over 200 pounds.

AVOID A BLOWUP

The possibility of an uncontrolled blowup is always present when diving with deep-sea gear, and is greatest when hand-jetting. Sand or silt in the water will collect in the exhaust valve, preventing it from opening. This condition need only exist for a few seconds without the diver's knowledge to send him hurtling to the surface in a dangerous uncontrolled blowup. When hand-jetting with deep-sea gear, frequently hit your chin button to keep your exhaust valve clear.

Keep your consciousness tuned to your buoyancy, and if you feel yourself getting light, shut down your air and clear your exhaust. If you should find yourself in a blowup, it is important to alert your tender, so that he can take up your slack to prevent you from falling back to the bottom if your suit ruptures.

If your blowup advances to the stage where you are spread-eagled (your dress is inflated to the degree that air pressure has the arms and legs of your dress extended rigidly so that you cannot bend them to reach your valves), you will be utterly helpless until you reach the surface and topside assistance.

AVOID CO_2 POISONING AND/OR ANOXIA

There is a greater possibility of carbon-dioxide poisoning and/or anoxia when using deep-sea gear than when mask-diving, because it is possible to breathe normally within the gear with your air supply turned off. For this reason, constant attention must be paid to your air flow, to assure adequate ventilation.

Keep your exhaust valve opened wide, use plenty of air, and purge your helmet frequently, especially when working hard. On the bottom, in deep-sea gear, the upper chest and the shoulders of your dress will be inflated, with the rest of the dress squeezed tightly against you. The incoming air will be circulating in the helmet proper but the air in the inflated part of your dress will be stagnant. Carbon dioxide is heavier than air and will sink to the lower part of your dress and will collect there.

Avoid keeping your head bent forward, with your forehead against the faceplate and your mouth below the neck ring, where you will be inspiring a portion of this bad air. CO_2 poisoning can be very rapid in this situation, and this is another reason why your jock strap or chafing gear suspenders should be properly adjusted so that your nose and mouth are well up into the helmet in the fresh-air stream.

Monitor Your Buoyancy. With practice, an amazing versatility can be developed in the use of hard-hat gear. By knowing your equipment and carefully controlling your buoyancy, it is possible to swim on the surface or under water, control your descents and ascents, and also achieve neutral buoyancy for work on risers or ship bottoms.

The most difficult positions to work in, and also the most potentially dangerous, in terms of blowup, are bending far over, or head-down. In either of these situations, the air in your suit will collect at your back or rush into your legs, increasing your buoyancy and tending to blow you up, feet-first. Weighted shoes or ankle weights are a big help in overcoming this tendency. Wear them if these positions of work are anticipated. Otherwise, constantly and consciously monitor your buoyancy and your equilibrium and be prepared to take corrective action, i.e., shut off your air and hit your chin button.

The Air-Control Valve. Most commercial weight belts weigh about 65 pounds, but in slack water and with experience you can get by with 45 or 50 pounds. Most commercial helmets have the air-control valve connected to a three-foot section of hose, called the whip. The control valve is attached to the left-hand breastplate ring with a snap hook.

West-coast abalone divers were the first to have the control valve mounted permanently on the helmet, and the "Jap" hat is now supplied with this desirable feature as standard. Any other helmet manufacturer will gladly customize his product and add this feature. Another fairly recent innovation in diving helmets is the air silencer, and with one, communications in a hard hat are outstanding.

PERSONAL PROBLEMS UNDER WATER

Urination. Urinating when dressed out in deep-sea gear has been a problem throughout diving history. Some diving equipment manufacturers naively offer a rubber urinal, similar to that worn by sufferers of various kidney diseases, to be strapped to the diver's leg inside the dress. They fail to realize that this part of the diver's leg gets squeezed by the water pressure, emptying the contents of the urinal. I have known many divers who habitually and unabashedly urinated in their dress, considering it one of the evils of the trade. The unsavoriness of this practice need not be elaborated. Morse Diving Company catalogs a brass urinal which is attached permanently on the dress at the crotch, that is simply a removable cap. Again, the pressure squeezing in on the dress in this area can arrange this rigid brass fitting in agonizing juxtaposition to delicate parts of your anatomy.

The late Herb Ammerman, a professional hard-hat diver all his life, showed me an ingenious trick for overcoming the urination problem. While standing on deck, he would soap one of his wrists, and with his tender holding onto the cuff, he would withdraw his arm from the sleeve. The standard No. 4 diving dress was voluminous enough to allow him to unzip his fly within the dress. His tender would then pass a narrow but very long olive jar down through the neck ring of the

Fig. 10. (Top) Kirby-Morgan "Hard Hat." Note the striking hammer, the diver's best friend. Photo: Kirby-Morgan Corp. (Bottom) C-Vu mask. Photo: C-Vu, Inc.

breastplate. When finished, Herb would pass the filled bottle back through the neck ring, and repeat the operation as many times as was required. I used this system for years with almost complete success. The only difficulty is in knowing when the bottle is full.

Faceplate Fogging. Another constant problem in hard-hat diving is faceplate fogging. Old pros overcome this by treating the faceplate with a variety of substances, such as vinegar, sputum, or tobacco juice. I used to keep a "Bull Durham" sack in my kit, specifically for this purpose, and before a dive my tender would moisten it slightly and rub it over the inside of my faceplate.

HARD-HAT EQUIPMENT

Maintaining hard-hat gear consists of suit washing, drying and patching, tightening the packing on the control valve and checking the non-return valve. Occasionally the seat on the exhaust valve has to be lightly ground and the spring replaced. The neck-ring gasket needs frequent dressing with neat's-foot oil.

Hard-hat equipment is manufactured and distributed in the U. S. by the Morse Diving Equipment Company, Sleeper Street, Boston, Massachusetts; Diving Equipment and Supply Company, Milwaukee, Wisconsin; and Craftsweld Equipment Company, Jackson Avenue, Long Island City, New York. Craftsweld purchased the patents of the old Schrader Diving Equipment Company and currently manufactures a really superior helmet. The "Jap" hat, manufactured by the Yokahama Diving Equipment Company of Japan, is distributed in this country by the J & J Diving Equipment Company, Pasadena, Texas. J & J can also supply the Japanese deep-sea diving dresses which are less expensive than and far superior to any manufactured in this country. The commercial diving division of U.S. Divers is now distributing a commercial hard hat which is identical to the "Jap" hat.

DIVING MASKS

Just prior to the outbreak of World War II, a Dutchman named Victor Berg developed a rubber hose-supplied compressed-air diving mask for use by pearl divers in the South Pacific. The Victor Berg mask was, to my knowledge, the world's first commercial compressed-air diving mask. It was manufactured by the Ohio Rubber Company, and issued to units of the United States Navy during the early 1940's and designated as shallow-water diving equipment. During the war, many Navy divers, unable to get a Victor Berg mask, modified the standard Mark IV gas mask for use in diving.

The advantages of a mask over hard-hat gear, in terms of diver maneuverability, speed in getting into the water, low initial expense, etc., were quickly recognized and led to the development of the Jack Brown mask, known to us today as the Desco mask. Thousands of

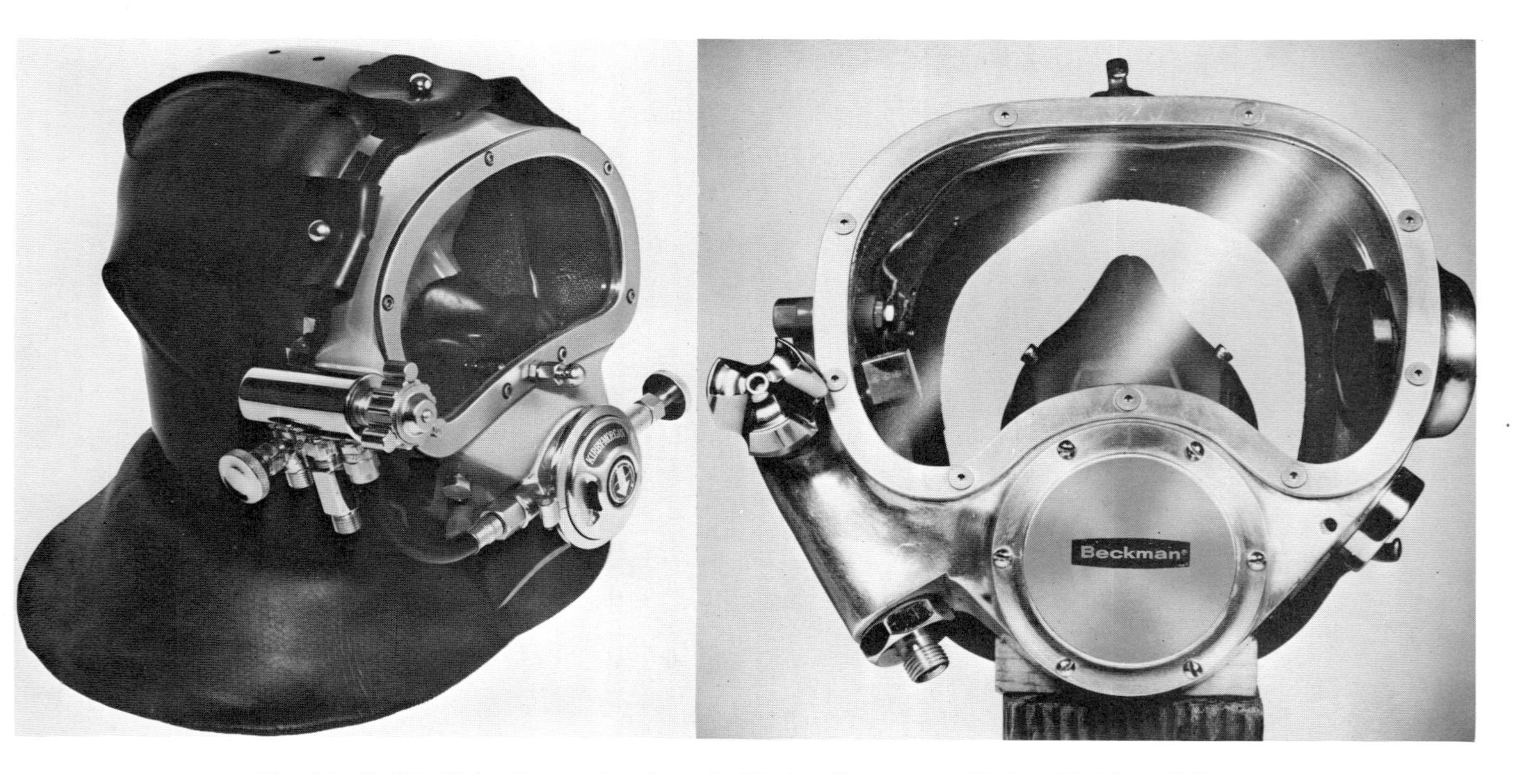

Fig. 11. (Left) Kirby-Morgan band mask. Photo: Commercial Diving Division, U.S. Divers. (Right) A Swindell, or Advanced diving mask. Note the oral-nasal mask for reduced dead space in the breathing circuit. Advanced Diving Equip. & Mfg. Inc. is now a subsidiary of Beckman Instruments, Inc. Photo: George Swindell and Beckman Instruments, Inc.

these masks were produced and distributed throughout the Navy. After the war, hundreds of Navy divers trained in the use of Desco lightweight gear entered the commercial diving industry and carried their preference for this kind of gear with them.

Some anonymous innovator discovered that a small speaker mounted in the mask provided adequate two-way communications when the diver temporarily shut off his air to eliminate the noise. With the development of the neoprene wet suit by Willard Bascom, telephone-equipped diving masks became truly effective diving gear and they were and are used extensively in all the offshore oil patches of the world.

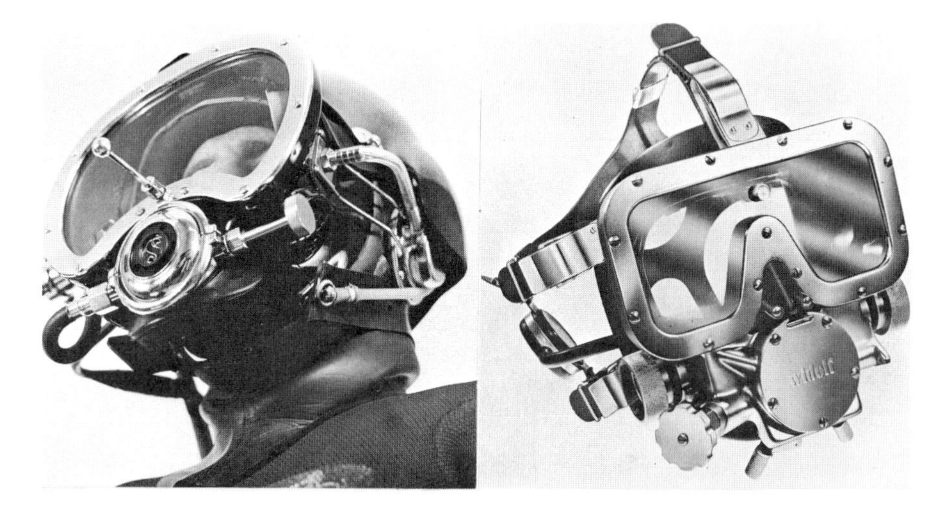

Fig. 12. (Left) Aquadyne mask. Note the extra ear piece for telephone communications, the sliding rod at the front of the mask to control the nose stopper, and the round knob for manually loading the demand regulator. Photo: General Aquadyne Incorporated. (Right) New Widolf mask for mixed-gas diving. Photo: Widolf Diving Equipment.

In 1948 or '49, Phil Widolf developed his rigid frame mask for abalone diving, and although this excellent piece of gear never enjoyed wide usage in the oil fields, it was the forerunner of all such masks manufactured today.

In an attempt to capitalize on the astounding commercial success of the Aqua-Lung® (a registered trademark of U.S. Divers Co.) in the rapidly developing sport of scuba diving, Scott Aviation Company produced its Hydro-pak and unwittingly supplied an important and much-needed piece of equipment to the burgeoning offshore petroleum diving industry. The Scott Hydro-pak was designed as scuba gear, to be used with a high-pressure cylinder of air, but instead of a mouthpiece, it used a full face-covering rubber mask with a demand regulator mounted on its side. Because a demand system uses only a fraction of the volume of air required with a free-flow system, it was possible to hook the mask to a surface supply hose and dive to relatively great depths with a very small portable compressor and a volume tank. From an average cost of about $2500, including compressor, to outfit

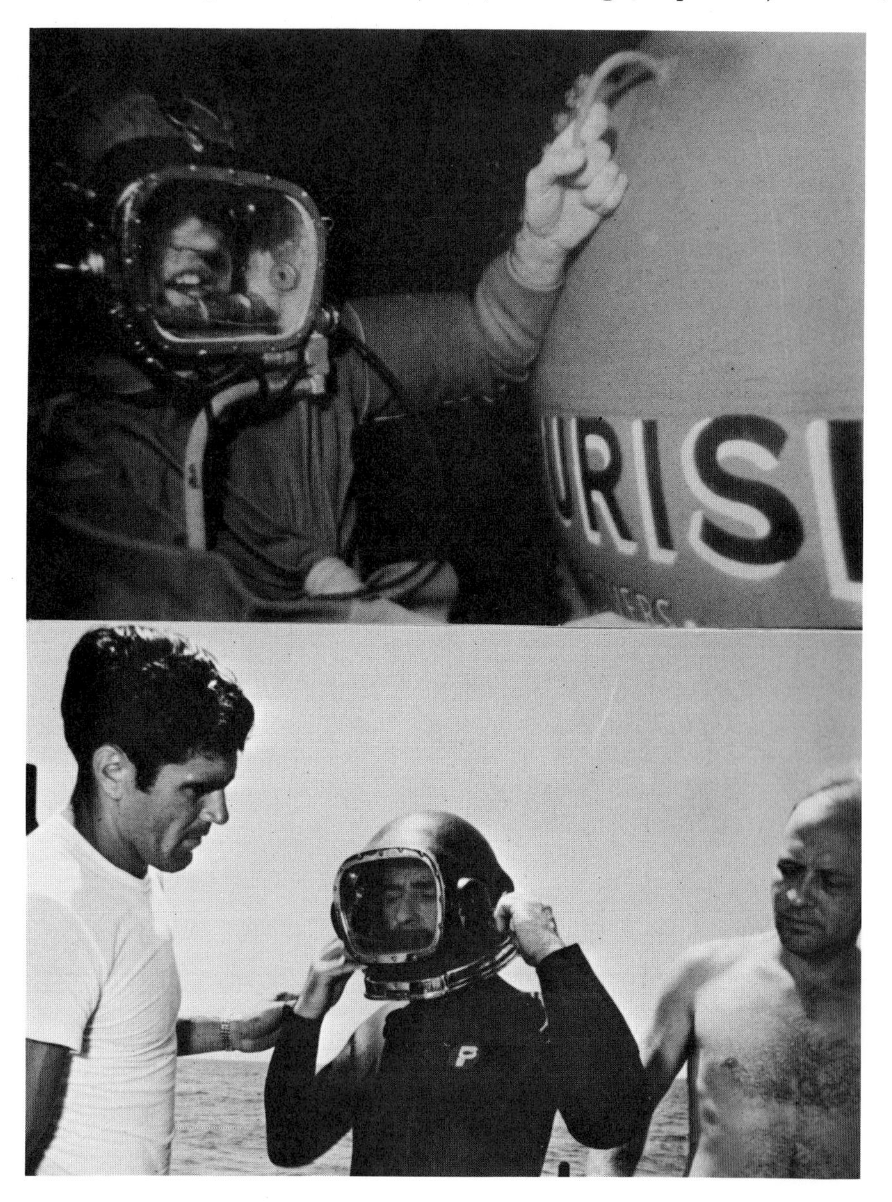

Fig. 13. (Top) Old-style Widolf mask, rigged for mixed-gas diving. Photo: Dan Wilson. (Bottom) The author in a Ratcliffe helmet. Divers John Becksted (right) and Earl Chacon (left). Photo: Paul J. Tzimoulis, *Skin Diver Magazine.*

a hard-hat diver, a free-lance diver with a Scott mask could then go into business with a beat-up Volkswagen and a capital outlay of about $500 for the rest of his diving equipment.

The Scott mask is still used extensively and it is one of the most comfortable masks ever manufactured. Its demand system is adequate, but for heavy work it is necessary to purge it often because of the large dead-air space in the mask. Many divers have customized the mask by adding a free-flow valve and a plastic faceplate with a bubble to hold a telephone speaker. The mask is then used as a free-flow mask while working, and the demand system still allows the diver to draw breath while talking or listening to the telephone. This feature is also excellent insurance in the event of a demand-regulator malfunction.

Desco soon produced a mask with a demand feature to compete with Scott, but it is, in my opinion, an inferior piece of equipment.

Normal-air produces a comfortable and easy-breathing demand mask, but I am very leery of the extremely thin plastic faceplate and the plastic parts of the regulator.

Draeger manufactures an excellent and comfortable demand mask, but the regulator is quite large and keeps getting knocked about in close quarters or in conditions of poor visibility.

The Kirby-Morgan Band mask was perhaps the first of the rigid plastic-frame diving masks to be mass-produced and it has found wide acceptance throughout the diving industry. Although high-priced, it is a fine piece of gear and combines both the free-flow and demand systems. It has an added advantage over the previously mentioned masks in that it is possible to manually adjust the demand regulator for the minimum possible breathing resistance at depth. It has an adjustable pad or stopper for blocking off the nostrils for easy ear-popping. This mask has been successfully used for short-duration helium dives in over 300 feet of water.

What is essentially the Widolf mask is now being manufactured by C-Vu, and it sells for less than $200.

George Swindell, of the Advanced Diving Equipment & Mfg. Co., is now marketing an excellent mask, built somewhat on the lines of a Widolf, but incorporating demand and free-flow systems. It also has a manually adjusted demand regulator. The Swindell mask is, to my knowledge, the only one with an air silencer, and it is the quietest mask on the market. It has an extra valve and hose fitting for a bailout bottle hookup, as does the Kirby-Morgan mask—an excellent feature.

Aquadyne is now manufacturing a mask identical in almost every detail to the Currin mask which Ocean Systems manufactured for in-house use shortly after the company was formed. This mask competes favorably with others on the market, but it is overpriced.

Widolf has once again entered the market, this time with a completely different mask, manufactured in several different models.

There is certainly a wide variety of masks from which to choose. Before buying one, because of the amount of money involved, try to dive the one of your choice.

Fig. 14. (Top) Savoie diving helmet, worn by diver Carl Bock. Photo: Lee Brown. (Bottom) The Miller helmet. Photo: Ben Miller.

HELMETS

Although the postwar metamorphosis of diving equipment was from helmets to masks, there were many divers trained in hard-hat diving who missed many of the advantages inherent in a helmet. First among these advantages is the unquestioned safety of the completely rigid head protection provided by a helmet. Second is the excellent two-way communications afforded without the necessity of shutting off the air or holding your breath. Third is the complete protection against ear infection. Last is the small but nonetheless vital reserve volume of air contained in the helmet which makes possible the rebreathing of helmet air for a few minutes during an ascent in the event of air-supply failure.

The first person to successfully combine the above features of a diving helmet with the simplicity of donning and ease of maneuverability of a mask was Joe Savoie of Boutte, Louisiana. The Savoie lightweight helmet was the forerunner of all the plastic helmets on the market today, and it is still one of the best. The Savoie rubber neck dam or seal has been copied and used by virtually every other plastic helmet manufacturer.

Most of the plastic helmets on the market can be attached by the neck ring to a Bell-Aqua or Bailey dry suit. When rigged this way, their operation is essentially the same as for deep-sea gear, with one exception. Most of these lightweight helmets do not have as large an exhaust valve as a deep-sea helmet, or a quick-dump valve, and, therefore, the possibility of uncontrolled blowup is much greater when using them with a dry suit. The Savoie helmet can be purchased with either a fixed or opening faceplate. I see no advantage in a helmet with an opening faceplate. It is just as easy and as fast to remove the entire helmet when the diver is on the surface, and much more comfortable for the diver.

This also applies to the deep-sea helmet. The Savoie helmet is fashioned from a motorcycle or race-driver's crash helmet and it fits closer than any other helmet, giving it a small internal volume and overall displacement. It is weighted, and further held onto the head by a chin strap that buckles before the helmet is locked to the neck ring. Because of its small size, it is probably the easiest helmet of all with which to swim.

Bob Ratcliffe built his first helmet, known as the "Rat Hat," shortly after Joe Savoie's came on the market. I own a Rat Hat, and prefer it to all the others, for some reasons practical and some indefinable. The Rat Hat is an esthetically satisfying piece of equipment. It is good looking and has the heft and rugged solidity of a first-class tool; something built to last. The communications, acoustics and air-noise level in my helmet are unqualifiedly the best I have ever encountered. The helmet came equipped with a simple lamb's wool air-noise damper, but it does its job remarkably well. My helmet was delivered with two Western Electric H-C-3 speakers and these lasted through more than three years of abusive operation. The Rat Hat, much like the Savoie helmet, attaches to the neck ring with two easily operated cam levers. The Rat Hat also anchors to the head with a chin strap, but it is outside of the neck dam. It has a built-in demand system for mixed-gas use and it is upon this feature that I am forced to direct some criticism.

The first stage of the regulator is mounted externally at the back of the helmet and the second stage is in a recessed pod inside the helmet proper. Breathing is through a mouthpiece at the base of the faceplate. With supply pressures in excess of 200 pounds, the regulator will flow free and there is no way to shut it off. The poorest aspect of this demand system is that exhaled gas is exhausted inside the helmet and if you do not wear nose clips, there is a very real danger of breathing in the high concentrations of CO_2 that collect in the helmet when it is being used on the demand circuit. My only other criticism is that the exhaust valve, a standard adjustable exhaust valve from a Desco mask, is a little small for the helmet, especially when used

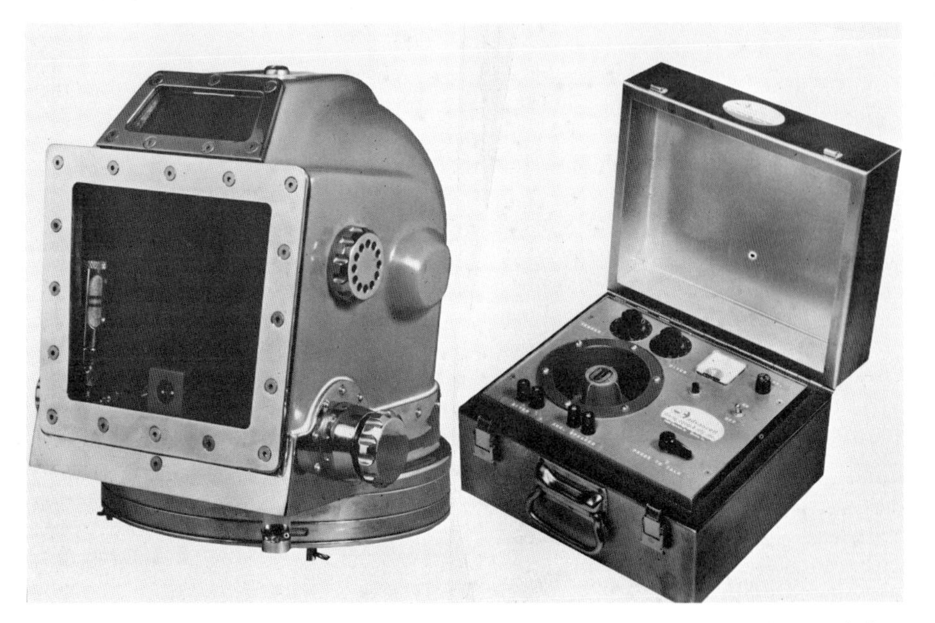

Fig. 15. (Left) Swindell, or Advanced diving air helmet, manufactured by Advanced Diving Equipment & Mfg., Inc. Photo: Commercial Diving Division, U.S. Divers. (Right) A Swindell, or Advanced diver's telephone. Photo: George Swindell and Beckman Instruments, Inc.

with a dry suit, and it does not have a chin button or quick-dump feature. With the exhaust valve in any but the wide-open position, an adequate supply of incoming air will tend to float the hat, causing it to ride hard against the chin strap. The Rat Hat is now owned by Cal Divers, and unfortunately is no longer manufactured for general sale.

The first David Clark helmets were an attempt to use space equipment for diving equipment and they were entirely unsuitable. They had a curved plastic faceplate and extremely delicate fittings. The current model is a tremendous improvement but it is still not a completely developed piece of gear and has many shortcomings. Its configuration is almost an exact copy of the Rat Hat, but it lacks the solidity and ruggedness of the latter. The Clark helmet is extremely light and not weighted; under water, it has a strong positive buoyancy. It relies on an adjustable jock strap and front and back kidney weights to hold it on the diver's head. Because of its light weight and great buoyancy, it is difficult to work in a head-down position.

The Swindell helmet, manufactured by the Advanced Diving Equipment & Mfg. Co., is a product of flawless construction and technical perfection, but I find it to be the most uncomfortable piece of diving gear that I have ever worn. The helmet attaches to the neck ring with one-eighth-turn interrupted threads similar to those used on conventional hard hats, but the neck ring lacks the leverage that a breastplate affords for the counterforce required to put it on and especially to take it off. The helmet is extremely awkward and difficult to handle, even for the tender. Although it is weighted, it also relies on a jock strap to hold it onto the diver. The weight, however, is in the exact top of the helmet, and in any but a vertical position, it puts an uncomfortable and sometimes painful strain on the neck muscles. The helmet has a very large interior volume and displacement and, unlike most other plastic helmets but exactly like a conventional hard hat, you turn your head inside the helmet rather than the helmet's turning when you pivot your neck. The Swindell non-flooding exhaust system is peerless.

The Aquadyne helmet is well made and comfortable to wear. It, too, is a positive buoyancy helmet, very lightweight, and requires a jock strap to hold it on. I find the exhaust valve to be overelaborate and it forms a bulky projection at the back of the hat. Also, the exhaust will not pass the volume of air through it that I feel most comfortable working with. The air silencer, lamb's wool stuffed into a narrow tube, is subject to clogging by moisture or oil, and it can cause an air-supply failure. The sealing mechanism between the helmet and the neck ring is difficult to operate and could be improved.

The Kirby-Morgan helmet is simply an extension of the mask to protect the top and back of the head. It has no neck ring and depends on the gasket around the diver's face for its seal. The helmet is negatively buoyant, and is held onto the head by a rubber spider similar to the one used on the mask, and all of its valves, controls and fittings are identical to those found on the mask. Both the Kirby-Morgan mask and helmet are provided with a manifold block and extra valve for use with a bailout bottle.

The Desco helmet is constructed of spun copper and bronze, little more than a conventional hard hat, modified for use with a wet suit. The only feature for which I can recommend it is its low price.

The Miller helmet is the latest entry into the lightweight helmet market, and it, too, is made completely of metal—in this case, a bronze casting. The earlier models were similar to the Kirby-Morgan helmet in that they depend on a face seal. Recent models use a separate cam-locked neck dam. It also has a demand and free-flow system, and is an excellent piece of gear.

All of these helmets are extremely expensive, ranging from $600 to $1,500. There is a genuine need for and possibility of producing a versatile lightweight helmet that would sell for around $500. The two-piece feature of helmet and mechanically attached neck ring is unnecessary in a simple air hat and it adds considerably to the cost. The neck dam can easily be attached directly to the helmet, with two tabs to pull it down into position while you are inserting your head into the helmet; or the neck dam could be secured around the neck with a waterproof zipper or an ordinary zipper backed with Velcro.

Note: Since this was written, the author has built a helmet conforming to the foregoing features called the Ski Hat; it is shown in Fig. 16.

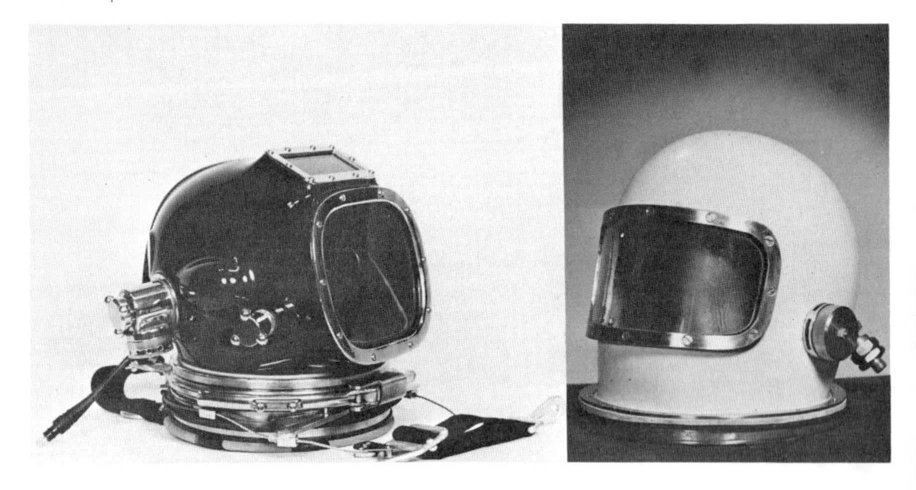

Fig. 16. (Left) Aquadyne air helmet. Photo by Brooks Institute; General Aquadyne Incorporated. (Right) The Ski Hat, diving helmet designed by the author. Photo: Art Lemane.

The Neck Dam. The neck dam, for the lightweight helmet that has one, is worn in one of two ways. If the dam opening is a little large or loose for you, the edge can be turned up and pushed up on the neck to give you an adequate seal. When worn this way, any water collecting in the resulting pocket of the neck dam around your neck will have to be purged through your exhaust valve. Another disadvantage of wearing the neck dam this way is that if your exhaust valve is not open enough, nor large enough, air pressure greater than the surrounding bottom pressure will build up in your hat, tending to blow it off your head. With the edge of the neck dam pointing down, any excess air will blow out the bottom of the dam and moisture or water in the hat can also be blown out this way. If the neck dam is moving up and down like a diaphragm with each inhalation and exhalation, your exhaust valve is closed too much and you do not have sufficient air coming into your helmet.

TELEPHONES

There are many diver's telephones on the market, and they all do their jobs with varying degrees of dependability and clarity. The best phone that I have ever used is the Landry Phone.

One important capability that no present-day telephones incorporate is that which allows the diver to be heard at all times. With today's telephones, when the "Press-to-Talk" switch is activated for the surface to talk to the diver, the diver cannot be heard until the switch

Fig. 17. The author with a Landry diver's telephone. Photo: Jack McKenney, *Skin Diver Magazine.*

is released. Frequently some long-winded engineer gets on the phone and the diver is unable to cut back in. This is a very dangerous situation when working with a crane. A phone should be built so that the diver can override the surface speaker and be heard at all times.

Remember when operating a telephone, that if your diver is wearing a Desco mask he will have to shut off his air and hold his breath to hear you. Break up your conversation into small segments and give him a chance to catch a breath every now and then.

WEIGHT BELTS

The only suitable weight belts on the market today are the Navy lightweight quick-release model supplied by Desco, and a new belt being manufactured by Ben Miller. Advanced Diving Equipment will also have a belt on the market soon.

A diver's weight belt should be easy to remove, certainly, but the diver should not be constantly preoccupied with the possibility of accidentally losing his belt. I have stated my objections to the type of weight belt used by scuba divers earlier. It is a simple matter to round up a good piece of leather or belting, a brass buckle from a saddle-cinch strap, some rivets, bronze or stainless steel rings, and fashion your own belt. Neat's-foot oil is a must for any leather equipment that will be used in salt water.

HOSE

For many years, the hose supplied by diving equipment manufacturers for use with mask or lightweight gear was nothing more than standard ¼-inch oxygen hose supplied in 50-foot lengths. This hose was easily kinked and not capable of withstanding too much pressure. Once again, an anonymous innovator discovered that the hose manufactured by Gates Rubber Company for use in crop-spraying operations made excellent diving hose. This hose, 100 RS, is now available in ⅜-inch I.D. and 300-foot unspliced lengths. It can withstand an internal pressure of 800 psi and it has a tensile strength of 1200 pounds. It is difficult to kink and with Aeroquip or Lenz fittings, it is unquestionably the best diving hose available. Diver's hose is generally made up in 300-foot lengths with a safety line, telephone cable and pneumohose. The combined tensile capacity of the hose, the phone cable and the pneumohose is so great that I believe the safety line to be redundant, except in the case of a rig made up of spliced sections.

The components of the hose assembly are generally taped every eight inches or so, but they should also be seized together every five or six feet with seine twine. When taping a hose, it is important that all components be in equal tension to avoid bights. Hose can be supplied made up in any desired lengths by the Advanced Diving Equipment & Mfg. Company.

All diving should be done with a pneumofathometer to insure accurate depth readings. This is simply an air supply operating through a valve, and a gauge reading out in feet of sea water attached to the pneumohose on your diving rig. In operation, the diver holds the end of the pneumohose on the bottom to check the depth or at chest level to determine decompression depths. The tender turns on the air supply long enough for air to flow freely out of the end of the pneumohose. When the air is shut off and the flow out of the end of the hose stops, the gauge will register the depth.

The diver's hose requires little maintenance, but it should be kept coiled when not in use, out of direct sunlight, and, of course, away from oil and grease. It should be kept well taped, and the ends should be cut back two or three inches at least once a year, and new fittings attached.

If your helmet, mask or telephones use single-pole terminals, rather than jacks, the ends of the communications wire should be soldered so that they can be easily inserted into the terminal holes. The ends of the communications cable, except for the bare wires, should be heavily daubed with rubber cement to prevent water from running down under the insulation, rotting it and corroding the wire.

COMPRESSORS

Although many different brands of compressor are used for diving, Quincy seems to dominate the market. Quincy compressors come in a full range of volume and pressure output, and these can be mounted in units with a variety of power sources. The importance of a dependable diving compressor does not have to be stressed. Buy the best available for your purposes. Change oil as well as filter elements frequently. Detergent lubricating oils should never be used. Lubricating oils with a viscosity less than SAE 30 should never be used, either, and the oil should never be filled over the "full" mark, because of the possibility of oil contaminating the air supply.

Diving air compressors should be used for diving only, and never to power pneumatic tools or for any other purpose. Remember, when setting up a job, that if you will be using a recompression chamber, you will need twice as much air. Always be sure that your compressor is set up on the job so that the hazard of carbon-monoxide poisoning, either from your compressor drive engine or from any other source, is eliminated. Complete diving compressor units including drive engine, filters, volume tanks and gauges can be supplied by Bolstad-Lister, Advance Diving Equipment & Mfg. Co., Equitable Equipment Co., and many others.

RECOMPRESSION CHAMBERS

A recompression chamber is a major piece of diving equipment and all of them are very expensive. They are required by unions and insurance companies on all diving jobs in over 100 feet of water. The

Army Corps of Engineers requires a chamber on any of their projects in 40 feet of water or more.

Chambers come in a variety of models, but they are all basically alike in configuration and operation. Chambers should be kept clean and painted. The hatch dogs should be kept free and lubricated with silicone or some other flash-resistant lubricant. The O-rings or gaskets

Fig. 18 (Top) Perry Submarine Builders Recompression Chamber. (Bottom) A diver's air compressor. Photo: Advanced Diving Equipment & Mfg. Co., Inc.

should be frequently replaced, and the hatch seats, if not stainless steel, should also be kept protected with silicone lubricant. Leaking valves should immediately be replaced or repaired. Whenever possible, chambers should be placed in the shade, in a position where objects cannot be dropped on them. When the chamber is not in use, the hatches should be dogged closed to prevent their being damaged in rolling seas. All gauges are delicate instruments, and when not in use, they should be covered and protected from the elements. Chamber gauges should have their read-outs in feet of sea water only, to eliminate the possibility of misinterpretation and overpressurization of the diver. Each compartment in a chamber should have two gauges so that they can be checked against each other for accurate readings. Chamber gauges should be frequently calibrated. Discourage the interior use of the chamber as a storage area, and do not use the top of the chamber for coiling hoses, etc.

All combustibles must be kept out of and away from a recompression chamber. Never smoke in a chamber, either under pressure or when the hatches are open. Remember, when operating a chamber, that it must be vented frequently. Nothing should ever be placed on or against the ports of a recompression chamber. The bulbs used in chamber lights should never be greater than 25 watts and these should always be enclosed in pressure-proof globes. If internal lamp globes are cracked, or in any way questionable, electric power must be disconnected from the chamber and hand lights used for illumination. Diver's battery-operated hand lights, such as Dacor or Ike lights, are acceptable for this use.

Whenever you are inside a recompression chamber under pressure, be sure to keep your body, clothing and any loose material away from the inlet and exhaust openings. The exhaust opening is especially dangerous because the high differential pressure is capable of sucking a part of your body out of the opening, causing severe injury.

Scott Aviation Company is currently marketing a system which dumps the exhalations from the oxygen-demand mask outside of the chamber, thereby greatly reducing the danger of fire. Before a diving job is commenced, check out the chamber completely, and bring it down on a dry run. Whenever possible, hook the chamber up to an auxiliary air supply.

Chambers are manufactured by the Advanced Diving Equipment & Mfg. Co., Equitable Equipment Company, Bolstad-Lister, Perry Submarine Company, and the Southern Tank Company. Galliazi chambers are distributed in the U.S. by Underseas Industries, Inc.

SCUBA EQUIPMENT

My first encounter with compressed-air scuba-diving equipment was in Southern California in 1949. It was with one of the first Aqua-Lungs®, at that time distributed by the Sea Net Manufacturing Co., of Los Angeles. I was in the abalone business and 90 percent of my underwater experience prior to that time had been with full deep-sea diving gear.

Being an avid snorkeler, I enjoyed the sensation of long-duration swims under water, without the need for surfacing, but it was difficult to get tanks charged, and compared to hose diving, the time period under water with an Aqua-Lung® was very short. I considered the Aqua-Lung® to be an enjoyable but expensive toy, and I completely deprecated its potential as a working tool.

During the early 1950's a substantial part of my income was derived from wheel jobs for tugboats and fishing trawlers in Boston harbor. Another fairly frequent chore was cleaning the suction strainers for the many lobster companies that ringed the harbor. By 1955, my calls for this type of work had virtually stopped. Tube suckers were doing for $15.00 or $20.00 the jobs for which I charged $85.00. At that time I bought scuba gear, and have used, enjoyed and profited by it in many different commercial applications, but, except for the summertime, it did not become a complete tool for me until the introduction of the wet suit.

I found the Pirelli and Bell Aqua dry suits wholly unsatisfactory with scuba gear. Dressed out in enough woolens to keep warm in New England waters, a short swim would completely exhaust me, and I generally got soaking wet anyway. The Bell Aqua and Bailey dry suits of today are useful cold-water gear when used with the light-weight helmets, but if I am going to wear a dry suit, I prefer to dress out in complete deep-sea gear.

The ultimate in cold-water diving comfort is the Diurene, or hot-water diving suit, as supplied by Dick Long's Skin Divers Unlimited. The entire system includes a boiler or hot-water heater, a pump, and neoprene or insulated hose. Hot water is continuously pumped down to the diver, creating an envelope of comfortable warm water inside the diving suit. The special hot-water suit with perforated tubes running down the arms, legs, and back is not entirely necessary, and if you have a supply of hot water and extra hose, the end of the hose can be tucked under your wet-suit jacket with amazingly comfortable results. I have spent as long as 11 hours in 50-degree water wearing a hot-water suit. Hot water from any source can be utilized, but, of course, you must be careful not to pump excessively hot water down to the diver. The diver must have a shutoff valve or a by-pass valve at his suit or be able to pull out the hose easily if he should be accidentally supplied with uncomfortably hot water.

There are many wet-suit manufacturers and the brand is the diver's personal choice. I have found the Bailey and Imperial suits best for my purposes. An ideal combination is a ⅛-inch vest under full shoulder-covering pants, topped off with a jacket. Zippers greatly simplify getting in and out of your wet suit, but a suit with zippers is not as warm as one without them. A good fit is usually possible only with a custom-tailored suit, but a well-fitting one is indispensable to realize the full potential for thermal protection offered by a wet suit.

As mentioned previously, the introduction of the Aqua-Lung® and other scuba gear gradually took over a share of the limited available commercial work, primarily because a man could get under water and perform simple work with a minimum investment of about $250.00.

But scuba gear never has replaced surface-supplied hose gear in the commercial diving industry, mainly because of the working diver's need for direct telephone communication with the surface. Scuba gear has limited use in the offshore oil industry, but it is ideal in one particular application—inspecting the pontoon or stinger of a pipe-lay barge. This will be covered fully in Chapter 9. There are many good books on scuba diving, so I will not elaborate on the techniques involved. If you anticipate using scuba gear, I strongly recommend that you enroll in an accredited NAUI, NASOS, or YMCA course.

The greatest impact of scuba gear on the commercial diving industry was the development and marketing of the many accessories aimed at the incredibly large sport-diving field, but that have found excellent application in commercial diving. Most important among these, as already mentioned, is the wet suit. Other items designed for the sport diver, but widely used by commercial offshore divers, are swim fins, diving knives, hand lights, life vests, wrist compasses and depth gauges, regulators and small cylinders for bailout rigs, and the Calypso or Nikonos 35-millimeter underwater camera. The commercial offshore diving industry is also indebted to sport diving for the many excellent divers who began their professional careers as sport divers.

DIVER'S KIT

An old carpenter's saying is, "A man is only as good as his tools." Although I don't agree with this completely, I do accept the inference that you must have good tools to do good work. As a professional carpenter must have a kit of good sharp tools to perform his job properly, so must the professional diver have his kit and carry it with him on every job. Some diving companies supply all the diver's gear, except for his personal wet suit, and other companies want the diver to supply his mask, hose, telephone, etc. Regardless of what gear the contractor is supplying, a good diver will always carry a number of articles with him, in addition to his wet suit.

A good kit should consist of a complete wet suit, in excellent condition, suitable for the water temperature you will be working in; a can of wet-suit cement; at least a square foot of patching material and a roll of wet-suit seam stripping; a tube of silicone lubricant for the zippers; a good weight belt with a selection of weights; neat's-foot oil; at least two diving knives; a pocket sharpening stone for the knives; coveralls; a pair of galoshes; swim fins; a complete set of tables; a hand light with spare bulb and battery; gloves; spare hose fittings; hose-to-pipe fittings; a spare quick-release hook; spare single-pole terminals; a bailout bottle with regulator, or a life jacket; and a couple of rolls of tape.

If you have found that you work better in one type of mask or helmet than you do in another, by all means invest in that type of gear, whether you will be paid rental for it or not. You will profit from it in the long run, by doing a better job and thereby building a good reputation. If you do provide your own helmet or mask, carry a spare non-return valve and spare telephone speakers. If you furnish your own telephone, carry spare batteries and fuses. You should also carry a small tool kit with you containing a couple of small adjustable wrenches, pliers, and a screwdriver. If your helmet or mask requires special tools, such as Allen wrenches or a Phillips screwdriver to completely disassemble it, carry these items with you.

A 3- or 4-pound short-handled striking hammer is an invaluable tool under water. As mentioned earlier, a foxhole shovel or a large gardener's trowel is especially handy for small digging jobs. I carry a few extra items, in case I have to do any burning. These include a couple of pairs of rubber lineman's gloves, spare collets and washers for the Craftsweld torch, spare washers for the Swindell torch, a No. 4 welding lens, and a 4-inch stiff-bladed putty knife. These items will be discussed in Chapter 7. This is not really as much gear as it seems, and quite often any one of these items could mean the difference between doing a professional job or holding up the project until the item can be found.

Your wet suit, fins, galoshes, coveralls, etc., will fit into a surplus sea bag or parachute bag. It is a simple matter to build a suitable chest to carry the other items, or you may find one in the nearest surplus outlet. Add to this list any additional items you deem necessary to do a good job. When you report on the job with all of your equipment in good condition and ready to go, and with spares to cover normal contingencies, you carry the mark of the professional.

Be completely familiar with your equipment. When you go out on a new job, don't dive into the bunk and wait until it is your turn in the water. Check out all of your equipment and be sure that it is operating. Check out the chamber and the air compressor; check out the job. Anticipate the job you will have to do and round up the tools and supplies that may be needed. Buy the finest equipment available. A list of diving equipment manufacturers and distributors is in the Appendix.

Chapter 5

DECOMPRESSION AND TREATMENT TABLES FOR COMPRESSED-AIR DIVING

Man has been breath-hold diving since the beginning of time, and he has been diving out of bells or buckets with an entrapped air supply from before the time of Christ.

Surface-supplied diving as we know it began about 1800 with the development of air pumps. During the next 100 years or so, an incredible amount of useful work was accomplished under water by thousands of divers who, almost to a man, were completely without the slightest understanding of the physiological laws they were daily transgressing.

It was generally known that some divers, after ascending, were afflicted with excruciating pains, paralysis and often death, but the cause was ascribed to bad air, weak constitutions or a variety of superstitious beliefs. Some Greek divers ate honey by the bucket as a preventative against decompression sickness, and treatments for it ranged from hot baths to cold packs with liberal use of analgesics and morphine. Many men survived years of diving without crippling effects, and even today, we occasionally encounter divers who consistently ignore the tables and seem to survive.

Boyle observed that bubbles would form in the blood and tissues of test animals after reduction of atmospheric pressure as early as 1670. In 1857, Bucquoy proposed the bubble theory as the principal cause of caisson disease or diver's palsy, and he suggested that the probable prevention of this illness was slow decompression. In 1869, a recommended decompression rate was one minute per meter of depth. In 1875, the first recompression chamber was built to treat caisson workers on the Brooklyn Bridge project. In 1878, the first compression and decompression schedules were published for caisson workers. These schedules specified a period of 25 minutes to compress a worker to three atmospheres pressure over surface, a workshift of four hours, followed by a decompression time of 40 minutes, with the pressure reduced at a uniform rate.

The U.S. Navy extreme exposure tables specify a total ascent time of 283 minutes for a four-hour exposure at 100 feet, so we can speculate that either men ain't what they used to be, or the mortality rate was high for compressed-air workers in the old days.

As early as 1883, Dr. Paul Bert, a distinguished French physiologist, and one of the foremost contributors to the field of compressed-air medicine, clearly understood the process of nitrogen absorption by the tissues during exposure to high air pressures. He recognized the beneficial effects of oxygen breathing for combating caisson disease and recommended its use *after* an ascent from elevated pressures.

As an indication of the poor communications of those times and the consequent delayed dissemination of knowledge, the U.S. Navy, as recently as 1905, published a diving manual without any decompression tables. The only indication of the awareness of any physiological hazards was the admonition to descend at the rate of one fathom a minute and to ascend at the same rate. The manual also stated that, "As a rule, diving should not be carried on much below twenty fathoms, although experienced divers have gone as deep as thirty-four fathoms."

It is curious that today, in the light of present knowledge, "experience," "constitution," and "machismo" are in some quarters still considered to be qualities which allow certain individuals either to ignore or merely pay lip service to the inflexible laws of physics as interpreted by the decompression tables.

Our current air-decompression tables derive from the first stage decompression tables for divers, computed by the eminent physiologist, Professor J. S. Haldane, under the auspices of the British Admiralty.

The basis for these first tables was Haldane's postulation that the absolute pressure on a diver could be reduced by one-half without the formation of bubbles, and although this is not absolutely or universally true, especially in deep water or with mixed gases, the theory survived practical experiments to depths of 35 fathoms. The Admiralty accepted these tables in 1907 with a maximum depth limit of 204 feet. This limit was imposed more by the limited capacity of the hand pumps in use at that time than by any other factor.

Starting with the solid foundation of Haldane's theories and tables, the U.S. Navy, after tens of thousands of practical and experimental dives, developed the current Standard Decompression Tables for Compressed Air Diving to 190 feet.

All of the tables and much of the remaining text in this chapter have been excerpted from the *U.S. Navy Diving Manual.* These tables are recognized throughout the world as the best and most reliable available. Adhere to them strictly for all compressed-air diving. The tables and procedures have been developed to provide safety from the hazards of decompression sickness and oxygen toxicity. Use them for all dives where air is the breathing medium, regardless of the type of diving equipment used.

As you know, a quantity of nitrogen is taken up by the body during every dive. The amount depends on the depth and duration of the dive. If the amount of nitrogen absorbed by the body during a dive exceeds a certain amount, this nitrogen will come out of solution and form bubbles in the blood and tissues. The formation of these bubbles is accompanied by pain and other symptoms and the condition, called decompression sickness, is dangerous and sometimes fatal. The rate at which various tissues absorb and give off nitrogen at different depths and for different times of exposure has been calculated within reasonable limits of accuracy, and these tables are the result of those calculations. The elimination of nitrogen from the body after a dive before bubbles are formed is called decompression. Decompression is accomplished by holding the diver at a series of diminishing depths, called decompression stops, occurring in ten-foot increments, for varying periods of time, as clearly specified in the tables.

U.S. NAVY STANDARD AIR DECOMPRESSION TABLE

The depths as listed in the first column represent the deepest depth achieved during the dive. They are listed in increments of ten feet, from 40 to 190 feet. For fractional depths, more than the listed ten-foot increments, use the next deeper table. For instance, for a dive to 42 feet, you would use the 50-foot table. For a dive to 98 feet, you would use the 100-foot table, and so on.

The second column lists the bottom times for each depth. Bottom time represents the time of the dive, measured from the time the diver leaves the surface at the beginning of his dive until he leaves the bottom at the start of his ascent.

The decompression schedule to use for any dive is that schedule that shows the deepest depth achieved during the dive and the full time of the dive, as explained above. If the bottom time falls between any two of the listed times in the table, use the next greater time listed.

Column 3, Time to First Stop, indicates the time it should take to pull the diver up to his first decompression stop. This is at a standard rate of 60 feet a minute for all depths and times listed in this table. If the rate of ascent for your particular dive turns out to be slower than that listed in the tables, and this delay occurs below fifty feet, add this extra time to the bottom time and decompress according to this new total bottom time.

For instance, for a dive to 80 feet for 80 minutes, it should take one minute to get the diver to his first stop at 20 feet. If the diver's hose became fouled, and it took him two minutes to clear it, at a depth of 60 feet, this extra two minutes would be added on to his bottom time. This would give him a new total bottom time of 82 minutes at 80 feet, and he would then have to be decompressed under the 90-minute schedule. If the delay in the ascent takes place above 50 feet, the extra time is added on to his first stop. In the above dive, if the diver spent the two minutes clearing his hose at the 40-foot level, his first stop at 20 feet would have to be increased from two to four minutes.

In cases where the ascent rate is exceeded, that is if the diver is pulled up in less time than that listed in the tables, the time should be made up in a stop ten feet deeper than the first stop listed in the tables. For instance, for a 50-minute dive to 150 feet, it should take two minutes to pull the diver up to his first stop at 30 feet. If he should inadvertently be pulled up in only one minute, the additional minute should be spent at a stop at 40 feet.

For dives in extremely cold water, or for dives where the diver has been working exceptionally hard, use the next deeper and next longer schedule. For instance, if the dive was to 130 feet for 40 minutes, decompress the diver on the schedule for a dive to 140 feet for 50 minutes.

You, as the diver, are the best judge of how cold you were, or how hard you worked. Do not be reluctant to tell your tender or your supervisor that you would like to be decompressed to a longer schedule. A few extra minutes in the water is always preferable to a long treatment for decompression sickness.

The next wide column in the table lists the depths of the decompression stops, from fifty to ten feet. The diver's chest should be kept at the depth of the stop. During rough weather, when the diver cannot be maintained at an accurate decompression depth because of waves and the rolling of the diving vessel, keep the diver deep so that his shallowest depth in the trough of the waves or at the top of the roll is always equal to or a few feet deeper than the listed decompression stop.

Some type of stage or hang-off bar should always be provided for the diver, especially during long decompressions. This is not a matter of the diver's comfort alone. The exertion and muscle tension required by the diver to grip a descending line for long periods of time can reduce the circulation to those muscles used, and cause bubbles to form. In addition to this, the pain caused by strained muscles can easily be confused with decompression sickness, and since all symptoms, regardless of how slight, must be treated, avoid the above possibility by keeping the diver comfortable and relaxed during his decompression stops. The ascent rate between stops is not critical for stops of 50 feet or less. Begin timing each stop when the diver's chest reaches the specified depth, and when the time for that depth, as listed in the tables, has expired, pull him up to the next stop. The tender should always inform the diver before lifting him to the next stop.

The total ascent column lists the time it should take to bring a diver from the bottom to the surface, including his decompression stops. The times are given in minutes and seconds, but in my opinion, the next longest minute is certainly close enough.

The last column lists the repetitive group designation for the time and depth of the particular dive. This letter designation is explained in the following table.

REPETITIVE DIVES

After any dive, even one for which you have been thoroughly decompressed, your body still contains an amount of nitrogen over and above what it normally carries at surface or sea-level atmospheric pressure. This is called "residual nitrogen," and it will take your body 12 hours to completely eliminate it and return to its normal surface tissue tension of nitrogen. This residual nitrogen has to be taken into account when figuring out the decompression for any successive dives. Any dive performed within 12 hours of a previous dive is a repetitive dive and must be made in accordance with the repetitive dive tables.

After the first dive, the diver notes the repetitive group designation letter listed in the far right-hand column of his decompression schedule in the Navy Standard Air Decompression Table. This letter designates a value for the amount of residual nitrogen left in his body when he reaches the surface. After a no-decompression dive, the diver can ascertain his repetitive group designation from the No-Decompression Limits and Repetitive Group Designation Table for No-Decompression Dives schedule. It is important to do this if repetitive dives are to be made, because the body will contain residual nitrogen, even after a no-decompression dive.

To determine the decompression necessary for a repetitive dive, you need the Standard Air Decompression Table, the Surface Interval Credit Table, and the Repetitive Dive Timetable.

The Surface Interval Credit Table is the real reason for the success and efficiency of the repetitive dive system. The diver continues to lose nitrogen while he is on the surface until he is completely desaturated. This desaturation requires 12 hours. To provide efficient decompression instructions, it is necessary to know the amount of nitrogen remaining in the tissues at the time a repetitive dive commences. The Surface Interval Credit Table provides that information.

The repetitive groups are measuring units, reflecting a value for the amount of residual nitrogen still contained in the tissues. In this table, the loss of nitrogen with the increasing length of the surface interval is shown in the change from one group to another. Details and an example of its use are given directly in the table. The Repetitive Dive Timetable lists the number of minutes at each depth that will build up the nitrogen partial pressure represented by each repetitive group. With the diver's known repetitive group designation, the system gives an arbitrary bottom time (the residual nitrogen time), that the diver must assume he has already completed, before he starts his repetitive dive. This arbitrary bottom time and the actual bottom time of the repetitive dive are added together and the decompression for the repetitive dive must then be according to the schedule shown in the Standard Air Decompression Table for the total of these combined times at the depth of the repetitive dive. Details and an example of its use are given directly on the table.

TABLE 1-10 FROM U.S. NAVY DIVING MANUAL

(This table is to be used only if no dives were made during the preceding 12 hours. If more than one dive is made during any 12-hour period, the decompression for the second and all succeeding dives must be calculated from the repetitive dive tables.)

U.S. Navy Standard Air Decompression Table

Depth (feet)	Bottom time (min)	Time to first stop (min:sec)	Decompression stops (feet)					Total ascent (min:sec)	Repetitive group
			50	40	30	20	10		
40	200						0	0:40	(*)
	210	0:30					2	2:40	N
	230	0:30					7	7:40	N
	250	0:30					11	11:40	O
	270	0:30					15	15:40	O
	300	0:30					19	19:40	Z
50	100						0	0:50	(*)
	110	0:40					3	3:50	L
	120	0:40					5	5:50	M
	140	0:40					10	10:50	M
	160	0:40					21	21:50	N
	180	0:40					29	29:50	O
	200	0:40					35	35:50	O
	220	0:40					40	40:50	Z
	240	0:40					47	47:50	Z

60	60							0	1:00	(*)
	70	0:50						2	3:00	K
	80	0:50						7	8:00	L
	100	0:50						14	15:00	M
	120	0:50						26	27:00	N
	140	0:50						39	40:00	O
	160	0:50						48	49:00	Z
	180	0:50						56	57:00	Z
	200	0:40					1	69	71:00	Z
70	50							0	1:10	(*)
	60	1:00						8	9:10	K
	70	1:00						14	15:10	L
	80	1:00						18	19:10	M
	90	1:00						23	24:10	N
	100	1:00						33	34:10	N
	110	0:50					2	41	44:10	O
	120	0:50					4	47	52:10	O
	130	0:50					6	52	59:10	O
	140	0:50					8	56	65:10	Z
	150	0:50					9	61	71:10	Z
	160	0:50					13	72	86:10	Z
	170	0:50					19	79	99:10	Z
80	40							0	1:20	(*)
	50	1:10						10	11:20	K
	60	1:10						17	18:20	L
	70	1:10						23	24:20	M
	80	1:00					2	31	34:20	N
	90	1:00					7	39	47:20	N
	100	1:00					11	46	58:20	O
	110	1:00					13	53	67:20	O
	120	1:00					17	56	74:20	Z
	130	1:00					19	63	83:20	Z
	140	1:00					26	69	96:20	Z
	150	1:00					32	77	110:20	Z

TABLE 1-10 (Cont.)

Depth (feet)	Bottom time (min)	Time to first stop (min:sec)	Decompression stops (feet)					Total ascent (min:sec)	Repetitive group
			50	40	30	20	10		
90	30						0	1:30	(*)
	40	1:20					7	8:30	J
	50	1:20					18	19:30	L
	60	1:20					25	26:30	M
	70	1:10				7	30	38:30	N
	80	1:10				13	40	54:30	N
	90	1:10				18	48	67:30	O
	100	1:10				21	54	76:30	Z
	110	1:10				24	61	86:30	Z
	120	1:10				32	68	101:30	Z
	130	1:00			5	36	74	116:30	Z
100	25						0	1:40	(*)
	30	1:30					3	4:40	I
	40	1:30					15	16:40	K
	50	1:20				2	24	27:40	L
	60	1:20				9	28	38:40	N
	70	1:20				17	39	57:40	O
	80	1:20				23	48	72:40	O
	90	1:10			3	23	57	84:40	Z
	100	1:10			7	23	66	97:40	Z
	110	1:10			10	34	72	117:40	Z
	120	1:10			12	41	78	132:40	Z
110	20						0	1:50	(*)
	25	1:40					3	4:50	H
	30	1:40					7	8:50	J
	40	1:30				2	21	24:50	L
	50	1:30				8	26	35:50	M
	60	1:30				18	36	55:50	N
	70	1:20			1	23	48	73:50	O
	80	1:20			7	23	57	88:50	Z

	90	1:20	--------	--------	12	30	64	107:50	Z
	100	1:20	--------	--------	15	37	72	125:50	Z
120--------------	15	--------	--------	--------	--------	--------	0	2:00	(*)
	20	1:50	--------	--------	--------	--------	2	4:00	H
	25	1:50	--------	--------	--------	--------	6	8:00	I
	30	1:50	--------	--------	--------	--------	14	16:00	J
	40	1:40	--------	--------	--------	5	25	32:00	L
	50	1:40	--------	--------	--------	15	31	48:00	N
	60	1:30	--------	--------	2	22	45	71:00	O
	70	1:30	--------	--------	9	23	55	89:00	O
	80	1:30	--------	--------	15	27	63	107:00	Z
	90	1:30	--------	--------	19	37	74	132:00	Z
	100	1:30	--------	--------	23	45	80	150:00	Z
130--------------	10	--------	--------	--------	--------	--------	0	2:10	(*)
	15	2:00	--------	--------	--------	--------	1	3:10	F
	20	2:00	--------	--------	--------	--------	4	6:10	H
	25	2:00	--------	--------	--------	--------	10	12:10	J
	30	1:50	--------	--------	--------	3	18	23:10	M
	40	1:50	--------	--------	--------	10	25	37:10	N
	50	1:40	--------	--------	3	21	37	63:10	O
	60	1:40	--------	--------	9	23	52	86:10	Z
	70	1:40	--------	--------	16	24	61	103:10	Z
	80	1:30	--------	3	19	35	72	131:10	Z
	90	1:30	--------	8	19	45	80	154:10	Z
140--------------	10	--------	--------	--------	--------	--------	0	2:20	(*)
	15	2:10	--------	--------	--------	--------	2	4:20	G
	20	2:10	--------	--------	--------	--------	6	8:20	I
	25	2:00	--------	--------	--------	2	14	18:20	J
	30	2:00	--------	--------	--------	5	21	28:20	K
	40	1:50	--------	--------	2	16	26	46:20	N
	50	1:50	--------	--------	6	24	44	76:20	O
	60	1:50	--------	--------	16	23	56	97:20	Z
	70	1:40	--------	4	19	32	68	125:20	Z
	80	1:40	--------	10	23	41	79	155:20	Z

TABLE 1-10 (Cont.)

Depth (feet)	Bottom time (min)	Time to first stop (min:sec)	Decompression stops (feet)					Total ascent (min:sec)	Repetitive group
			50	40	30	20	10		
150	5						0	2:30	C
	10	2:20					1	3:30	E
	15	2:20					3	5:30	G
	20	2:10				2	7	11:30	H
	25	2:10				4	17	23:30	K
	30	2:10				8	24	34:30	L
	40	2:00			5	19	33	59:30	N
	50	2:00			12	23	51	88:30	O
	60	1:50		3	19	26	62	112:30	Z
	70	1:50		11	19	39	75	146:30	Z
	80	1:40	1	17	19	50	84	173:30	Z
160	5						0	2:40	D
	10	2:30					1	3:40	F
	15	2:20				1	4	7:40	H
	20	2:20				3	11	16:40	J
	25	2:20				7	20	29:40	K
	30	2:10			2	11	25	40:40	M
	40	2:10			7	23	39	71:40	N
	50	2:00		2	16	23	55	98:40	Z
	60	2:00		9	19	33	69	132:40	Z
	70	1:50	1	17	22	44	80	166:40	Z
170	5						0	2:50	D
	10	2:40					2	4:50	F
	15	2:30				2	5	9:50	H
	20	2:30				4	15	21:50	J
	25	2:20			2	7	23	34:50	L
	30	2:20			4	13	26	45:50	M

	40	2:10	----	1	10	23	45	81:50	O
	50	2:10	----	5	18	23	61	109:50	Z
	60	2:00	2	15	22	37	74	152:50	Z
	70	2:00	8	17	19	51	86	183:50	Z
180	5	----	----	----	----	----	0	3:00	D
	10	2:50	----	----	----	----	3	6:00	F
	15	2:40	----	----	----	3	6	12:00	I
	20	2:30	----	----	1	5	17	26:00	K
	25	2:30	----	----	3	10	24	40:00	L
	30	2:30	----	----	6	17	27	53:00	N
	40	2:20	----	3	14	23	50	93:00	O
	50	2:10	2	9	19	30	65	128:00	Z
	60	2:10	5	16	19	44	81	168:00	Z
190	5	----	----	----	----	----	0	3:10	D
	10	2:50	----	----	----	1	3	7:10	G
	15	2:50	----	----	----	4	7	14:10	I
	20	2:40	----	----	2	6	20	31:10	K
	25	2:40	----	----	5	11	25	44:10	M
	30	2:30	----	1	8	19	32	63:10	N
	40	2:30	----	8	14	23	55	103:10	O
	50	2:20	4	13	22	33	72	147:10	Z
	60	2:20	10	17	19	50	84	183:10	Z

*See table 1–11 for repetitive groups in no-decompression dives.

There is one exception to the table. It occasionally occurs when the repetitive dive is to the same depth as, or to a greater depth than the initial dives, and when the surface interval is short. Because of the necessity to account for the greatest exposure within a group, the arbitrary bottom time assigned may be greater than the sum of the actual bottom times of the previous dives. In such a case (if the repetitive dive is to the same depth as, or greater depth than, the previous dive or dives), add the actual bottom time of the previous dive to the actual bottom time of the repetitive dive, and decompress to this deeper schedule.

Whenever in doubt about a repetitive dive, or if you should find yourself without the necessary tables, add the times of the two dives together and decompress to the schedule for the deepest dive. When one repetitive dive is to be followed by another, the procedure for selecting the proper decompression schedule for the first repetitive dive is repeated. That is, at the end of the repetitive dive, the repetitive group designation shown at the far right of the Standard Air Decompression Table at the schedule for that dive is noted, and the second or next repetitive dive is computed from this letter designation.

Personally, I am not convinced of the safety of making more than one repetitive dive within any 12-hour period, even when the tables are strictly adhered to. Though I am at a loss to give you a sound physiological explanation for this opinion, which I admit is strictly visceral, I can state, nevertheless, that the practice of continuous repetitive diving should be avoided, if for no other reason than because it creates a situation where the divers on the job are not receiving the amount of rest or sleep they require for performing their arduous work.

Repetitive dives should not be made with a surface interval of less than ten minutes. If the necessity for such a dive does arise, the total bottom times of the two dives will have to be added together and decompression made to the depth of the deepest dive.

The U.S. Navy Standard Air Decompression Table for Exceptional Exposures includes only schedules of decompression for exceptional or emergency cases. Great demands are imposed upon the diver's endurance by emergencies necessitating use of this table. Therefore, complete assurance of success of the decompression schedules, detailed in this table, is impossible. Plan all routine dives according to the schedules as set forth in the U.S. Navy Standard Air Decompression Tables and resort to the Extreme Exposure Tables only in the case of dire emergencies, or when a long exposure is unavoidable, as in the case of a diver fouled or trapped on the bottom. Repetitive dives should never be made after an exceptional exposure. A diver so exposed should not be permitted to dive again for at least 24 hours.

SURFACE DECOMPRESSION

The practice of surface decompression using a recompression chamber began and was developed from the operational necessity of getting divers out of the water before their proper decompression was completed, because of accident or severe weather. Surface decompression was routinely used by British divers as early as 1914, during the salvage of the *Empress of Ireland,* in water depths of 170 to 190 feet. The early British practice of surface decompression differed from the routines currently used in the United States in that the diver was brought rapidly to the surface from whatever depth he was diving in, and then put immediately into a recompression chamber. He was then recompressed to the depth of his dive and kept there for five minutes. After this, he was brought up to his first stop, and then decompressed according to the standard table.

The first large-scale operational use of surface decompression procedures by the U.S. Navy was during the salvage of the submarine S-51, in 1925, and the practice was resorted to because of the extremely cold water and fast currents at the wreck.

The Surface Decompression Table Using Air is used in any situation where the normal decompression in the water is impracticable, and when oxygen is not available. This table is longer than the Standard Air Table, because it generally repeats the 20-foot water stop in the chamber. Its principal advantage is the comfort and security of the diver during a long decompression. The ascent rate for this table is 60 feet per minute, in the water, between stops, and when surfacing from the chamber.

Surface decompression using oxygen holds many advantages for the diver, the principal one being much shorter decompression schedules for most dives. Surface decompression using oxygen is now a standard practice on almost every commercial diving job in water depths of over 100 feet.

The Navy Surface Decompression Table Using Oxygen has not been recently recomputed, and there are some discrepancies in it concerning the limits of allowable exposure. However, the Navy considers it to be safe in its present form. The one disadvantage in this table is that it only goes down to 170 feet, and many commercial air dives are conducted to 200 feet. Most divers and diving supervisors, in the absence of published oxygen decompression tables beyond 170 feet, will use the much longer Standard Air Decompression Tables, and decompress in the water.

The British Navy uses oxygen for surface decompression for compressed air dives down to 200 feet, and perhaps in the near future, the U.S. Navy will extend their O_2 tables to this depth.

TABLE 1-11 FROM U.S. NAVY DIVING MANUAL

(FORMERLY TABLE 1–6, 1963 DIVING MANUAL)

No-decompression limits and repetitive group designation table for no-decompression air dives

Depth (feet)	No-decompression limits (min)	Repetitive groups (air dives)														
		A	B	C	D	E	F	G	H	I	J	K	L	M	N	O
10	----	60	120	210	300	----	----	----	----	----	----	----	----	----	----	----
15	----	35	70	110	160	225	350	----	----	----	----	----	----	----	----	----
20	----	25	50	75	100	135	180	240	325	----	----	----	----	----	----	----
25	----	20	35	55	75	100	125	160	195	245	315	----	----	----	----	----
30	----	15	30	45	60	75	95	120	145	170	205	250	310	----	----	----
35	310	5	15	25	40	50	60	80	100	120	140	160	190	220	270	310
40	200	5	15	25	30	40	50	70	80	100	110	130	150	170	200	----
50	100	----	10	15	25	30	40	50	60	70	80	90	100	----	----	----
60	60	----	10	15	20	25	30	40	50	55	60	----	----	----	----	----
70	50	----	5	10	15	20	30	35	40	45	50	----	----	----	----	----
80	40	----	5	10	15	20	25	30	35	40	----	----	----	----	----	----
90	30	----	5	10	12	15	20	25	30	----	----	----	----	----	----	----
100	25	----	5	7	10	15	20	22	25	----	----	----	----	----	----	----
110	20	----	----	5	10	13	15	20	----	----	----	----	----	----	----	----
120	15	----	----	5	10	12	15	----	----	----	----	----	----	----	----	----
130	10	----	----	5	8	10	----	----	----	----	----	----	----	----	----	----
140	10	----	----	5	7	10	----	----	----	----	----	----	----	----	----	----
150	5	----	----	5	----	----	----	----	----	----	----	----	----	----	----	----
160	5	----	----	----	5	----	----	----	----	----	----	----	----	----	----	----

170	5	-----	-----	-----	5	-----	-----	-----	-----	-----	-----	-----	-----	-----	-----	-----
180	5	-----	-----	-----	5	-----	-----	-----	-----	-----	-----	-----	-----	-----	-----	-----
190	5	-----	-----	-----	5	-----	-----	-----	-----	-----	-----	-----	-----	-----	-----	-----

Instructions for Use

I. No-decompression limits:

This column shows at various depths greater than 30 feet the allowable diving times (in minutes) which permit surfacing directly at 60 feet a minute with no decompression stops. Longer exposure times require the use of the Standard Air Decompression Table (table 1–10).

II. Repetitive group designation table:

The tabulated exposure times (or bottom times) are in minutes. The times at the various depths in each vertical column are the maximum exposures during which a diver will remain within the group listed at the head of the column.

To find the repetitive group designation at surfacing for dives involving exposures up to and including the no-decompression limits: Enter the table on the ***exact or next greater depth*** than that to which exposed and select the listed exposure time ***exact or next greater*** than the actual exposure time. The repetitive group designation is indicated by the letter at the head of the vertical column where the selected exposure time is listed.

For example: A dive was to 32 feet for 45 minutes. Enter the table along the 35-foot-depth line since it is next greater than 32 feet. The table shows that since group D is left after 40 minutes' exposure and group E after 50 minutes, group E (at the head of the column where the 50-minute exposure is listed) is the proper selection.

Exposure times for depths less than 40 feet are listed only up to approximately 5 hours since this is considered to be beyond field requirements for this table.

TABLE 1-12 FROM U.S. NAVY DIVING MANUAL

(FORMERLY TABLE 1-7, 1963 DIVING MANUAL)

Surface Interval Credit Table for air decompression dives

[Repetitive group at the end of the surface interval (air dive)]

Repetitive group at the beginning of	Z	O	N	M	L	K	J	I	H	G	F	E	D	C	B	A
Z	0:10 0:22	0:23 0:34	0:35 0:48	0:49 1:02	1:03 1:18	1:19 1:36	1:37 1:55	1:56 2:17	2:18 2:42	2:43 3:10	3:11 3:45	3:46 4:29	4:30 5:27	5:28 6:56	6:57 10:05	10:00 12:00*
O		0:10 0:23	0:24 0:36	0:37 0:51	0:52 1:07	1:08 1:24	1:25 1:43	1:44 2:04	2:05 2:29	2:30 2:59	3:00 3:33	3:34 4:17	4:18 5:16	5:17 6:44	6:45 9:54	9:55 12:00*
N			0:10 0:24	0:25 0:39	0:40 0:54	0:55 1:11	1:12 1:30	1:31 1:53	1:54 2:18	2:19 2:47	2:48 3:22	3:23 4:04	4:05 5:03	5:04 6:32	6:33 9:43	9:44 12:00*
M				0:10 0:25	0:26 0:42	0:43 0:59	1:00 1:18	1:19 1:39	1:40 2:05	2:06 2:34	2:35 3:08	3:09 3:52	3:53 4:49	4:50 6:18	6:19 9:28	9:29 12:00*
L					0:10 0:26	0:27 0:45	0:46 1:04	1:05 1:25	1:26 1:49	1:50 2:19	2:20 2:53	2:54 3:36	3:37 4:35	4:36 6:02	6:03 9:12	9:13 12:00*
K						0:10 0:28	0:29 0:49	0:50 1:11	1:12 1:35	1:36 2:03	2:04 2:38	2:39 3:21	3:22 4:19	4:20 5:48	5:49 8:58	8:59 12:00*
J							0:10 0:31	0:32 0:54	0:55 1:19	1:20 1:47	1:48 2:20	2:21 3:04	3:05 4:02	4:03 5:40	5:41 8:40	8:41 12:00*
I								0:10 0:33	0:34 0:59	1:00 1:29	1:30 2:02	2:03 2:44	2:45 3:43	3:44 5:12	5:13 8:21	8:22 12:00*
H									0:10 0:36	0:37 1:06	1:07 1:41	1:42 2:23	2:24 3:20	3:21 4:49	4:50 7:59	8:00 12:00*

	G	0:10 0:40	0:41 1:15	1:16 1:59	2:00 2:58	2:59 4:25	4:26 7:35	7:36 12:00*
the surface interval from previous dive		F	0:10 0:45	0:46 1:29	1:30 2:28	2:29 3:57	3:58 7:05	7:06 12:00*
			E	0:10 0:54	0:55 1:57	1:58 3:22	3:23 6:32	6:33 12:00*
				D	0:10 1:09	1:10 2:38	2:39 5:48	5:49 12:00*
					C	0:10 1:39	1:40 2:49	2:50 12:00*
						B	0:10 2:10	2:11 12:00*
							A	0:10 12:00*

Instructions for Use

Surface interval time in the table is in *hours* and *minutes* (7:59 means 7 hours and 59 minutes). The surface interval must be at least 10 minutes.

Find the *repetitive group designation letter* (from the previous dive schedule) on the diagonal slope. Enter the table horizontally to select the surface interval time that is exactly between the actual surface interval times shown. The repetitive group designation for the *end* of the surface interval is at the head of the vertical column where the selected surface interval time is listed. For example, a previous dive was to 110 feet for 30 minutes. The diver remains on the surface 1 hour and 30 minutes and wishes to find the new repetitive group designation: The repetitive group from the last column of the 110/30 schedule in the Standard Air Decompression Tables is "J." Enter the surface interval credit table along the horizontal line labeled "J." The 1-hour-and-30-minute surface interval lies between the times 1:20 and 1:47. Therefore, the diver has lost sufficient inert gas to place him in group "G" (at the head of the vertical column selected).

*NOTE.—Dives following surface intervals of *more* than 12 hours are not considered repetitive dives. *Actual* bottom times in the Standard Air Decompression Tables may be used in computing decompression for such dives.

TABLE 1-13 FROM U.S. NAVY DIVING MANUAL

(FORMERLY TABLE 1–8, 1963 DIVING MANUAL)

Repetitive dive timetable for air dives

Repetitive groups	Repetitive dive depth (ft) (air dives)															
	40	50	60	70	80	90	100	110	120	130	140	150	160	170	180	190
A	7	6	5	4	4	3	3	3	3	3	2	2	2	2	2	2
B	17	13	11	9	8	7	7	6	6	6	5	5	4	4	4	4
C	25	21	17	15	13	11	10	10	9	8	7	7	6	6	6	6
D	37	29	24	20	18	16	14	13	12	11	10	9	9	8	8	8
E	49	38	30	26	23	20	18	16	15	13	12	12	11	10	10	10
F	61	47	36	31	28	24	22	20	18	16	15	14	13	13	12	11
G	73	56	44	37	32	29	26	24	21	19	18	17	16	15	14	13
H	87	66	52	43	38	33	30	27	25	22	20	19	18	17	16	15
I	101	76	61	50	43	38	34	31	28	25	23	22	20	19	18	17
J	116	87	70	57	48	43	38	34	32	28	26	24	23	22	20	19
K	138	99	79	64	54	47	43	38	35	31	29	27	26	24	22	21
L	161	111	88	72	61	53	48	42	39	35	32	30	28	26	25	24
M	187	124	97	80	68	58	52	47	43	38	35	32	31	29	27	26
N	213	142	107	87	73	64	57	51	46	40	38	35	33	31	29	28
O	241	160	117	96	80	70	62	55	50	44	40	38	36	34	31	30
Z	257	169	122	100	84	73	64	57	52	46	42	40	37	35	32	31

The bottom times listed in this table are called "residual nitrogen times" and are the times a diver is to consider he has *already* spent on bottom when he *starts* a repetitive dive to a specific depth. They are in minutes.

Enter the table horizontally with the repetitive group designation from the Surface Interval Credit Table. The time in each vertical column is the number of minutes that would be required (at the depth listed at the head of the column) to saturate to the particular group.

For example: The final group designation from the Surface Interval Credit Table, on the basis of a previous dive and surface interval, is "H." To plan a dive to 110 feet, determine the residual nitrogen time for this depth required by the repetitive group designation: Enter this table along the horizontal line labeled "H." The table shows that one must *start* a dive to 110 feet as though he had already been on the bottom for 27 minutes. This information can then be applied to the Standard Air Decompression Table or No-Decompression Table in a number of ways:

(1) Assuming a diver is going to finish a job and take whatever decompression is required, he must add 27 minutes to his actual bottom time and be prepared to take decompression according to the 110-foot schedules for the sum or equivalent single dive time.

(2) Assuming one wishes to make a quick inspection dive for the minimum decompression, he will decompress according to the 110/30 schedule for a dive of 3 minutes or less (27+3=30). For a dive of over 3 minutes but less than 13, he will decompress according to the 110/40 schedule (27+13=40).

(3) Assuming that one does not want to exceed the 110/50 schedule and the amount of decompression it requires, he will have to start ascent before 23 minutes of actual bottom time (50−27=23).

(4) Assuming that a diver has air for approximately 45 minutes bottom time and decompression stops, the possible dives can be computed: A dive of 13 minutes will require 23 minutes of decompression (110/40 schedule), for a total submerged time of 36 minutes. A dive of 13 to 23 minutes will require 34 minutes of decompression (110/50 schedule), for a total submerged time of 47 to 57 minutes. Therefore, to be safe, the diver will have to start ascent before 13 minutes or a standby air source will have to be provided.

TABLE 1-14 FROM U.S. NAVY DIVING MANUAL

(FORMERLY TABLE 1-9, 1963 DIVING MANUAL)

U.S. Navy Standard Air Decompression Table for exceptional exposures

Depth (ft)	Bottom time (min)	Time to first stop (min:sec)	Decompression stops (feet)													Total ascent time (min:sec)
			130	120	110	100	90	80	70	60	50	40	30	20	10	
40	360	0:30													23	23:40
	480	0:30													41	41:40
	720	0:30													69	69:40
60	240	0:40												2	79	82:00
	360	0:40												20	119	140:00
	480	0:40												44	148	193:00
	720	0:40												78	187	266:00
80	180	1:00												35	85	121:20
	240	0:50											6	52	120	179:20
	360	0:50											29	90	160	280:20
	480	0:50											59	107	187	354:20
	720	0:40										17	108	142	187	455:20
100	180	1:00										1	29	53	118	202:40
	240	1:00										14	42	84	142	283:40
	360	0:50									2	42	73	111	187	416:40
	480	0:50									21	61	91	142	187	503:40
	720	0:50									55	106	122	142	187	613:40
120	120	1:20										10	19	47	98	176:00
	180	1:10									5	27	37	76	137	284:00
	240	1:10									23	35	60	97	179	396:00
	360	1:00								18	45	64	93	142	187	551:00
	480	0:50							3	41	64	93	122	142	187	654:00
	720	0:50							32	74	100	114	122	142	187	773:00

140	90	1:30									2	14	18	42	88	166:20
	120	1:30									12	14	36	56	120	240:20
	180	1:20								10	26	32	54	94	168	386:20
	240	1:10							8	28	34	50	78	124	187	511:20
	360	1:00						9	32	42	64	84	122	142	187	684:20
	480	1:00						31	44	59	100	114	122	142	187	801:20
	720	0:50					16	56	88	97	100	114	122	142	187	924:20
170	90	1:50								12	12	14	34	52	120	246:50
	120	1:30						2	10	12	18	32	42	82	156	356:50
	180	1:20					4	10	22	28	34	50	78	120	187	535:50
	240	1:20					18	24	30	42	50	70	116	142	187	681:50
	360	1:10				22	34	40	52	60	98	114	122	142	187	873:50
	480	1:00			14	40	42	56	91	97	100	114	122	142	187	1007:50
200	5	3:10													1	4:20
	10	3:00												1	4	8:20
	15	2:50											1	4	10	18:20
	20	2:50											3	7	27	40:20
	25	2:50											7	14	25	49:20
	30	2:40										2	9	22	37	73:20
	40	2:30									2	8	17	23	59	112:20
	50	2:30									6	16	22	39	75	161:20
	60	2:20								2	13	17	24	51	89	199:20
	90	1:50					1	10	10	12	12	30	38	74	134	324:20
	120	1:40				6	10	10	10	24	28	40	64	98	180	473:20
	180	1:20		1	10	10	18	24	24	42	48	70	106	142	187	685:20
	240	1:20		6	20	24	24	36	42	54	68	114	122	142	187	842:20
	360	1:10	12	22	36	40	44	56	82	98	100	114	122	142	187	1058:20
210	5	3:20													1	4:30
	10	3:10												2	4	9:30
	15	3:00											1	5	13	22:30
	20	3:00											4	10	23	40:30
	25	2:50										2	7	17	27	56:30
	30	2:50										4	9	24	41	81:30
	40	2:40									4	9	19	26	63	124:30
	50	2:30								1	9	17	19	45	80	174:30

TABLE 1-14 (Cont.)

Depth (ft)	Bottom time (min)	Time to first stop (min:sec)	Decompression stops (feet)													Total ascent time (min:sec)
			130	120	110	100	90	80	70	60	50	40	30	20	10	
220	5	3:30													2	5:40
	10	3:20												2	5	10:40
	15	3:10											2	5	16	26:40
	20	3:00										1	3	11	24	42:40
	25	3:00										3	8	19	33	66:40
	30	2:50									1	7	10	23	47	91:40
	40	2:50									6	12	22	29	68	140:40
	50	2:40								3	12	17	18	51	86	190:40
230	5	3:40													2	5:50
	10	3:20											1	2	6	12:50
	15	3:20											3	6	18	30:50
	20	3:10										2	5	12	26	48:50
	25	3:10										4	8	22	37	74:50
	30	3:00									2	8	12	23	51	99:50
	40	2:50								1	7	15	22	34	74	156:50
	50	2:50								5	14	16	24	51	89	202:50
240	5	3:50													2	6:00
	10	3:30											1	3	6	14:00
	15	3:30											4	6	21	35:00
	20	3:20										3	6	15	25	53:00
	25	3:10									1	4	9	24	40	82:00
	30	3:10									4	8	15	22	56	109:00
	40	3:00								3	7	17	22	39	75	167:00
	50	2:50							1	8	15	16	29	51	94	218:00
250	5	3:50												1	2	7:10
	10	3:40											1	4	7	16:10
	15	3:30										1	4	7	22	38:10
	20	3:30										4	7	17	27	59:10
	25	3:20									2	7	10	24	45	92:10
	30	3:20									6	7	17	23	59	116:10

	40	3:10								5	9	17	19	45	79	178:10
	60	2:40					4	10	10	10	12	22	36	64	126	298:10
	90	2:10		8	10	10	10	10	10	28	28	44	68	98	186	514:10
260	5	4:00												1	2	7:20
	10	3:50											2	4	9	19:20
	15	3:40										2	4	10	22	42:20
	20	3:30									1	4	7	20	31	67:20
	25	3:30									3	8	11	23	50	99:20
	30	3:20								2	6	8	19	26	61	126:20
	40	3:10							1	6	11	16	19	49	84	190:20
270	5	4:10												1	3	8:30
	10	4:00											2	5	11	22:30
	15	3:50										3	4	11	24	46:30
	20	3:40									2	3	9	21	35	74:30
	25	**3:30**								**2**	**3**	**8**	**13**	**23**	**53**	**106:30**
	30	**3:30**								**3**	**6**	**12**	**22**	**27**	**64**	**138:30**
	40	**3:20**							**5**	**6**	**11**	**17**	**22**	**51**	**88**	**204:30**
280	**5**	**4:20**												**2**	**2**	**8:40**
	10	**4:00**										**1**	**2**	**5**	**13**	**25:40**
	15	**3:50**									**1**	**3**	**4**	**11**	**26**	**49:40**
	20	**3:50**									**3**	**4**	**8**	**23**	**39**	**81:40**
	25	**3:40**								**2**	**5**	**7**	**16**	**23**	**56**	**113:40**
	30	**3:30**							**1**	**3**	**7**	**13**	**22**	**30**	**70**	**150:40**
	40	**3:20**						**1**	**6**	**6**	**13**	**17**	**27**	**51**	**93**	**218:40**
290	**5**	**4:30**												**2**	**3**	**9:50**
	10	**4:10**										**1**	**3**	**5**	**16**	**29:50**
	15	**4:00**									**1**	**3**	**6**	**12**	**26**	**52:50**
	20	**4:00**									**3**	**7**	**9**	**23**	**43**	**89:50**
	25	**3:50**								**3**	**5**	**8**	**17**	**23**	**60**	**120:50**
	30	**3:40**							**1**	**5**	**6**	**16**	**22**	**36**	**72**	**162:50**
	40	**3:30**						**3**	**5**	**7**	**15**	**16**	**32**	**51**	**95**	**228:50**

TABLE 1-14 (Cont.)

Depth (ft)	Bottom time (min)	Time to first stop (min:sec)	Decompression stops (feet)													Total ascent time (min:sec)
			130	120	110	100	90	80	70	60	50	40	30	20	10	
300	5	4:40												3	3	11:00
	10	4:20										1	3	6	17	32:00
	15	4:10									2	3	6	15	26	57:00
	20	4:00								2	3	7	10	23	47	97:00
	25	3:50							1	3	6	8	19	26	61	129:00
	30	3:50							2	5	7	17	22	39	75	172:00
	40	3:40						4	6	9	15	17	34	51	90	231:00
	60	3:00		4	10	10	10	10	10	14	28	32	50	90	187	460:00

U.S. Navy Standard Air Decompression Table for exceptional exposures—Continued

Extreme exposures—250 and 300 ft

Depth (ft)	Bottom time (min)	Time to first stop (min:sec)	Decompression stops (feet)																				Total ascent time (min:sec)
			200	190	180	170	160	150	140	130	120	110	100	90	80	70	60	50	40	30	20	10	
250	120	1:50							5	10	10	10	10	16	24	24	36	48	64	94	142	187	684:10
	180	1:30					4	8	8	10	22	24	24	32	42	44	60	84	114	122	142	187	931:10
	240	1:30					9	14	21	22	22	40	40	42	56	76	98	100	114	122	142	187	1109:10
300	90	2:20					3	8	8	10	10	10	10	16	24	24	34	48	64	90	142	187	693:00
	120	2:00			4	8	8	8	8	10	14	24	24	24	34	42	58	66	102	122	142	187	890:00
	180	1:40	6	8	8	8	14	20	21	21	28	40	40	48	56	82	98	100	114	122	142	187	1168:00

The ascent rate when using the Surface Decompression Table Using Oxygen differs from the other tables. Ascent is at a rate of 25 feet per minute from the bottom to the first stop, one minute between stops and one minute for the final 30-foot ascent to the surface. The final ascent from 40 feet in the chamber is made in two minutes, but many operators routinely stretch this time to five minutes. Oxygen is breathed during this final ascent in the chamber. The one section of the tables that does not reflect current practice has to do with Column 5, regarding the period of time that the diver must breathe oxygen at the 40-foot level. If this specified period is longer than 20 minutes, it is now routinely broken up into two periods with a five-minute air-breathing period between them. This is done to reduce the possibility of oxygen poisoning, and some people currently believe that the longest uninterrupted period of breathing oxygen should only be 15 minutes. Thus, if the schedule for a particular dive indicates in Column 5 an oxygen breathing time of 26 minutes, this should be broken up into two periods of 13 minutes, with a five-minute interval of air breathing between them. These five-minute air-breathing periods are not considered as part of the decompression, and these times are in addition to the total decompression times as shown in the table. Oxygen decompression and the operation of recompression chambers are covered in Chapters 2-4, but the subject is important enough to be reviewed.

The principal dangers of oxygen decompression are oxygen poisoning and fire. To eliminate the possibility of fire, the chamber must be kept clean and free of all unnecessary articles. The diver's and tender's clothes must be checked to be sure there are no matches, cigarette lighters or other fire-producing or combustible articles in them. All oil and grease must be kept out of and away from the chamber. The diver should be checked carefully before entering the chamber, because quite often he will have picked up oil or grease on his dress during his work. If this has happened, strip the diver of his diving suit or dress before he enters the chamber. The chamber light bulbs should never be larger than 25 watts, and the lights and all other electrical equipment must be in pressure-proof containers. If there is any question whatsoever concerning the repair or suitability of any electrical equipment, including the lights, power to the chamber must be disconnected. A battery-operated hand light (such as the Ike light) may be used in this case, or a light can be beamed in through one of the ports.

The Navy recommends the placement of a bucket of sand or water in the chamber for fire prevention purposes, and although I have never seen this done in the commercial diving field, I think it is a very good idea. The chamber light must never be covered with clothing or any other substance for the purpose of shading. This can and has caused fatal chamber fires. The chamber should be hooked up to an emergency air supply. All mechanical components of the chamber,

TABLE 1-27 FROM U.S. NAVY DIVING MANUAL

(FORMERLY TABLE 1-18, 1963 DIVING MANUAL)

Surface decompression table using air

Depth (ft)	Bottom time (min)	Time to first stop (min:sec)	Time at water stops (min)				Chamber stops (air) (min)		Total ascent time (min:sec)
			30	20	10		20	10	
40	230	0:30			3	NOT TO EXCEED 3 MINUTES AND 30 SECONDS		7	14:30
	250	:30			3			11	18:30
	270	:30			3			15	22:30
	300	:30			3			19	26:30
50	120	:40			3			5	12:40
	140	:40			3			10	17:40
	160	:40			3			21	28:40
	180	:40			3			29	36:40
	200	:40			3			35	42:40
	220	:40			3			40	47:40
	240	:40			3			47	54:40
60	80	:50			3			7	14:50
	100	:50			3			14	21:50
	120	:50			3			26	33:50
	140	:50			3			39	46:50
	160	:50			3			48	55:50
	180	:50			3			56	63:50
	200	:40		3			3	69	80:10
70	60	1:00			3			8	16:00
	70	1:00			3			14	22:00
	80	1:00			3			18	26:00
	90	1:00			3			23	31:00
	100	1:00			3			33	41:00
	110	:50		3			3	41	52:20
	120	:50		3			4	47	59:20
	130	:50		3			6	52	66:20
	140	:50		3			8	56	72:20
	150	:50		3			9	61	78:20
	160	:50		3			13	72	93:20

Depth (ft)	Bottom time (min)	Time to first stop (min:sec)	Time at water stops (min)						Chamber stops (air) (min)		Total ascent time (min:sec)
			50	40	30	20	10		20	10	
	170	:50				3		TIME ON SURFACE	19	79	106:20
80	50	1:10					3			10	18:10
	60	1:10					3			17	25:10
	70	1:10					3			23	31:10
	80	1:00				3			3	31	42:30
	90	1:00				3			7	39	54:30
	100	1:00				3			11	46	65:30
	110	1:00				3			13	53	74:30
	120	1:00				3			17	56	81:30
	130	1:00				3			19	63	90:30
	140	1:00				26			26	69	126:30
	150	1:00				32			32	77	146:30
90	40	1:20					3			7	15:20
	50	1:20					3			18	26:20
	60	1:20					3			25	33:20
	70	1:10				3			7	30	45:40
	80	1:10				13			13	40	71:40
	90	1:10				18			18	48	89:40
	100	1:10				21			21	54	101:40
	110	1:10				24			24	61	114:40
	120	1:10				32			32	68	137:40
	130	1:00			5	36			36	74	156:40
100	40	1:30					3			15	23:30
	50	1:20				3			3	24	35:50
	60	1:20				3			9	28	45:50
	70	1:20				3			17	39	64:50
	80	1:20				23			23	48	99:50
	90	1:10			3	23			23	57	111:50
	100	1:10			7	23			23	66	124:50
	110	1:10			10	34			34	72	155:50
	120	1:10			12	41			41	78	177:50
	30	1:40					3			7	15:40
	40	1:30				3			3	21	33:00
	50	1:30				3			8	26	43:00
	60	1:30				18			18	36	78:00
	70	1:20			1	23			23	48	101:00
	80	1:20			7	23			23	57	116:00

TABLE 1-27 (Cont.)

Depth (ft)	Bottom time (min)	Time to first stop (min:sec)	Time at water stops (min)						Chamber stops (air) (min)		Total ascent time (min:sec)
			50	40	30	20	10		20	10	
110	90	1:20			12	30		TIME ON SURFACE NOT TO EXCEED 3 MINUTES AND 30 SECONDS	30	64	142:00
	100	1:20			15	37			37	72	167:00
120	25	1:50					3			6	14:50
	30	1:50					3			14	22:50
	40	1:40				3			5	25	39:10
	50	1:40				15			15	31	67:10
	60	1:30			2	22			22	45	97:10
	70	1:30			9	23			23	55	116:10
	80	1:30			15	27			27	63	138:10
	90	1:30			19	37			37	74	173:10
	100	1:30			23	45			45	80	189:10
130	25	2:00					3			10	19:00
	30	1:50				3			3	18	30:20
	40	1:50				10			10	25	51:20
	50	1:40			3	21			21	37	88:20
	60	1:40			9	23			23	52	113:20
	70	1:40			16	24			24	61	131:20
	80	1:30		3	19	35			35	72	170:20
	90	1:30		8	19	45			45	80	203:20
140	20	2:10					3			6	15:10
	25	2:00				3			3	14	26:30
	30	2:00				5			5	21	37:30
	40	1:50			2	16			16	26	66:30
	50	1:50			6	24			24	44	104:30
	60	1:50			16	23			23	56	124:30
	70	1:40		4	19	32			32	68	161:30
	80	1:40		10	23	41			41	79	200:30
150	20	2:10				3			3	7	19:40
	25	2:10				4			4	17	31:40
	30	2:10				8			8	24	46:40
	40	2:00			5	19			19	33	82:40
	50	2:00			12	23			23	51	115:40
	60	1:50		3	19	26			26	62	142:40
	70	1:50		11	19	39			39	75	189:40
	80	1:40	1	17	19	50			50	84	227:40

160	20	2:20				3		TIME ON SURFACE NOT TO EXCEED 3 MINUTES AND 30 SECONDS	3	11	23:50
	25	2:20				7			7	20	40:50
	30	2:10			2	11			11	25	55:50
	40	2:10			7	23			23	39	98:50
	50	2:00		2	16	23			23	55	125:50
	60	2:00		9	19	33			33	69	169:50
	70	1:50	1	17	22	44			44	80	214:50
170	15	2:30				3			3	5	18:00
	20	2:30				4			4	15	30:00
	25	2:20			2	7			7	23	46:00
	30	2:20			4	13			13	26	63:00
	40	2:10		1	10	23			23	45	109:00
	50	2:10		5	18	23			23	61	137:00
	60	2:00	2	15	22	37			37	74	194:00
	70	2:00	8	17	19	51			51	86	239:00
180	15	2:40				3			3	6	19:10
	20	2:30			1	5			5	17	35:10
	25	2:30			3	10			10	24	54:10
	30	2:30			6	17			17	27	74:10
	40	2:20		3	14	23			23	50	120:10
	50	2:10	2	9	19	30			30	65	162:10
	60	2:10	5	16	19	44			44	81	216:10
190	15	2:50				4			4	7	22:20
	20	2:40			2	6			6	20	41:20
	25	2:40			5	11			11	25	59:20
	30	2:30		1	8	19			19	32	86:20
	40	2:30		8	14	23			23	55	130:20
	50	2:20	4	13	22	33			33	72	184:20
	60	2:20	10	17	19	50			50	84	237:20

NOTE.—The ascent rates in this table are 60 feet per minute to the first stop, between stops and to the surface in the water and in the chamber. The descent rate in the chamber is also 60 feet per minute. The total ascent time may be shortened only by shortening the surface interval.

TABLE 1-26 FROM U.S. NAVY DIVING MANUAL

(FORMERLY TABLE 1-17, 1963 DIVING MANUAL)

Surface decompression table using oxygen

1	2	3	4				5	6	7	8
Depth (ft, gage)	Bottom time (min) [1]	Time to first stop or surface [2]	Time (min) breathing air at water stops (ft) [3]				Surface interval [4]	Time at 40-foot chamber stop (min) on oxygen [5]	Surface [6]	Total decompression time (min: sec) [7]
			60	50	40	30				
70	52	2:48	0	0	0	0	NOT TO EXCEED 5 MINUTES	0	CHAMBER TO SURFACE WHILE BREATHING OXYGEN	2:48
	90	2:48	0	0	0	0		15		23:48
	[8] 120	2:48	0	0	0	0		23		31:48
	150	2:28	0	0	0	0		31		39:48
	180	2:48	0	0	0	0		39		47:48
80	40	3:12	0	0	0	0		0		3:12
	70	3:12	0	0	0	0		14		23:12
	85	3:12	0	0	0	0		20		29:12
	100	3:12	0	0	0	0		26		35:12
	[8] 115	3:12	0	0	0	0		31		40:12
	130	3:12	0	0	0	0		37		46:12
	150	3:12	0	0	0	0		44		53:12
90	32	3:36	0	0	0	0		0		3:36
	60	3:36	0	0	0	0		14		23:36
	70	3:36	0	0	0	0		20		29:36
	80	3:36	0	0	0	0		25		34:36
	[8] 90	3:36	0	0	0	0		30		39:36
	100	3:36	0	0	0	0		34		43:36
	110	3:36	0	0	0	0		39		48:36
	120	3:36	0	0	0	0		43		52:36
	130	3:36	0	0	0	0		48		57:36
100	26	4:00	0	0	0	0		0		4:00
	50	4:00	0	0	0	0		14		24:00
	60	4:00	0	0	0	0		20		30:00
	70	4:00	0	0	0	0		26		36:00
	[8] 80	4:00	0	0	0	0		32		42:00
	90	4:00	0	0	0	0		38		48:00
	100	4:00	0	0	0	0		44		54:00
	110	4:00	0	0	0	0		49		59:00

	120	4:00	0	0	0	0	SURFACE INTERVAL	53	2-MINUTE ASCENT FROM 40 FEET IN	63:00
110	22	4:24	0	0	0	0		0		4:24
	40	4:24	0	0	0	0		12		22:24
	50	4:24	0	0	0	0		19		29:24
	60	4:24	0	0	0	0		26		36:24
	[8] 70	4:24	0	0	0	0		33		43:24
	80	3:12	0	0	0	1		40		51:12
	90	3:12	0	0	0	2		46		58:12
	100	3:12	0	0	0	5		51		66:12
	110	3:12	0	0	0	12		54		76:12
120	18	4:48	0	0	0	0		0		4:48
	30	4:48	0	0	0	0		9		19:48
	40	4:48	0	0	0	0		16		26:48
	50	4:48	0	0	0	0		24		34:48
	[8] 60	3:36	0	0	0	2		32		44:36
	70	3:36	0	0	0	4		39		53:36
	80	3:36	0	0	0	5		46		61:36
	90	3:12	0	0	3	7		51		72:12
	100	3:12	0	0	6	15		54		86:12
130	15	5:12	0	0	0	0		0		5:12
	30	5:12	0	0	0	0		12		23:12
	40	5:12	0	0	0	0		21		32:12
	50	4:00	0	0	0	3		29		43:00
	[8] 60	4:00	0	0	0	5		37		53:00
	70	4:00	0	0	0	7		45		63:00
	80	3:36	0	0	6	7		51		75:36
	90	3:36	0	0	10	12		56		89:36
140	13	5:36	0	0	0	0		0		5:36
	25	5:36	0	0	0	0		11		22:36
	30	5:36	0	0	0	0		15		26:36
	35	5:36	0	0	0	0		20		31:36
	40	4:24	0	0	0	2		24		37:24
	45	4:24	0	0	0	4		29		44:24
	50	4:24	0	0	0	6		33		50:24
	[8] 55	4:24	0	0	0	7		38		56:24
	60	4:24	0	0	0	8		43		62:24
	65	4:00	0	0	3	7		48		70:00
	70	3:36	0	2	7	7		51		79:36
150	11	6:00	0	0	0	0		0		6:00
	25	6:00	0	0	0	0		13		25:00
	30	6:00	0	0	0	0		18		30:00

See footnotes at end of table.

TABLE 1-26 (Cont.)

	35	4:48	0	0	0	4	SURFACE INTERVAL NOT TO EXCEED 5 MINUTES	23	2-MINUTE ASCENT FROM 40 FEET IN CHAMBER TO SURFACE WHILE BREATHING OXYGEN	38:48
	40	4:24	0	0	3	6		27		48:24
	45	4:24	0	0	5	7		33		57:24
	[8] 50	4:00	0	2	5	8		38		66:00
	55	3:36	2	5	9	4		44		77:36
160	9	6:24	0	0	0	0		0		6:24
	20	6:24	0	0	0	0		11		23:24
	25	6:24	0	0	0	0		16		28:24
	30	5:12	0	0	0	2		21		35:12
	35	4:48	0	0	4	6		26		48:48
	40	4:24	0	3	5	8		32		61:24
	[8] 45	4:00	3	4	8	6		38		73:00
170	7	6:48	0	0	0	0		0		6:48
	20	6:48	0	0	0	0		13		25:48
	25	6:48	0	0	0	0		19		31:48
	30	5:12	0	0	3	5		23		44:12
	35	4:48	0	4	4	7		29		57:48
	[8] 40	4:24	4	4	8	6		36		72:24

[1] Time interval in minutes from leaving the surface to leaving the bottom.

[2] Time of ascent in minutes and seconds to the first stop or to the surface at a rate of 25 feet per minute.

[3] Water stops: Time spent at tabulated stops using air. If no water stops are required, use a 25-foot-per-minute rate of ascent to the surface. When water stops are required, use a 25-foot-per-minute rate of ascent to the first stop. Take an additional minute between stops. Use 1 minute for the ascent from 30 feet to the surface.

[4] Surface interval: The surface interval shall not exceed 5 minutes and is composed of the following elements:

a. Time of ascent from the 30-foot water stop to the surface (1 minute).

b. Time on the surface for landing the diver on deck and undressing (not to exceed 3 minutes and 30 seconds).

c. Time of descent in the recompression chamber from the surface to 40 feet (about 30 seconds).

[5] During the period of oxygen breathing, the chamber shall be ventilated unless an oxygen-elimination system is used.

[6] Surfacing: Oxygen breathing during this 2-minute period shall follow without interruption the period of oxygen breathing tabulated in col. 6.

[7] Total decompression time in minutes and seconds. This time includes:

a. Time of ascent from the bottom to the first stop at 25 feet per minute, col. 3.

b. Sum of tabulated water stops, col. 4.

c. One minute between water stops.

d. The surface interval, col. 5.

e. Time at 40 feet in the recompression chamber, col. 6.

f. Time of ascent, an additional 2 minutes, from 40 feet to the surface, col. 7.

The total decompression time may be shortened only by decreasing the time required to undress the diver on deck.

[8] These are the optimum exposure times for each depth and represent for the average diver the best balance of safety, length of work period, and amount of useful work. Exposure beyond these limits of time is permitted only under special conditions.

valves, hatch seals, dogs, etc., must be in good repair. Pressure gauges must be calibrated. The chamber ports must be kept clean and unobstructed, and they should never be rapped on for the purpose of signaling. The chamber operator must observe the chamber occupant at all times, especially when he is breathing oxygen. The chamber must be ventilated frequently, at least for two minutes out of every five when oxygen is being breathed. The operator must time all oxygen-breathing periods accurately. The diver must not be allowed to fall asleep when breathing oxygen. I recommend that all chambers be hooked up with a compressed air by-pass on the oxygen breathing circuit, so that if a man goes into convulsions while wearing the mask, the breathing medium can be shifted from oxygen to compressed air by the operator.

Both the diver and the operator must be familiar with oxygen poisoning symptoms, i.e., twitching, especially of facial muscles, dizziness, nausea, blurring of vision, and convulsions. If any of these symptoms are detected, the oxygen mask must be removed immediately, and the chamber well ventilated. If symptoms do not stop immediately the pressure in the chamber should be reduced by ten feet. When symptoms disappear, the diver can be brought down to the original depth, and the decompression continued under the longer Surface Decompression Table using air. In this situation, any decompression time previously spent on the oxygen table will be disregarded, and the diver will be decompressed to the full schedule for his dive on the Surface Decompression Table Using Air.

Decompression sickness and embolism have been covered in Chapter 2, but they are also important enough to justify a brief review.

DECOMPRESSION SICKNESS

Decompression sickness is the condition that results after a dive or exposure to elevated pressures, when the nitrogen or other inert gas (such as helium) which was absorbed by the blood during exposure to high pressure is not eliminated from the body by adequate decompression. This nitrogen or other inert gas forms bubbles in the blood and tissues of the body, and these gas bubbles cause a variety of dangerous and harmful symptoms. This condition has been variously called caisson disease, the bends, the screws, the niggles or just plainly, "a hit." The proper name for this condition is decompression sickness, and this name should always be used to avoid confusion and ambiguity. The symptoms of decompression sickness depend on the location and size of the bubble or bubbles that develop. These symptoms consist of pain in the joints, muscles or bones when a bubble locates in one of these structures. Bubble formation in the brain can produce blindness, dizziness, paralysis, or even unconsciousness and convulsions. When the spinal cord is affected, paralysis or loss of feeling in some part of the body can occur. Bubbles in the lungs can cause asphyxia or choking. Skin bubbles produce itching, rash, or both. Unusual fatigue or exhaustion after a dive is probably also due to bubbles, but the location of bubbles causing such symptoms is not known. Many other symptoms can be caused by bubbles in unusual locations. Decompression sickness which affects the central nervous system (brain, or spinal cord) or lungs can produce serious disabilities, and may even threaten life, if not treated promptly and properly. When other areas, such as joints, are affected, the condition may produce excruciating pain, and may lead to local damage if not treated.

Prevention of decompression sickness is generally accomplished by following the decompression tables correctly. Even when the tables are used correctly, however, unusual conditions, either in the diver or in connection with the dive, will produce a small percentage of decompression sickness cases. To be absolutely safe under all possible circumstances, the decompression time specified would have to be far in excess of that normally needed. On the other hand, under ideal circumstances, some individuals can ascend safely in less time than the tables specify. This fact must not be taken to mean that the tables contain an unnecessarily large safety factor. As a matter of fact, the tables generally represent the minimum decompression time that will permit average divers to surface safely from normal working dives without an unacceptable incidence of decompression sickness.

Factors in the diver which apparently favor the development of decompression sickness, even when the tables are correctly followed, include age, obesity, excessive fatigue, loss of of sleep, alcoholic indulgence and its aftereffects, various illnesses, and anything which contributes to generally poor physical condition and poor circulatory efficiency. Unusually heavy exertion and extremes of temperature during the dive can have unfavorable effects. Heavy work at depths speeds up the circulation and increases the uptake of inert gas. Exercise during decompression, although it hastens the elimination of inert gas from some tissues, often increases the incidence of decompression sickness. Anything that impedes blood flow in any part of the body during decompression can favor bubble formation. Keeping a leg or an arm in a cramped position is an example of a condition which would impede blood flow.

The treatment for decompression sickness is immediate recompression and decompression, according to the most appropriate treatment table. The sooner treatment is initiated, the more effective it will be. Treat any unusual signs or symptoms in a diver by immediate recompression.

AIR EMBOLISM

Air embolism is another serious divers' illness that can only be successfully treated by recompression. Air embolism is most frequently caused by divers holding their breath during the ascent. When this occurs, the pressure in the lungs becomes greater than the surrounding water pressure, and after the lungs expand to their maximum capacity, they rupture, and air is forced into the blood vessels that pass through the lungs, or into the surrounding tissues. Air embolism can also be caused by recent respiratory diseases, such as pneumonia or severe chest colds. In this situation, pockets of air are trapped in the lungs during the dive by mucus or phlegm, and these pockets are not freely vented during the ascent, causing an air embolism. Serious and even fatal air embolisms can occur during an ascent from very shallow depths. Any diver manifesting difficulties or any unusual signs or symptoms after a dive too short or too shallow to have contracted the bends, must be assumed to be suffering from air embolism, and he must be recompressed immediately. Symptoms of air embolism will usually develop within seconds of surfacing, and can include weakness, dizziness, paralysis, visual disturbances, chest pains, etc. An unconscious diver must be assumed to be suffering from air embolism.

The signs that may be observed in a diver suffering from air embolism, progressing from less serious to more serious, are bloody or frothy sputum, staggering, evidence of confusion or difficulty in seeing, paralysis or weakness in the extremities, collapse, unconsciousness, convulsions, cessation of breathing. Note that the onset may be so sudden that none but the most serious signs can be seen.

The treatment of air embolism consists of recompression in a recompression chamber. This treatment reduces the size of the bubble, and may permit resumption of normal circulation of blood in the brain. Recompress the patient without delay to a depth of 165 feet. Descend at the maximum rate that is within the capability of the tender or tenders to equalize. (The normal descent rate of 25 feet a minute, used during treatment for decompression sickness, does not apply to the treatment of air embolism.) If the tenders have difficulty equalizing, descent must continue regardless. After reaching 165 feet, follow Treatment Table 3, 4, 5A, or 6A, whichever is indicated by the response of the patient. If he completely recovers within 15 minutes, use Table 5A. If he completely recovers within 30 minutes, use Table 3 or Table 6A. If he does not completely recover within 30 minutes, use Table 4. Be extremely watchful for any evidence of recurrence of symptoms during ascent.

Use oxygen where permitted by the treatment table, but discontinue if there is evidence that oxygen breathing is producing lung irritation (pain or coughing) or difficulty in breathing. Note that a helium-oxygen mixture can be used at any time during treatment with Table 3 or 4.

In any case of air embolism or related accidents, breathing may cease. In this case, artificial respiration must be started in addition to immediate recompression. Shock must also be treated when it exists.

Air embolism is an extreme emergency, and the victim may die or suffer permanent brain damage unless he receives immediate recompression and proper treatment.

Author's Note: There are a number of serious situations and pathological conditions that may develop during diving operations including, but not limited to, mediastinal or subcutaneous emphysema, pneumothorax, etc. The diagnosis and treatment of these and other conditions are not within the scope of this book, nor generally among the capabilities of the average diver. Further information on the medical aspects of diving can be found in the *U.S. Navy Diving Manual* and in other books listed in the Bibliography.

In my opinion, as a diver lacking formal medical training, your responsibilities towards a seriously afflicted diving mate, include and are limited to the following: the administration of first aid (a Red Cross First-Aid Training Course is an asset to every working diver), artificial respiration, in the event of drowning or cessation of breathing; immediate recompression in a chamber to the depth of relief, or 165 feet in any serious or doubtful situation; the immediate summoning of qualified medical help.

Following are notes on the treatment of an unconscious diver, on artificial respiration, on first aid, and on recompression, concluding with Treatment Tables, 1, 1A, 2, 2A, 3, 4, 5, 6, 5A and 6A.

These notes and tables have been taken word-for-word out of the *U.S. Navy Diving Manual.* The only change is the deletion of references to other tables and to other parts of the *U.S. Navy Diving Manual.* This was done because the various tables in the 1963 *U.S. Navy Diving Manual* were given certain letter and number designations for identifying purposes, and although these tables remain substantially unchanged in the 1970 issue of the manual, the numbers and letters designating them have been changed. To avoid confusion, any references made to the various tables is done by using the full written title, such as: U.S. Navy Standard Air Decompression Table.

TREATMENT OF AN UNCONSCIOUS DIVER

1. If the diver is not breathing, start mouth-to-mouth or manual artificial respiration at once (see note a).
2. Recompress promptly (see note d).
3. Examine for injuries and other abnormalities; apply first aid and other measures as required. (Secure medical help as soon as possible.)

NOTES

Artificial respiration:

a. Shift to a mechanical resuscitator if one is available and working properly, but never wait for it. Always start the mouth-to-mouth or manual methods first.

b. Continue artificial respiration by some method without interruption until normal breathing resumes or the victim is pronounced dead. Continue on the way to the chamber and during recompression. (Do not use oxygen deeper than 60 feet in the chamber.)

Recompression:

c. Remember that an unconscious diver may have air embolism or serious decompression sickness even though some other accident seems to explain his condition.

d. Recompress unless—
   1. The victim regains consciousness and is free of nervous system symptoms before recompression can be started.
   2. The possibility of air embolism or decompression sickness can be ruled out without question.
   3. Another lifesaving measure is absolutely required and makes recompression impossible.

e. Try to reach a recompression chamber no matter how far it is.

f. Treat according to the treatment tables, depending on response. Remember that early recovery under pressure never rules out the need for adequate treatment.

ARTIFICIAL RESPIRATION

1. Start artificial respiration immediately, whenever a man is not breathing due to drowning or any other cause.
   a. Never wait for a mechanical resuscitator.
   b. Delay only to stop serious bleeding (if possible, have another person tend to such measures while you start artificial respiration).
   c. Send another person for a doctor or other competent aid.
2. Before starting, remove victim from the cause of his trouble, but do not waste time moving him any further than necessary.
3. Get on with artificial respiration. Leave details to others or try to get them done quickly between cycles.
   a. Recheck position of victim:
      1. In position for mouth-to-mouth resuscitation.
      2. Head slightly lower than feet if possible, especially in drowning.
      3. Chin pulled toward operator.
   b. Recheck airway:
      1. Remove froth, debris, or other material.
      2. See that tongue stays forward; have someone hold it if it draws back (you can run a safety pin through tongue if necessary).

3. If artificial respiration does not move any air, there is an obstruction. Strangulation must be overcome.

c. Loosen any tight clothing—collar, belt, etc.
d. Keep victim warm.
e. Check pulse. Combat shock.

4. Continue artificial respiration without interruption. (Minimum time is 4 hours unless victim revives or is pronounced dead by a doctor.)
   a. Do not apply too much back pressure. (A strong operator can crack ribs of small victim.)
   b. If you become tired, let another operator take over. Do not break rhythm during shift.
   c. Watch carefully for signs of return of natural breathing movements. If they appear, time your movements to assist them.
   d. Shift to a mechanical resuscitator if one is available, ready, and operating properly.
   e. If victim starts breathing for himself, watch him carefully. Resume artificial respiration if he stops or if movements become too feeble.
5. If victim revives, continue care:
   a. Keep him lying down.
   b. Remove wet clothes; keep him warm.
   c. Give nothing by mouth until fully conscious.
   d. Attend to any injuries.
   e. Be sure he is seen promptly by a doctor.

Note: If victim has been under water with any kind of breathing apparatus, he may have air embolism. This can seldom be ruled out in an unconscious diver, whether he is breathing or not, and recompression should be given if any doubt exists. Do not delay artificial respiration. Give it by some method on way to chamber and during recompression.

FIRST AID

Proper first aid can make the difference between life and death. Every diver should have a good knowledge of first aid, and a First Aid Manual should be kept handy wherever diving is done. This table is only a reminder of some vital points.

1. If nature of injury is not certain, check victim over quickly but carefully.
   a. Is he breathing?
   b. Is he bleeding?
   c. Any broken bones?
   d. Any sign of head injury?
2. Start artificial respiration if breathing has stopped.
3. Stop bleeding. If bleeding is very heavy, do this before anything else:
   a. Try direct pressure with snug bandage.
   b. Use "pressure points."
   c. Apply tourniquet only as a last resort.

4. If victim is a diver, consider possible need for immediate recompression.
5. Combat shock.
   a. Know its signs:
      1) Paleness
      2) Skin cold and moist
      3) Weak, rapid pulse
      4) Fainting
   b. Remember that shock is a serious danger in almost any injury or severe illness. Take steps to prevent or treat it:
      1) Keep victim flat (head slightly lower than rest of body—except with head injury or if this causes trouble breathing).
      2) Keep warm by covering.
      3) Try to calm him; do what you can to lessen pain.
      4) If conscious, able to swallow, not vomiting, and with no abdominal injury, give as much shock solution as victim will take (1 teaspoonful table salt and ½ teaspoonful baking soda per quart of water).
   c. If shock is present, give plasma or plasma-substitute intravenously if possible.
6. Take immediate action in poisoning or chemical burns.
   a. In poisoning:
      1) If victim is conscious and poison is not a corrosive one, get him to vomit.
      2) Dilute poison in stomach (but give nothing by mouth if unconscious), and repeat vomiting.
      3) Determine nature of poison, give proper antidote.
   b. In chemical burns:
      1) Flush with large quantities of water.
      2) Avoid strong "neutralizers."
7. Send for medical help, or get victim to hospital or dispensary, in anything but minor conditions.
   a. If another person is present, send him at once for medical assistance.
   b. Do not move victim unless you can do it properly (see 8).
8. Handle any injured person with care.
   a. If victim must be moved, use stretcher (or improvise one). Transfer him to it with as little movement as possible. Use special precautions with possible back or neck injuries.
   b. Splint broken bones temporarily on the spot.
9. Cover wounds and burns.
   a. Avoid handling; do not try to clean or disinfect (let the doctor do this).
   b. Use sterile dressing (or cleanest cloth available) and apply bandage over it.
10. In head injuries:
   a. Keep patient lying down and quiet.
   b. Secure medical attention even if injury seems slight.

11. In convulsions:
    a. Put something soft between teeth.
    b. Try to prevent injury, but do not restrain movements.
12. In collapse in hot surroundings:
    a. Check for signs of heat stroke.
        1) Skin hot and dry.
        2) Pulse rapid but full.
        3) High body temperature.
    b. If signs are present,
        1) Get medical assistance.
        2) Take immediate steps to lower body temperature.

NOTES ON RECOMPRESSION

1. General considerations:
    a. Follow the treatment tables accurately.
    b. Permit no shortening or other alterations of the tables except on the advice of a doctor who has been trained in diving medicine, or in an extreme emergency.
2. Rate of descent in the chamber:
    a. The normal descent rate is 25 feet per minute.
    b. If serious symptoms are present: rapid descent is desirable.
    c. If pain increases on descent: stop, resume at a rate tolerated by the patient.
3. Treatment depth:
    a. Go to the full depth indicated by the table required.
    b. Do not go beyond 165 feet, except on the decision of a doctor who has been trained in diving medicine.
4. Examination of the patient.
    a. If no serious symptoms are evident and pain is not severe, examine the patient thoroughly before treatment.
    b. If any serious symptom is noted, do not delay recompression for examination or for determining depth of relief.
    c. If Treatment Tables 5, 6, 5A, or 6A are used, a doctor trained in diving medicine and a qualified medical attendant must always accompany the patient in the chamber during treatment.
    d. In "pain only" cases, make sure that relief is complete within 10 minutes at 60 feet on oxygen if Table 5 is used. If not, Table 6 may be used. If Table 1 is used, make sure that complete relief has been reported before reaching 66 feet.
    e. On reaching treatment depth, examine the patient as completely as possible to detect—
        1) Incomplete relief.
        2) Any symptoms overlooked.

        Note: At the very least, have the patient stand and walk the length of the chamber, if this is at all possible.

f. Recheck the patient before leaving the treatment depth.
g. Ask the patient how he feels before and after coming to each stop, and periodically during long stops.
h. Do not let the patient sleep through changes of depth or for more than an hour at a time at any stop. (Symptoms can develop or recur during sleep.)
i. Recheck the patient before leaving the last stop.
j. During treatment make sure that the patient can obtain all the things that he needs, such as food, liquids, and any other items that he might require.

5. Patient getting worse:
   a. Never continue ascent if the patient's condition is worsening.
   b. Treat the patient as a recurrence during treatment (see 6).
   c. Consider the use of helium-oxygen as a breathing medium for the patient (see 8).

6. Recurrence of symptoms:
   a. During treatment:
      1) Recompress to depth of relief (but never less than 30 feet or deeper than 165 feet, except on decision of a doctor trained in diving medicine).
      2) If a doctor, trained in diving medicine, is available and the depth of relief is less than 60 feet, recompress to 60 feet and treat on Table 6.
      3) If a doctor, trained in diving medicine, is not available or the depth of relief is greater than 60 feet, complete the treatment according to Table 4; i.e., remain at depth of relief for 30 minutes and complete remaining stops of Table 4.
      4) If recurrence involves serious symptoms not previously present, take the patient to 60 feet and treat on Table 6 or take the patient to 165 feet and treat on Table 4.
   b. Following treatment:
      1) Recompress to 60 feet and use Table 6, if a doctor, trained in diving medicine, is available.
      2) If the depth of relief is less than 30 feet, recompress the patient to 30 feet and decompress from the 30-foot stop according to Table 3.
      3) If the depth of relief is deeper than 30 feet, keep the patient at depth of relief for 30 minutes and decompress according to Table 3.
      4) If the original treatment was on Table 5 or 6, use Table 6. If the original treatment was on Table 5A or 6A, use 6, 6A, or Table 4. If the original treatment was on Table 3, use Table 6, 6A or **Table 4**.

5) Examine the patient carefully to be sure no serious symptom is present. If the original treatment was on Table 1 or 2, appearance of a serious symptom requires full treatment on Table 6, 3, or 4.

c. Using oxygen treatment tables during or following treatment:

1) Table 6 can be lengthened by an additional 25 minutes at 60 feet (20 minutes on oxygen and 5 minutes on air) or an additional 75 minutes at 30 feet (15 minutes on air and 60 minutes on oxygen), or both. Table 6A can be lengthened in the same manner.
2) If relief is not complete at 60 feet or if the patient's condition is worsening, the additional time above may be used or the patient can be recompressed to 165 feet and treated on Table 2, 2A, 3, or 4 as appropriate.

7. Use of oxygen:

a. Use oxygen wherever permitted by the treatment tables unless the patient is known to tolerate oxygen poorly.

b. If a doctor trained in diving medicine is available, he may recommend the use of oxygen for patients who are known to tolerate oxygen poorly.

c. Take all precautions against fire.

d. Tend carefully, being alert for such symptoms of oxygen poisoning as—

1) Twitching of the face and lips.
2) Nausea.
3) Dizziness and vertigo.
4) Vomiting.
5) Convulsions.
6) Anxiety.
7) Confusion.
8) Malaise or excessive tiredness.
9) Restlessness and irritability.
10) Changes in vision as blurring or narrowing of the visual field.
11) Incoordination.
12) Tremors of the arms and legs.
13) Numbness or tingling of the fingers or toes.
14) Fainting.
15) Spasmodic breathing.

e. Know what to do in the event of a convulsion:

1) Halt ascent.
2) Remove mask at once.
3) Maintain depth.
4) Protect the convulsing patient from injury, but do not restrain or forcefully oppose the convulsive movements.
5) Use a padded mouth bit to protect the tongue of a convulsing patient.

6) If the patient is not convulsing, have him hyperventilate with chamber air for a few breaths.

f. If oxygen breathing must be interrupted:
   1) On Table 1, proceed on Table 1A.
   2) On Table 2, proceed on Table 2A.
   3) On Table 3, continue on Table 3, using air.
   4) On Table 5, 6, 5A, or 6A, allow 15 minutes after the reaction has entirely subsided and resume the schedule at the point of its interruption.
   5) On Table 5, if the reaction occurred at 60 feet, upon arrival at the 30-foot stop, switch to the schedule of Table 6.

g. At the doctor's discretion, oxygen breathing may be resumed at the 40-foot stop. If oxygen breathing is resumed, complete treatment as follows:
   1) Resuming from Table 1A: breathe oxygen at 40 feet for 30 minutes and at 30 feet for 1 hour.
   2) Resuming from Table 2A: breathe oxygen at 40 feet for 30 minutes and at 30 feet for 2 hours.
   3) In both cases, then surface in 5 minutes, still breathing oxygen.
   4) Resuming from Table 3: breathe oxygen at 40 feet for 30 minutes and at 30 feet for the first hour, and then finish the treatment with air.

8. Use of helium-oxygen:
   a. Helium-oxygen mixtures in a ratio of about 80:20 can be used instead of air (not in place of oxygen) in all types of treatment and at any depth.
   b. The use of helium-oxygen mixtures is especially desirable in any patient who—
      1) Has serious symptoms which fail to clear within a short time at 165 feet.
      2) Has a recurrence of symptoms or otherwise becomes worse at any stage of treatment.
      3) Has any difficulty in breathing.

9. Tenders:
   a. A qualified tender must be in the chamber at all times.
   b. The tender must be alert for any change in the condition of the patient, especially during oxygen breathing.
   c. The tender must breathe oxygen if he has been with the patient throughout treatment using Table 1 or 2.
      1) On Table 1, breathe oxygen at 40 feet for 30 minutes.
      2) On Table 2, breathe oxygen at 30 feet for 1 hour.
   d. A tender in the chamber only during the oxygen-breathing part of Table 1 or 2 gains a safety factor by breathing oxygen for 30 minutes of the last stop, but it is not essential. Tenders

may breathe oxygen during the use of Table 3 or 4 at depths of 40 feet or less.

e. When Tables 5, 6, 5A or 6A are used, the tender normally breathes air throughout. However, if the treatment is a repetitive dive for the tender or if Tables 6 or 6A are lengthened, the tender must breathe oxygen during the last 30 minutes of ascent from 30 feet to the surface.

f. Anyone entering the chamber and leaving before completion of the treatment must be decompressed according to standard diving tables.

g. Personnel outside the chamber must specify and control decompression of anyone leaving the chamber and must review all decisions concerning treatment or decompression made by personnel (including the doctor) inside the chamber.

10. Ventilation of the chamber:

a. All ventilation will be continuous and the volumes specified are measured at the chamber pressure.

b. If ventilation must be interrupted for any reason, the time will not exceed 5 minutes in any 30-minute period. When the ventilation is resumed, twice the volume of ventilation will be used for twice the time interrupted and then the basic ventilation will be used again.

c. When air or a helium-oxygen mixture is breathed, provide 2 cubic feet per minute for a man at rest and 4 cubic feet per minute for a man who is not at rest, such as a tender actively taking care of a patient.

d. When oxygen is breathed, provide 12.5 cubic feet per minute for a man at rest and 25 cubic feet per minute for a man who is not at rest. When these ventilation rates are used, no additional ventilation is required for personnel breathing air. These ventilation rates apply only to the number of people breathing oxygen.

e. The above rules apply to all chambers that do not have facilities to monitor the oxygen concentration in the chamber. Chambers that can monitor oxygen concentration may use intermittent ventilation so that the oxygen concentration in the chamber does not exceed 22.5 percent. This ventilation also requires no additional ventilation for personnel breathing air.

f. If an oxygen-elimination system is used for oxygen breathing, the ventilation rate required for air breathing may be used and applies to all personnel, whether or not the oxygen-elimination system is used to obtain the correct ventilation rate.

11. First aid:
    a. First aid may be required in addition to recompression. Do not neglect it.
12. Recompression in the water:
    a. Recompression without a chamber is difficult and hazardous. Except in grave emergencies, seek the nearest chamber even if it is at a considerable distance.
    b. If water recompression must be used, and the diver is conscious and able to care for himself:
        1) Use the deep-sea diving rig if available.
        2) Follow treatment tables as closely as possible.
        3) Maintain constant communication.
        4) Have a standby diver ready and preferably use a tender with the patient.
    c. If the diver is unconscious or incapacitated, send another diver down with him to control his valves and otherwise assist him.
    d. If a lightweight diving outfit or scuba must be used, keep at least one diver with the patient at all times. Plan carefully for shifting rigs or cylinders. Have an ample number of tenders topside and at intermediate depths.
    e. If depth is inadequate for full treatment according to the tables:
        1) Take the patient to maximum available depth.
        2) Keep him there for 30 minutes.
        3) Bring him up according to Table 2A. Do not use stops shorter than those of Table 2A.
13. The most frequent errors related to treatment:
    a. Failure of the diver to report symptoms early.
    b. Failure to treat doubtful cases.
    c. Failure to treat promptly.
    d. Failure to treat adequately.
    e. Failure to recognize serious symptoms.
    f. Failure to keep the patient near the chamber after treatment.
14. ALWAYS KEEP THE DIVER CLOSE TO THE CHAMBER FOR AT LEAST 6 HOURS AFTER TREATMENT. (Keep him for 24 hours unless very prompt return can be assured.)

TABLE 1-30 FROM U.S. NAVY DIVING MANUAL

(FORMERLY TABLE 1-21, 1963 DIVING MANUAL)

Treatment of decompression sickness and air embolism

Stops	Bends—pain only		Serious symptoms	
Rate of descent—25 feet per minute. Rate of ascent—1 minute between stops.	Pain relieved at depths less than 66 feet. Use table 1A if O_2 is not available......	Pain relieved at depths greater than 66 feet. Use table 2A if O_2 is not available...... If pain does not improve within 30 minutes at 165 feet, the case is probably not bends. Decompress on table 2 or 2A.	Serious symptoms include any one of the following: 1. Unconsciousness. 2. Convulsions. 3. Weakness or inability to use arms or legs. 4. Air embolism. 5. Any visual disturbances. 6. Dizziness. 7. Loss of speech or hearing. 8. Severe shortness of breath or chokes. 9. Bends occurring while still under pressure.	
			Symptoms relieved within 30 minutes at 165 feet. Use table 3	Symptoms not relieved within 30 minutes at 165 feet. Use table 4

Pounds	Feet	Table 1	Table 1A	Table 2	Table 2A	Table 3	Table 4
73.4	165	....	....	30 (air)	30 (air)	30 (air)	30 to 120 (air)
62.3	140	....	....	12 (air)	12 (air)	12 (air)	30 (air)
53.4	120	....	....	12 (air)	12 (air)	12 (air)	30 (air)
44.5	100	30 (air)	30 (air)	12 (air)	12 (air)	12 (air)	30 (air)
35.6	80	12 (air)	12 (air)	12 (air)	12 (air)	12 (air)	30 (air)
26.7	60	30 (O_2)	30 (air)	30 (O_2)	30 (air)	30 (O_2) or (air)	6 hr (air)
22.3	50	30 (O_2)	30 (air)	30 (O_2)	30 (air)	30 (O_2) or (air)	6 hr (air)
17.8	40	30 (O_2)	30 (air)	30 (O_2)	30 (air)	30 (O_2) or (air)	6 hr (air)
13.4	30		60 (air)	60 (O_2)	2 hr (air)	12 hr (air)	First 11 hr (air) Then 1 hr (O_2) or (air)
8.9	20	5 (O_2)	60 (air)		2 hr (air)	2 hr (air)	First 1 hr (air) Then 1 hr (O_2) or (air)
				5 (O_2)			
4.5	10		2 hr (air)		4 hr (air)	2 hr (air)	First 1 hr (air) Then 1 hr (O_2) or (air)
Surface			1 min (air)		1 min (air)	1 min (air)	1 min (O_2)

Time at all stops in minutes unless otherwise indicated.

MINIMAL RECOMPRESSION, OXYGEN BREATHING METHOD FOR TREATMENT OF DECOMPRESSION SICKNESS AND AIR EMBOLISM (TABLES 5 & 6)

Table 5

Depth (Ft.)	Time (Min.)	Breathing Media	Total Elapsed Time (Min.)
60	20	O_2	20
60	5	Air	25
60	20	O_2	45
60-30	30	O_2	75
30	5	Air	80
30	20	O_2	100
30	5	Air	105
30-0	30	O_2	135

Table 6

Depth (Ft.)	Time (Min.)	Breathing Media	Total Elapsed Time (Min.)
60	20	O_2	20
60	5	Air	25
60	20	O_2	45
60	5	Air	50
60	20	O_2	70
60	5	Air	75
60-30	30	O_2	105
30	15	Air	120
30	60	O_2	180
30	15	Air	195
30	60	O_2	255
30-0	30	O_2	285

Instructions for Use of Tables 5 and 6

1. Choice of Table. The short (135-minute) schedule of Table 5 is used for treatment of pain only bends if all pain is completely relieved within 10 minutes of reaching 60 feet. The long (285-minute) schedule of Table 6 is used for all serious symptoms, for recurrence, or if pain is not completely resolved after 10 minutes at 60 feet.

2. Oxygen Breathing. Commence O_2 breathing prior to descent. Descent time is not counted as time at 60 feet. If oxygen intolerance develops, discontinue the oxygen until 15 minutes after the reaction has entirely subsided, then resume the schedule at its point of interruption.

3. Descent. Normal rate of descent is 25 feet per minute. A more rapid descent is desirable if more serious symptoms are present.

4. Ascent. Ascent is continuous at 1 foot per minute. Do not compensate for slowing of the rate by subsequent acceleration. Do compensate if the rate is exceeded. If necessary halt ascent and hold depth while ventilating the chamber.

5. Relief Not Complete. If relief is not complete at 60 feet, proceed with Table 6 and observe patient's condition closely for any change, lengthen the schedule if thought necessary, or compress to 165 feet and treat patient on Table 2, 2A, 3 or 4 as appropriate.

6. Recurrence. If symptoms recur or if new symptoms appear during course of treatment with Table 5 or 6, return to 60 feet and retreat the patient on Table 6.

7. Lengthened Treatment. Table 6 can be lengthened by an additional 25 minutes at 60 feet (20 minutes O_2—5 minutes air) or an additional 75 minutes at 30 feet (15 minutes air—60 minutes O_2) or both.

8. Medical Evaluation. Before making a recommendation the responsible medical officer should carefully consider:

a. The diagnosis and exact condition of the patient.
b. The nature of any defect remaining.
c. The diving schedule which precipitated his injury and the magnitude of the omitted decompression, if any.
d. The time intervals elapsed between the end of the patient's last dive, the onset of injury, and the commencement of treatment.
e. The circulo-pulmonary condition of the patient and the status of his inert gas exchange.
f. The presence of other medical conditions which might complicate treatment or necessitate later transfer to a hospital.
g. Adjuvant medical treatment which might be of benefit.

9. Serious Symptoms. Unconsciousness, convulsions, weakness or inability to use arms or legs, air embolism, any visual disturbances, dizziness, loss of speech or hearing, chokes, bends under pressure.

10. Tender. Tender routinely breathes chamber air. If the treatment schedule is lengthened or if the treatment constitutes a repetitive dive for the tender, he must breathe oxygen for the final 30 minutes of ascent from 30 feet to the surface.

OXYGEN BREATHING METHOD FOR TREATMENT OF AIR EMBOLISM

Table 5A

Depth (Ft.)	Time (Min.)	Breathing Media	Total Time (Min.)
165	15*	Air	15
165-60	4	Air	19
60	20	O_2	39
60	5	Air	44
60	20	O_2	64
60-30	30	O_2	94
30	5	Air	99
30	20	O_2	119
30	5	Air	124
30-0	30	O_2	154

*Total time will vary as function of this stop. Medical attendant should take enough time to accomplish a thorough physical examination, since the ensuing treatment is based on patient's physical status.

Table 6A

Depth (Ft.)	Time (Min.)	Breathing Media	Total Time (Min.)
165	30	Air	30
165-60	4	Air	34
60	20	O_2	54
60	5	Air	59
60	20	O_2	79
60	5	Air	84
60	20	O_2	104
60	5	Air	109
60-30	30	O_2	139
30	15	Air	154
30	60	O_2	214
30	15	Air	229
30	60	O_2	289
30-0	30	O_2	319

Instructions for Use of Tables 5A and 6A

1. Recompression to 165 feet should be accomplished as rapidly as possible (usually less than 1 minute).

2. Total time at 165 feet will vary with the clinical status of the patient. The medical attendant should take the time to make a thorough physical appraisal of the patient. If all major symptoms and signs are gone before 15 minutes total bottom time, proceed to 60 feet at 25 feet per minute on air and begin oxygen as in Table 5A.

3. If serious or major symptoms or signs persist beyond 15 minutes, but show signs of moderating within 30 minutes (total bottom time), proceed to 60 feet at 25 feet per minute and begin oxygen as in Table 6A.

4. Should serious symptoms and signs persist beyond 30 minutes at 165 feet without moderation, begin treatment on Table 4.

5. Recurrence. If symptoms recur or if new symptoms appear during course of treatment with Table 5A or 6A, return to 60 feet and re-treat patient according to Table 6A.

6. Relief Not Complete. If relief is not complete at 60 feet, proceed with Table 6A and observe patient's condition closely for any change, lengthen the schedule if thought necessary or compress to 165 feet and treat patient on Table 4.

7. Lengthened Treatment. Table 6A can be lengthened by an additional 25 minutes at 60 feet (20 minutes O_2—5 minutes air) or an additional 75 minutes at 30 feet (15 minutes air—60 minutes O_2) or both.

8. Oxygen Toxicity. Should oxygen intolerance develop during the course of treatment, discontinue the oxygen until 15 minutes after the reaction has entirely subsided, then resume the schedule at its point of interruption. If the reaction occurred at 60 feet on Table 5A, upon arrival at 30 feet switch to Table 6A.

9. Tenders. Inside tenders routinely breathe air; however, if treatment constitutes a repetitive dive for a tender, he must breathe O_2 from 30 feet to the surface.

10. Serious Symptoms. Unconsciousness, convulsions, major paralysis or weakness.

11. Follow-up. Patients with air embolism on leaving the chamber should be routinely admitted for observation and given a thorough medical examination, including X-ray, carefully checking for pneumothorax. Patients should be released to duty only if completely asymptomatic and medical clearance is indicated by examination.

Chapter 6

RIGGING

There is virtually no underwater activity that does not require the application of good rigging practice. The diver seldom makes a move without using a rope, knot, cable, sling, shackle, or hook. His diving hose and safety line, if properly made up, is an excellent example of careful rigging. It incorporates an eye splice, seizing and square knots. Fancy or decorative knots also serve a utilitarian purpose when used on the diving hose.

Three-strand Turk's-heads tied with colored plastic line make excellent and attractive depth markers on the hose, one for 10 feet, two for 20 feet, and so on. A short four-strand round sennit makes a fine lanyard for the safety quick-release hook. There are literally thousands of knots, and learning how to tie them is a gratifying and useful hobby. See the Bibliography for a list of fine books covering the subject.

The few knots, hitches, bends and splices illustrated in this chapter are some of the most useful and most frequently used. Learn to tie them all with your eyes closed. One or another of them will serve you in many underwater situations requiring the use of rope or cordage.

Some simple terminology will be helpful.

A *knot* is a series of loops, turns and locking crosses in a line, which when drawn tight will hold its form in the line.

A *hitch* is a knot tied around an object, and will generally come apart when the object is removed.

A *bend* is a knot used for tying two ropes together.

Whipping is tape or small line around the end of a rope to keep it from fraying.

Seizing is twine or small line for binding two ropes or other objects together.

The *standing part* of the rope is that stationary part around which the end is passed to form a knot.

The *bitter end* is the end of the rope with which the knot is tied.

The *bight* of a line can be either an uncrossed arc formed by any part of the rope, or that portion of the rope used to form a knot in the middle of a rope, without recourse to the ends.

The simplest knot of all is the *overhand knot* which is used to keep an unwhipped rope from fraying, or to give a rope end more surface for heavy pulling, such as the end of a bucket rope.

USEFUL KNOTS AND HITCHES

The *reef* or *square knot* is unquestionably the handiest and the most frequently tied knot in the entire world. With the second half of the knot tied with two short bights, we have the universal shoelace knot. Even so, most people tie the square knot correctly only by accident, usually ending up with a granny knot. A square knot, well drawn up, will not come loose, but a granny knot will.

There is a very simple verbal formula for coming up with a perfect square knot every time—right over left, left over right. Most knots are easier to tie if the line is held at the start with the palms up and the thumbs pointing outward. For the square knot, the formula means

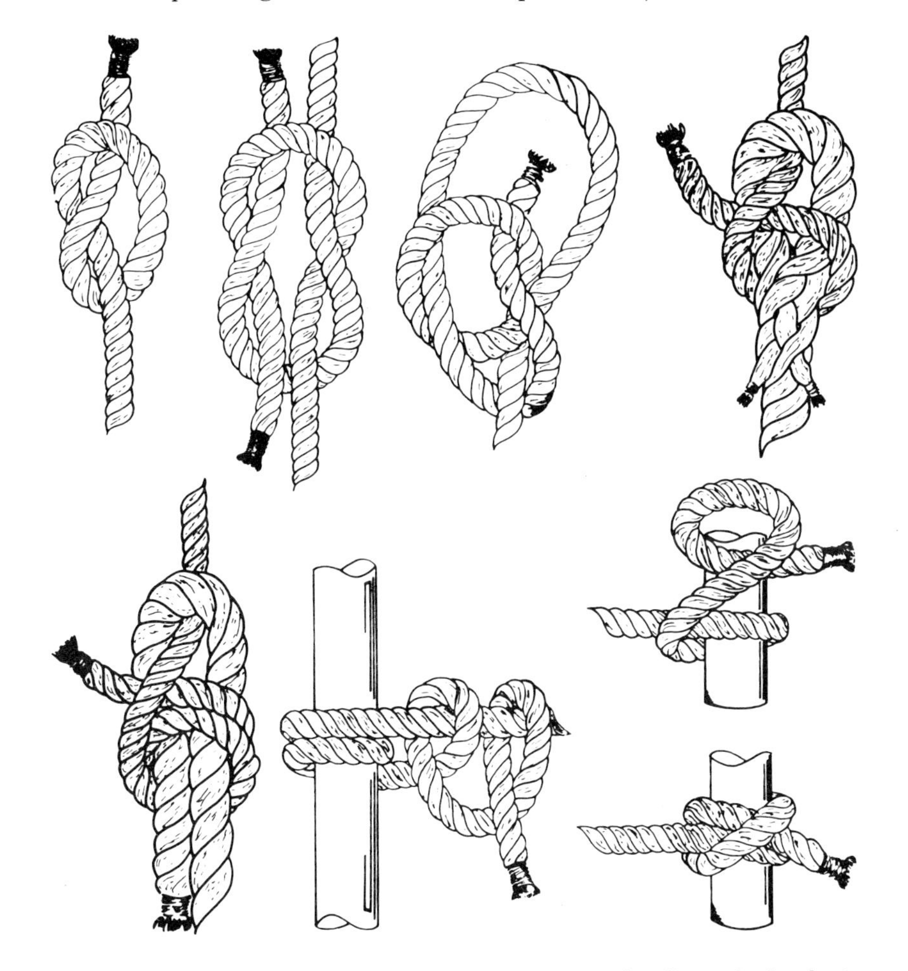

Fig. 19. (Top, l. to r.) Overhand knot; square knot; bowline; single sheet or becket bend. (Bottom, l. to r.) Double sheet or becket bend; round turn and two half hitches; clove hitch. Drawings by Art Herman.

the end of the rope in the right hand is passed over the standing part in the left hand and brought around to form an overhand knot. This end is now farthest to the left, and using the same end it is passed over the end on the right side and brought back up to form another overhand knot. A proper square knot is easy to identify in that both ends will lie in the same plane.

A square knot is excellent for fastening two ends of a line together or for a binder knot.

The *bowline* is another useful and universal knot. It is perfect for tying a quick, secure loop or eye in the end of a line. Bowlines seldom jam, even under heavy strain, and they are usually easy to untie.

The quickest way to tie a bowline is to lay the line across your palms with the end in your right hand. Grab the line, three or four inches from the end, between the thumb and forefinger of your right hand. Turn your hand over, facing down, and cross the end over the standing part in your left hand. Grab the standing part at the point of cross-over, with your thumb holding the cross firmly and roll your right hand forward, forming a half hitch, with the end of the line up through it. Pass the end of the line behind the standing part and then down through the half hitch. Hold the tail and the right leg of the loop in one hand and pull on the standing part to draw the knot tight. With most knots, and with the bowline in particular, it is important to leave enough bitter end of line protruding from the knot so that it will not slip out as the knot is drawn tight.

The best way to join two lines together, especially lines of different sizes, is with the *sheet* or *becket bend*. This bend can be doubled for greater security.

Form a tight U with one end of the lines to be joined and hold it in your left hand. Take the end of the other line with your right hand, pass it into the U from the underside, cross over the right side of the U, then back behind both legs and around to the front. Now pass the end over the left leg, then under the center strand and out over the right leg. To double it, just continue around the U in the same way for one more turn.

The easiest and fastest way to tie a rope to any object—a pile, a pipe, a ring, etc.—is with a *round turn* and *two half hitches*. This is simply passing the end of the rope around the object and then taking two half hitches around the standing part of the rope. Whenever tying this hitch, always take two round turns around the object. This lessens the strain on the half hitches, making them easier to untie, and it also makes the knot more secure.

The *clove hitch* is a simple, easy-to-tie hitch with a thousand uses. It is especially useful for attaching a line to a smooth object when you don't want the line to slide sideways. Tied with the end tucked back under the last round turn, it makes an excellent, easily-untied hitch for sending tools down to a diver. The clove hitch can be instantly tied in the bight of a line by pulling two half hitches together. The clove hitch is excellent for securing a line to a hook under water when the hook cannot be moused, with an additional half hitch around the standing part.

Without qualification, the handiest, most secure hitch I have ever tied is the *constrictor knot.*

I first learned to make this knot from the *Ashley Book of Knots*, and in over 20 years of diving and associated marine activities, I have never seen anyone else tie it. This is unfortunate, because it is an

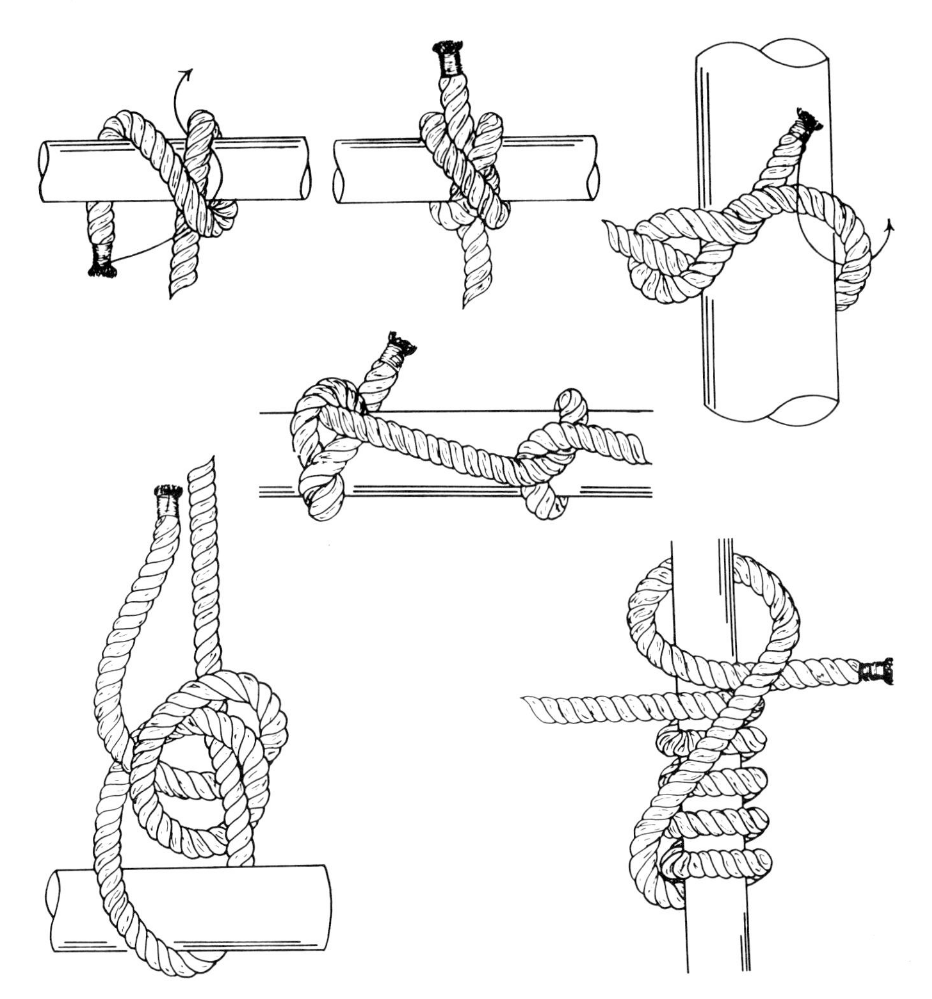

Fig. 20. (Top, l. to r.) Constrictor knot—two views; timber hitch. (Center) Half hitch. (Bottom, l. to r.) Midshipman's hitch; rolling hitch. Drawings by Art Herman.

ideal knot for securing tools and for many other purposes. Unlike the clove hitch, it will not loosen up when the strain comes from alternating directions. When tightly drawn up, this knot is very difficult to untie, but it can easily be tied with a slip loop.

This knot is perfect for such things as slippery hammer handles. It is also excellent for anchoring the end of a line to a pipe or rod before a series of half hitches is applied, for a vertical lift, in an application similar to that for which a timber hitch and half hitch might be used.

The *timber hitch* and a series of half hitches is quick and easy to tie and is used for a vertical lift, or for towing long objects, such as pipes, rods, pilings, etc. Be sure that the last half hitch is close to the forward or upper end of the object.

The *midshipman's hitch* is another excellent but seldom used hitch and it can be used to secure a loop in a line around an object while pulling a strain on the line. Pull the line around a pipe or stanchion, holding a strain on the line. Pass a half hitch around the standing part and then follow the half hitch around with a second one, right on top of the first, pulling hard on the end of the line to jam the knot. The knot will not slip with a strain on the standing part. One more half hitch must be passed around the standing part to secure the end.

The *rolling hitch* is perhaps the most secure hitch to use as a stopper on large rope, chain, cable, or any slippery cylindrical surface. The more turns applied, the greater the friction and therefore the security of this hitch. Each turn should be applied carefully with the slack taken out of the turns as you apply them. The pull should be in the direction opposite from the last hitch.

COILING ROPE

Quite often a diver will have to take a coil of small rope below with him as a search line, a measuring line, or whatever. If you are not going to spend half your dive untangling a hopelessly snarled mess, then you will have to spend a few minutes making up and securing a proper coil.

If it can possibly be avoided, long lengths of natural-fiber rope—manila, sisal, hemp, etc.—should never be given to a diver fresh out of the coil without first being wetted and then stretched with a healthy strain. If this is not done the rope will contract into an impossible twisted mass. This is not the case with synthetic fiber ropes, such as nylon or polypropylene, but you must constantly bear in mind that these ropes are very slippery and all knots tied with them should be drawn up as tight as possible, and further secured with an extra half hitch or two. An excellent way to safety a knot in synthetic rope is to open up the strands of the standing part after the knot is tied, and tuck the bitter end through.

The coil shown in Fig. 21 is easy to put together and easy to undo under water. With the bitter end tied to a ring on your weight belt, both hands are free until you are ready for it.

WHIPPING

Rope of any kind, natural or synthetic, should never be taken under water or sent to a diver without whipped ends. Without whipping, the strands of a rope will quickly unwind under water and the individual fine threads and yarns will cause an annoying and sometimes dangerous entanglement with almost anything they come in contact with.

A *lineman's whipping,* which is nothing more than three or four tight wraps of tape around the rope about an inch from the end, is perfectly acceptable for anything short of permanent rigging and applying one to a rope is no great imposition on any member of the crew. Lost time under water due to fooling around with cows' tails and twisted lumps of rope is impossible to calculate but the dollar loss to the industry from these two simple-to-correct causes is astronomical.

THE DIVER'S KNIFE

Most people, including a number of divers, believe the purpose of a diver's knife is to fight sharks, and this fact is substantiated by the very poor design, for cutting purposes, of almost every so-called diver's knife on the market. The primary reason a diver carries a knife is to cut cordage, either in the execution of his work or to clear himself if fouled. A good cutting knife for cordage must be thin-bladed, wedge-shaped, wide, and, of course, razor-sharp.

As mentioned earlier, the stainless-steel folding rigger's knife, often used by yachtsmen, makes an excellent diving knife. The marlinespike is excellent for starting or tightening small shackles. These knives do not have stainless-steel springs, and if they are not rinsed in fresh water and oiled after every dive, the springs soon deteriorate and crack.

The Case rigger's sheath knife is another exceptionally good diver's knife but a different sheath from the one supplied must be fashioned. The handle is wood and should be kept oiled or varnished. This knife is handiest when secured to your belt or sheath with a short lanyard.

A knife for cutting rope is best sharpened on a rough stone only. The rough stone leaves a microscopic saw edge that displaces the fine fibers during the cutting stroke, preventing a friction bind on the blade when cutting through thick rope. (Extremely heavy rope is easiest to cut if you roll the rope, cutting most of the way through one strand at a time, making the final cut through the center.) A diver should always take the time to cut away any extraneous ropes or lines from his work area to prevent fouling.

If a rope with frayed or loose ends is sent to you, take the time to clean up the end. If the line does not have to be passed through a small opening, an overhand knot pulled tight will suffice. For large diameter rope, or rope that must be poked through a hole, cut away a foot or so and unwind several yarns. Four or five turns around the rope end with these yarns secured with a square knot will make an adequate temporary whipping.

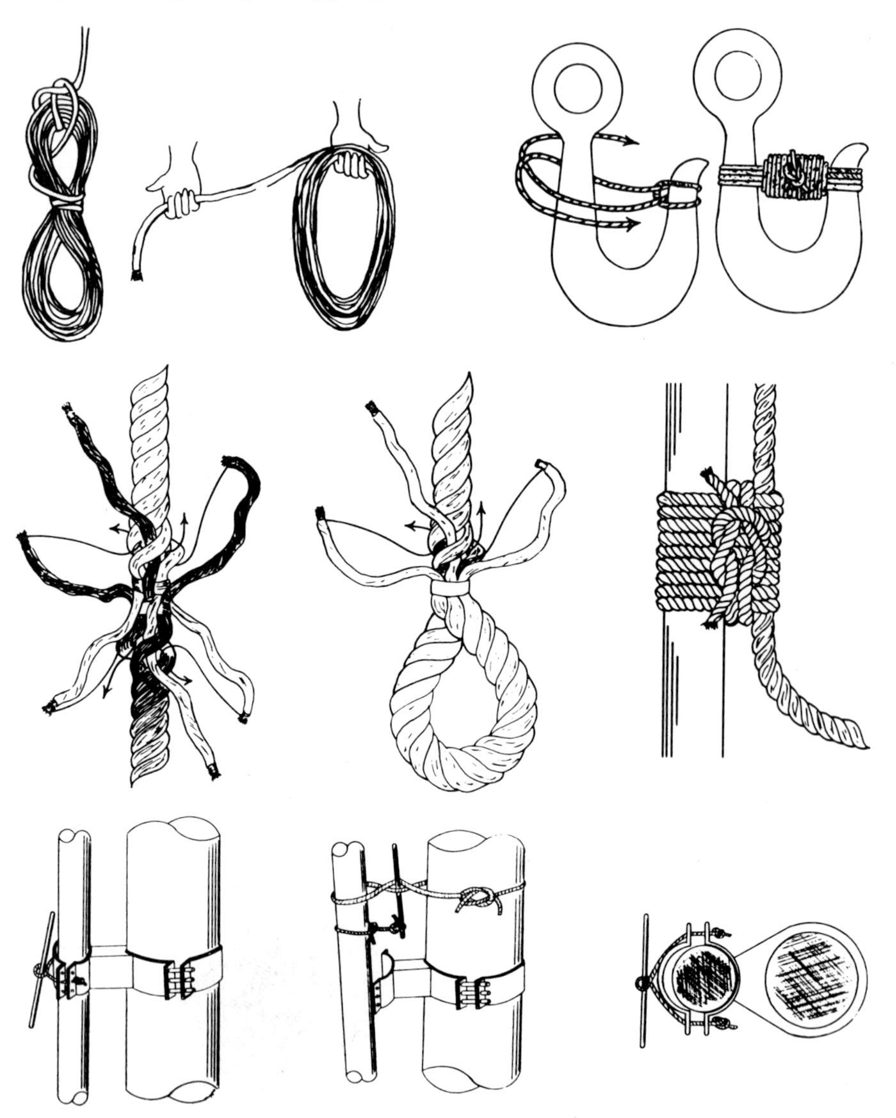

Fig. 21. (Top, l. and r.) Coil of line and mousing a hook. (Center, l. to r.) Short splice; eye splice; seizing hose to life line. (Bottom) Spanish windlass—three views. Drawings by Art Herman.

SPANISH WINDLASS

A Spanish windlass is a method of applying power to a rope by means of a bar or lever. There are several types of Spanish windlasses. The two shown in Fig. 21 can be used to advantage in a number of underwater situations.

Wind the bar until you have pulled the clamp or small riser in place. The end of the bar can be tied off to hold the strain until the bolts are in place.

MOUSING A HOOK

Whenever a sling is sent to a diver, or whenever a diver has to put a sling on a hook under water, the hook should always be moused if it does not have a mechanical safety. This is to prevent the sling eye from coming off the hook if the load is slacked or if the barge is rolling.

The simplest mousing is made by doubling up a piece of small stuff about two feet long. Pass the doubled legs through the loop as shown in Fig. 21 and loop the noose around the bill and the shank of the hook. Pull the noose tight around the back of the hook; cross the strands on either side of the bill; take several turns in opposite directions around the bill and the shank; and then make four or five frapping turns around the rope, ending with a square knot.

SEIZING

Seizing is used to join two or more lines together. For the diving hose and life line it is best done with cotton seine twine. Seizing, if it is worth doing at all, must be pulled very tight. In a 300-foot diving hose seized every five feet or so, there are 60 or more seizings. If you do not wear leather gloves, the first joint of the little finger of both hands will be sawed open after only eight or ten seizings.

Cut your seine twine into 2-foot lengths. Tuck a 6-inch tail between the life line and hose, and hold it firmly with your left hand. With your right hand, wrap the seine twine tightly around the hose and the safety line five or six times, making sure the successive turns ride close beside each other, but without crossing. Pass the working end between the hose and line and make several turns with both ends going in opposite directions, longitudinally between the hose and life line. End with a square knot. After the square knot is drawn tight, trim your ends but not too short; leave about an inch. Cover the seizing with three or four tight wraps of tape.

When tying the square knot with tension on the line, if someone is not standing by with a convenient index finger to hold down on the first overhand knot, take two turns with the first overhand knot. This will hold the tension, allowing you to tie the locking overhand knot without having the seizing slack off.

When making up a hose for seizing and taping, all the components—hose, telephone cable, pneumohose and life line—must be stretched out for at least a run of 50 feet ahead of the work area. This will eliminate twisting tensions in the assembly. All components must be nearly slack or under slight equal tension to prevent bights in the completed hose. The most common error in making up a hose is pulling too much tension into the life line while seizing and taping. When the tension is slacked, the hose ends up looking like a dragon's back because of all the bights. Put a good strain on the safety line to stretch it before you make it up; then pull a slight tension in all the components and tie them off. Test-tape eight or ten feet. Slack it off and see how it lies. When you have your tensions adjusted properly, seize and tape your hose.

SPLICING

It is unlikely that you will ever need to splice either rope or wire rope under water; however, the *eye splice* and the *short splice* for rope are very handy to know and will be used frequently in setting up your various rigging projects.

For the eye splice, put a few wraps of tape or a temporary whipping around your rope eight or more inches from the end, and unwind the strands. Form a loop with the end of the rope, the size of the desired eye, and fan out the three tails. Lay the tails on top of the standing part and hold them firmly with your left hand, thumb on top. Lay the center tail straight down the rope, the left tail pointing left and the right tail pointing right. Grip the standing part with your right hand and twist your hands in opposite directions, opening up the strands. Tuck the center tail under the topmost strand of the standing part, going from right to left. Cross over this strand with the left-hand tail and tuck it under the next strand, going from right to left. Roll the splice away from you and pass the last strand over the top of the next strand from left to right, and then back under the same strand from right to left. This completes the first tuck. No two tails should end up coming out from under the same strand. Pull each tail tight and apply two or more additional tucks, going over one strand and under the next. All tucks will be made working the tails from right to left. When working the splice, pull the slack out of each tail evenly.

When the splice is completed, roll it vigorously between the palms of your hands or under your shoe to round it off. Trim the ends, but not too short; the length of the ends should be equal to the diameter of the rope.

With Manila or natural fiber rope, three tucks are sufficient. Because synthetic rope is so slippery, put four or five tucks in your splice and finish it off with tape or whipping at the bottom of the splice for greater security. It will be much easier to tie your splices if a few wraps of tape are put around the ends of your individual strands to keep the fibers and yarns from fuzzing up. The ends of synthetic ropes can be fused by holding them into an open flame for a few seconds. Let the material cool and solidify before trying to work it.

For the short splice, put some tape or temporary whipping eight or more inches from the ends of the two lines to be joined and unwind the strands down to this point.

Join or "marry" the two ends with alternating tails against the standing parts, heading in opposite directions. Take one complete tuck in the right-hand rope, working the tails over one and under one, and then reverse the rope and take a complete tuck in the other end. Pull the tucks up tight and add two additional tucks to each rope. Finish off the same as an eye splice.

ROPE STRENGTHS

It is well to know the strength of the various materials and fittings used for rigging. Refer to the table for Manila rope.

Safe Load for New Manila Rope—3-Strand

SIZE			SIZE		
Circumference in Inches	Diameter in Inches	Safe Load Pounds	Circumference in Inches	Diameter in Inches	Safe Load Pounds
3/4	1/4	120	3	1	1,800
1-1/8	3/8	270	3-1/2	1-1/8	2,400
1-1/2	1/2	530	4-1/2	1-1/2	3,700
2-1/4	3/4	1,080	5-1/2	1-13/16	5,300
2-3/4	13/16	1,300	6	2	6,200

Note: For sisal rope, decrease weight of load by one-third or use rope one size larger than indicated in the table.

Nylon is about twice as strong as Manila rope of the same size. The various polyester fiber ropes or ropes of combinations of synthetic fibers have strengths ranging from that of Manila to that of nylon.

Knots and splices reduce the strength of rope. The eye splice and short splice retain about 85 percent of the strength of the rope, while the efficiency of a bowline is 53 percent and that of a square knot is only 43 percent.

A rule of thumb for Manila rope up to one inch in diameter is: The safe load in tons equals the diameter in inches, squared.

WIRE ROPE CLIPS AND SLINGS

Splicing wire rope is the purview of the professional rigger and will not be discussed here. However, you will be required to put an eye in wire rope occasionally, and this can be done with either a *Molly Hogan* (also known as a *quick eye* or *Flemish eye*) or with cable clips.

For the Molly Hogan, quick eye, or Flemish eye, put a soft wire whipping around the wire rope about three times the length of the desired eye from the end. Split your wire rope down the middle into two sections and unwind them down to the whipping. Form a loop of the desired size crossing the two sections at the top and lay the tails in opposite directions back into the wire rope, taking care to twist the tails as necessary so they lay up perfectly. After the eye is formed, bring the two tails together below the throat and lay them up into a single cable. Finish it off with a wire rope clip below the throat. The Molly Hogan is approximately 70 percent efficient.

Number of Clips and Distance Between Clips Needed for Safety

Diameter of Rope (Ins.)	Number of Clips	Distance Between Clips
1/4-3/8	3	2-1/4"
7/16-5/8	4	3-3/4"
3/4-1-1/8	5	6-3/4"
1-1/4-1-1/2	6	9"
1-5/8-1-3/4	7	10-1/2"

Safe Load in Pounds for New Improved Plow Steel Wire Rope Slings

Under Different Loading Conditions

6 Strands of 19 Wires, Hemp Center

Size Diameter in Inches	Single Wire Rope Sling Vertical Lift	Sling or 2 Wire Ropes—Used at 60° Angle	Sling or 2 Wire Ropes—Used at 45° Angle	Sling or 2 Wire Ropes—Used at 30° Angle
3/8	2,500	4,300	3,600	2,500
1/2	4,300	7,400	5,800	4,300
5/8	6,600	11,400	9,400	6,600
3/4	9,400	16,200	13,000	9,400
7/8	12,800	22,100	17,400	12,800
1	16,000	27,700	23,200	16,000
1-1/8	21,200	36,700	29,700	21,200
1-1/4	26,000	45,000	36,200	26,000
1-3/8	31,400	54,300	43,500	31,400
1-1/2	37,000	64,000	52,200	37,000

Note: Avoid angles of less than 45°.

When making an eye using wire rope clips, the U-bolts must always be on the dead end of the line and the saddles on the live or standing part. Alternating U-bolts or placing U-bolts on the standing part will dangerously weaken the cable because they crush the strands. Cable clips must be set up as tightly as possible with a wrench. After the cable has a strain on it, tighten the U-bolts again to take up the slack caused by the cable stretching.

A good way to remove all the slack when tightening the clips is to rap the back of the U-bolt with a hammer a few times while tightening the nuts. Whenever possible, use a steel thimble in any eye, in soft rope or wire rope. This greatly increases the strength of the eye. Do not form an eye by half-hitching wire rope and stopping the end with a wire rope clip. This type of eye is less than 50 percent efficient. A clipped eye is 80 percent efficient.

The safe load for plow steel cable is eight times the diameter in inches, squared.

HOOKS AND CHAIN SLINGS

Open hooks on cable or chain slings are not generally used for underwater rigging because of the ever-present chance of their coming unhooked. I do feel that there are a number of instances when a snap hook or a mechanical safety hook would be a timesaving alternative to shackles, especially for light loads and sending tools to a diver. There are several rules that must be observed when using hooks, either open or safety.

On vertical lifts with choked slings, the bill of the hook must always point down. When using two slings with hooks for lifting objects with pad eyes, the hook bills must point outward, away from the center of the load.

When using two slings with hooks for picking up a long load from two points, such as a joint of pipe, the load must be balanced as evenly as possible between the slings and the hook bills must again point outward, away from the center of the load. If the slings are used as chokers around the pipe, two turns should be taken before hooking back on the standing part. Sprung hooks should never be used.

Short chains with a safety hook on one end and a grab hook on the other can be used to advantage on a number of underwater rigging applications. Chain has a much better grip on cylindrical surfaces than cable and should be used whenever slipping would be undesirable, for instance when rigging a snatch block to a structure brace or diagonal for a sideways pull on a riser. After taking two or three round turns on the brace, the grab hook can be attached to the chain to give you the fairest lead. The length of a chain with a grab hook can be readily adjusted also. If the slip hook does not have a safety latch, mouse it as shown in Fig. 22.

It is quite often necessary to impart a rolling force to a pipe, as in lining up flange bolt holes, and a chain is much more effective for this than a cable. Wherever rigging passes over rough surfaces or sharp corners or edges, chain should be used. A rolling hitch tied with chain is an excellent stopper for cable. Again, the more turns, the more secure is the hitch. Chains should be frequently inspected for sprung or stretched links and for cracks.

Chains made of standard links and used for hoisting or similar purposes must not be subjected to loads greater than shown in the table.

Safe Load in Pounds for New Wrought Iron Chain Slings

Diameter of Link Stock in Inches	Single Chain Vertical Lift	Sling or 2 Chains—Used at 60° Angle	Sling or 2 Chains—Used at 45° Angle	Sling or 2 Chains—Used at 30° Angle
1/4	1,000	1,800	1,500	1,000
3/8	2,300	4,100	3,300	2,300
1/2	4,200	7,300	6,000	4,200
5/8	6,600	11,400	9,300	6,600
3/4	9,500	16,500	13,500	9,500

Note: Avoid angles of less than 45°.

A chain is no stronger than its weakest link. It should be discarded when it shows evidence of having been stretched. Stretching can be distinguished by: (1) small checks or cracks in the links; (2) the links binding on each other; (3) the links showing elongation.

SHACKLES

Safe Load in Pounds

Dropped Forged Steel, Weldless

Diameter of Pin (Ins.)	Max. Width Between Eyes (Ins.)	Safe Working Load (Lbs.)	Diameter of Pin (Ins.)	Max. Width Between Eyes (Ins.)	Safe Working Load (Lbs.)
1/4	3/8	560	1-1/4	1-7/8	16,000
3/8	9/16	1,400	1-3/8	2	20,000
1/2	11/16	2,700	1-1/2	2-1/8	24,000
5/8	13/16	3,600	1-5/8	2-1/4	28,000
3/4	1-1/16	5,600	1-3/4	2-3/4	32,000
7/8	1-1/4	7,800	2	2-3/4	36,000
1	1-1/2	10,400	2-1/4	3-1/4	46,000
1-1/8	1-5/8	13,200	2-1/2	4	56,000

All shackle pins must be straight and all pins of the screw-pin type must be screwed in all the way. If width between the eyes is greater than shown in the table, *the shackle has been overstrained and must not be used.*

The two types of shackles most often found offshore are the screw-pin shackle and the safety shackle (Fig. 22). The safety shackle has a number of distinct advantages for the diver and should always be used whenever a choice is possible. With a safety shackle, once the pin is rammed home, the load is reasonably secure and the diver will not have to support the weight or the bending tension of heavy slings while he is screwing on the nut.

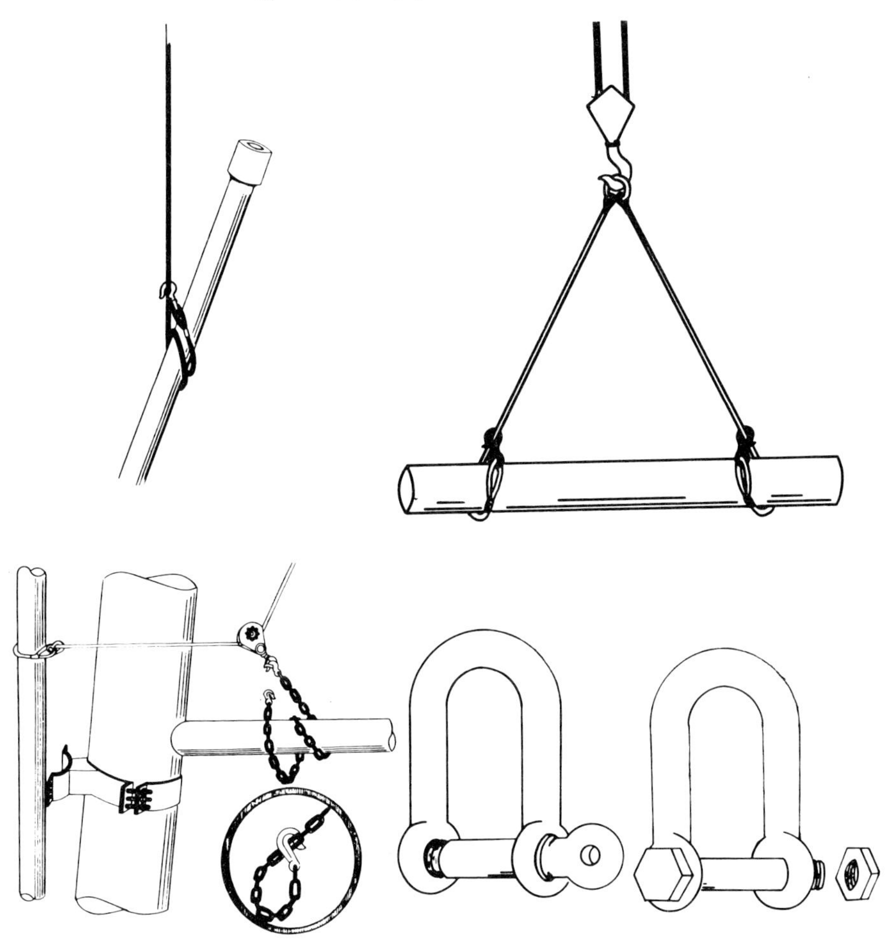

Fig. 22. (Top) Hooks must point down on vertical lifts; bridle hooks must point outboard on horizontal lifts. (Bottom) Chain and grab hook used to position snatch block; screw-pin shackle; safety shackle. Drawings by Art Herman.

With a screw-pin shackle, the diver must keep the weight of the sling off the pin until it is completely screwed home. This is a very difficult job when trying to shackle a heavy sling around a vertical object, and doubly difficult when the sling location is up off the bottom, or where the diver has no solid place on which to stand. Screw-pin shackles are also more difficult to safety, requiring several turns of wire through the eye of the pin and around a leg of the shackle. A safety shackle is positively secured from coming undone by simply poking a wire or a welding rod through the hole in the pin behind the nut and bending it to keep it from dropping out. A screw-pin shackle must have all the threads engaged to meet its full strength potential. Whenever shackles of any type are used with slings, the pin should always be in the eye to keep the pin from rolling and possibly coming undone. Sprung shackles should never be used.

Shackles should never be sent to a diver unless the threads are clean and easily screwed up by hand. When lowering slings with shackles to a diver, don't screw the pin or the nut up tight. Leave it a few threads loose so the diver can take it apart quickly.

When you are going to make a dive, either to fasten or unfasten shackles, take the proper tools with you. You will need a pair of side-cutting pliers to remove any safety wire or welding rod, a hammer to start the nut on a safety shackle, and a marlinespike or a spud wrench to loosen or tighten large screw pin shackles.

When you are rigging with shackles, visualize what will happen when the desired strain is applied. Hook up your shackle so that the pin can be readily inserted and removed without jamming. When rigging with slings and shackles, especially from a barge subject to sudden movement from swells, keep your fingers and hands clear. If you should find that you have passed a turn or two of the sling around your hose as well as the object on which you are working, thereby shackling yourself to the load, it won't be the first time it has happened. Keep your hose and yourself clear of the rigging. You are dealing with tremendous forces, capable of crippling and maiming in an unguarded instant.

Whenever possible, take an extra round turn with your sling. This reduces the strain on the shackle and it also helps to prevent the sling from slipping longitudinally on your load. Whenever passing heavy slings, or slings in awkward positions, tie a piece of soft line into the sling eye. In the case of a large-diameter pipe, you can poke the soft line under the pipe, cross over the top and pull your soft line until you have pulled the sling eye through. Then, hop up on top of the pipe and pull the sling eye up to you. You can stop the soft line off on the standing part (two round turns, two half hitches) and the sling eye will be held in place while you manipulate the shackle. In this particular instance, remove the shackle from the sling eye before trying to drag it under the pipe.

WORKING WITH LIFTS, SLINGS, CRANES, ETC.

When trying to impart a moving force to an object—for instance, picking up an end of pipe to line up two flanges, rather than lifting it free from the bottom—use a heavy nylon sling between your cable or chain choker and the crane hook, especially in rough seas. The stretching nylon makes an excellent shock absorber between the rolling barge and the pipeline. The nylon must, of course, be large enough to withstand the expected loads.

Whenever possible, as when installing risers, shift your lifting power source, winches, air tuggers, etc., to a permanent or stationary platform, to eliminate motion in your rigging caused by barge roll and heave.

Whenever working with any topside power source—tugger, winches, derricks, etc.—the diving telephones must be in excellent working order.

The telephone operator must be instructed to stay off the phone whenever power is applied or cables are moving. I have more than once given the order to take a strain with a rig and then watched in horror as my elaborately planned rigging was ripped out by the roots while some imbecile was chattering away over the telephone, preventing me from giving the order to stop.

You must give your orders for heaving up or slacking off clearly and unambiguously. When the rigging is in motion, don't talk or say anything between the order to pick up and stop.

Whenever you are ordering a lift, try to estimate the distance of travel. Convey this topside by saying "Pick up six inches and stop." "Pick up two feet and stop." Moving in short increments eliminates situations such as the above, and it also gives you a chance to observe your rigging as it is being tightened up. When you are using more than one cable from the surface, have them clearly identified. The crane or derrick will usually have one lifting cable rove through multiple sheaves, called the load block, and another single cable called the whip. Specify clearly which one you want in motion. In the case of air tuggers, you may have three or four cables to work with. Identify them with numbers 1, 2, 3, and so on, and mark them with corresponding tape or small line markers at the end you will be working with.

When taking a lift with a crane, you will want the crane hook centered as nearly as possible over your load. When you are on the bottom, try to picture your position relative to the center pin of the crane. Give your directions with this position in mind. If you want the load block moved away from the center pin, boom down, bearing in mind that the block will have to be picked up simultaneously to keep it off the bottom. If you want the load block to move towards the center pin, you must boom up, bearing in mind that the block will have to be slacked off to keep it in reach. For swinging to one side or the other, give your orders for a swing to port or starboard of the barge to avoid confusion with the crane operator.

Make all boom movements slowly and in short increments. Depending on the depth of water, the tip of the boom will reach a position along a line of movement several seconds before the load block. Give your order to move, then stop and allow the block to steady down, then move again, if necessary.

Before taking a heavy lift, move well away from the rigging in case anything parts. Don't allow a heavy or uncertain lift to precede you to the surface. If the load cannot be swung far enough away so that you are absolutely in the clear, the lift will have to wait until you get on deck.

Keep your fingers away from the sheaves of snatch blocks and load blocks. There are many odd-fingered divers who can attest to the wisdom of this statement.

HAND SIGNALS FOR DIRECTING A CRANE BOOM

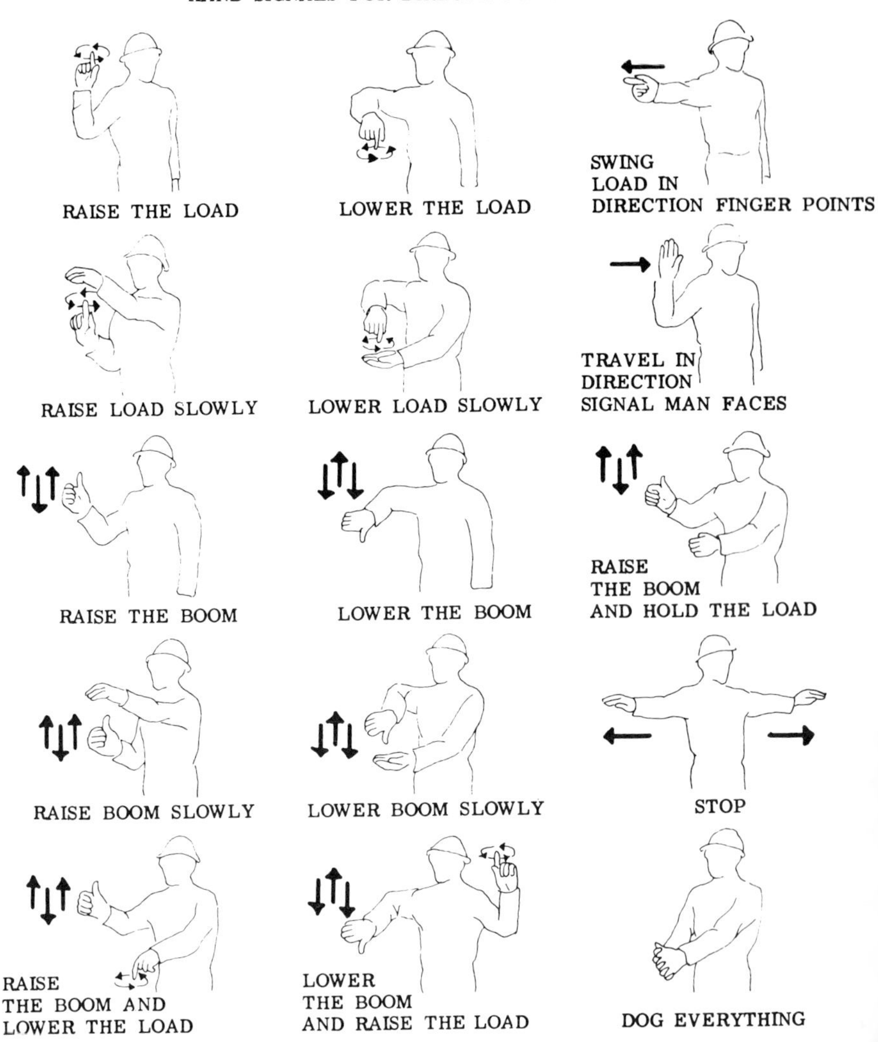

Fig. 23. Hand signals for directing a crane boom. Suggested by a National Safety Council poster. Drawing by Sue Zinkowski.

CRANE SIGNALS

When divers will be working in relays on a given project, it is important that an agreement be reached among all of them as to the rigging procedure to be followed, especially in deep water.

Proper rigging is time-consuming but essential. It is always worth the effort to plan and to rig properly from the start. Remember that the straightest line of force is the most efficient. If you are going to pick up, have your lift line directly above the load. If you want to move something laterally, rig your snatch block to pull from as close to 90 degrees as is possible under the circumstances. Learn proper crane signals. If there are any regional or trade variations on the ones shown in Fig. 23, learn the signals that are being used.

Make all signals clearly and positively. If you haven't made up your mind which signal you want to give, keep your hands by your side until you have decided. It is up to you to be sure that you are in the crane operator's line of sight. It must be determined beforehand who is going to tag, or signal, the crane operator. He can serve only one master.

LIFTING AND PULLING DEVICES

Chain falls, together with a number of different types of lever-operated, hand-powered lifting or pulling devices, such as Coffin Jacks, come-alongs and ratchet hoists, are important rigger's tools and they all have many applications in underwater work, especially for exerting lateral pulls and for balancing awkward loads, such as heavy riser clamps.

They are all subject to occasional idiosyncratic behavior, especially when it comes to reversing the direction of travel. Examine the particular one you will be using before it is put in the water, and be sure you know how to reverse it. All of these devices have many moving parts, and on-the-job maintenance is usually difficult. It is best to keep them in a drum of diesel oil when not actually in use, and at the end of the job the mechanic should completely disassemble the unit for proper cleaning and lubrication.

Spur-gear chain hoists have a fiber friction disc or shoe to hold the load or brake it when lowering. This fiber loses some of its friction efficiency when it is soaked in either water or fuel oil, and the unit should therefore be used at less than its rated capacity under water. The discs or shoes should be frequently replaced.

A cheater pipe should never be used on come-alongs or other lever-operated lift-pulls. Using one will only spring the hooks or damage the unit. If you are unable to jack the load with the handle supplied, it will be necessary to get a larger unit, or use several of them to do the job. Never use any of these devices with sprung hooks or chains. Always be sure the unit works in both directions and that the chains are not kinked or frozen, before they are ever put in the water.

Rig properly and rig safely.

... "come along" or a ratchet chain hoist. Photo: Divcon, Inc.

Chapter 7

BURNING AND WELDING UNDER WATER

Severing metal under water is one of the most frequent and important tasks that a diver performs. There are always cables to be burned away from tugboat and ship propellers, and burning sheet-piling and H-beam piling on cofferdams and bridge foundations constitutes perhaps the bulk of a construction diver's work.

For the oil-field diver, there are damaged pipelines to cut in preparation for repairs, jacket-leg piles to be cut in structure salvage and skirt piles to cut for structure installation, in addition to a thousand other jobs that require a cutting torch. A good diver must be a good burner.

GAS CUTTING

Torch cutting of iron and steel, on the surface or under water, is accomplished by the burning or oxidizing of the steel to be severed along a narrow line called the kerf. A small point of the steel to be cut is heated to the molten state by one of several methods and then oxygen is directed at that point. The oxygen combines chemically with the molten steel, instantly converting it into various gases and chemical compounds, literally burning it. In addition, the pressure of the oxygen stream drives the converted material, or slag, out of the cut or kerf. With a gas torch, the flame is angled or inclined slightly in the direction of the desired cut, thereby preheating the metal immediately ahead of the cut and permitting a continuous cutting operation.

Gas-torch cutting is a very refined industrial process, capable of producing smooth, even cuts in steel 20 inches and more in thickness.

Underwater gas-torch cutting uses acetylene or hydrogen for the fuel gas. Acetylene is very unstable at pressures above 15 psi and so it is only used in depths less than 25 feet. It is possible to use hydrogen at relatively great depths, but the hydrogen flame is not as hot as acetylene.

The technique of burning under water is exactly the same as burning on the surface. Underwater gas torches are modified slightly from surface-use torches in that they have an additional valve and a skirt around the tip through which compressed air is blown to create a bubble or artificial atmosphere, within which the flame burns. It is possible to burn with Map gas, a highly refined form of acetylene, with a standard surface torch and a special Map gas tip.

Although gas-torch cutting under water is possible and even practical it is seldom employed in the offshore oil business. There are several reasons for this; the dominant ones are the logistic difficulties in supplying the numerous and bulky cylinders of special gases required and the high degree of technical skill demanded of the diver to operate a gas torch properly under water. Because of its very limited usage, gas-torch cutting under water will not be elaborated upon.

If, however, you have a desire to include competent underwater gas-torch cutting of steel among your technical abilities, first learn all you can about gas-torch cutting on the surface. On every pipeline job there will be ample opportunity to practice on scrap and the welding foreman should be delighted to instruct you and help you with your burning.

When you have mastered this skill topside, whatever additional technical information you need can be found in the Navy Manual, *Underwater Cutting and Welding* and in several other books listed in the Bibliography.

OXY-ARC CUTTING

Almost all underwater cutting of steel in the offshore oil fields is done by the oxy-arc process.

This process uses the tremendous heat (about 6500°) of an electric arc to preheat the metal. As with gas cutting, when the steel is heated a high-pressure jet of oxygen is introduced to the molten steel to complete the cutting process. Unlike gas cutting, where varying periods of time (depending on the thickness of the steel) are required to heat the steel sufficiently for burning to take place, the oxy-arc torch begins burning almost instantaneously upon the striking of the arc and the introduction of oxygen. This is because of the concentration of the greater heat of the arc on a very tiny area of the steel to be cut. As with gas cutting, oxy-arc cutting can be and should be continuous. Compared to gas-torch cutting, oxy-arc cutting leaves a very rough and ragged edge.

Equipment required for oxy-arc cutting is a welding machine, a supply of high-pressure oxygen, an oxygen pressure regulator, welding and ground leads, a ground clamp, oxygen hose, a single-pole knife switch of sufficient size and capacity to carry the currents required, a supply of tubular steel electrodes, an oxy-arc cutting torch, a broad-knife stiff-blade scraper, sledgehammer, rubber gloves, and in some cases, a light welding lens.

In an emergency situation, oxy-arc cutting can be effected with surprisingly low currents. However, proper, fast and economical cutting can only be done with a machine capable of producing 300 or more amps. Thicker steel requires higher amperage, as do longer leads. It is possible to cut with either AC or DC machines, but DC is preferable because of its lower potential for dangerous shock and reduced corrosive action on the torch and metal parts of the diving gear. When using AC currents, any metallic parts of the diving mask or helmet, such as the chin button, that will come in contact with the diver must be insulated with electrical tape. The welding-machine frame should be grounded to the barge or work vessel.

The machine heat settings and the polarity of the current are critical to fast, efficient cutting. Higher heat translates into faster cutting. Welding machines of over 400 amps are rare around the oil field, and so I would recommend that the machine be set to its maximum for any cutting of steel over ½-inch thick. Higher heat also results in faster rod consumption, but this is compensated by faster cutting.

Straight polarity is best for efficient underwater oxy-arc burning in that most of the heat is in the work, not in the electrode. It also results in less corrosion of the torch.

The polarity of welding machines will sometimes change due to an electrical phenomenon. If no welding foreman is present to assure you of the correct polarity of the machine you intend to use, you can determine this for yourself by a simple test.

With the knife switch open and the leads hooked up to the machine, insert a rod in the torch and also clamp a rod in the ground clamp. Have someone help you by holding either the torch or the ground, and put the tips of both rods into a bucket of water so that they are about an inch apart. Bubbles will flow from the rod connected to the negative or minus pole, while hardly any will flow from the other rod. *Remember: Burn with straight polarity, electrode negative.*

Because underwater burning is done with much higher O_2 pressure than that required for surface burning, most standard O_2 regulators cannot be adjusted to provide sufficient pressures for burning at even moderate depths. When rigging up for a burning job, be sure you have a regulator that will deliver O_2 at a supply pressure at least 100 pounds greater than the water pressure at the depth of your work.

Most treatments of oxy-arc burning with which I am familiar recommend varying O_2 pressures ranging from 20 pounds over ambient for ¼-inch plate to 75 pounds over ambient for 1-inch plate. My experiences indicate an O_2 pressure of 60 pounds over ambient for anything up to ½-inch steel and up to 125 pounds over ambient for 1¼-inch steel. This may at first seem a bit prodigal with expensive O_2, but it is well justified by fast, efficient cutting, especially in deep water where diving time is so precious.

PREPARATIONS FOR EFFICIENT BURNING

The torch and ground leads must be completely insulated for their entire length, both on the surface and under water. Current leakage under water through cracked or cut leads will only mean insufficient heat at the cut with consequent slow, uncertain and inefficient burning. Current loss as a result of poorly insulated leads is a very difficult source of torch malfunction to track down. Sections of lead that are uninsulated and come in contact with any grounded metal surface will quickly melt, severing the lead. Long cutting leads are very heavy and difficult to lower overboard and retrieve, so be sure the leads are properly and completely insulated before they are put in the water. Any cracks or cuts must be repaired with rubber cement and rubber or electrical tape.

For burning in deep water, loss of current because of the length of the leads must be compensated for by heavier cable in the leads and by increased current. For burning in depths over 200 feet, the leads should be 4/0 welding cable. If long leads are made up of spliced shorter sections, the connecting lugs of the splices must be scraped shiny before they are bolted together to assure maximum electrical contact. The bolt tension must also be sufficient to assure proper contact. All splices must be properly insulated and, in addition, they should be strengthened by a section of strong rope, seized to the cables on both sides of the splice so that no strain comes on the splice when the cable is lowered or retrieved.

The torch lead should be made up with an oxygen hose in much the same way a diving hose is made up. It should be seized with seine twine about every five feet and well taped, without bights, about every eight or ten inches. Efficient burning requires an adequate volume of O_2 at the torch head and leads over 100 feet long should be made up with 3/8-inch I.D. hose to assure proper volume. Splices in the oxygen hose should also be strengthened in the same way and for the same reasons as the cable. Hose that is difficult to kink must be used for obvious reasons.

Salt water is an excellent conductor of electricity and it is therefore possible to strike an arc on a metal object under water with the ground pole of the welding machine grounded only to the steel of the work barge or ship. Because of this, many people believe that it is possible to burn without a ground leading directly to the work. This is pure fallacy. For proper, efficient, fast, economical burning, the electrical circuit must be completed by a positive mechanical ground to your work. If some part of the object to be burned, a structure leg, a pipeline riser, a portion of a sunken wreck, etc., protrudes above the water surface, it is acceptable to attach the ground to that part, if there is a continuous metal connection between the place to which your ground clamp is attached and the place where the burning is to be done. The metal at the point of attachment of the ground clamp must be scraped clean so that it is free of paint, marine growth, or any other substance that will impede the free flow of electrical current. The ground clamp must be tightly secured with ample slack in the cable to prevent it from being accidentally jerked free.

If the object to be burned is completely under water, then the ground clamp must be carried down, along with the torch, and secured to the object. Remember, the ground lead must be attached to the positive pole of the welding generator. The ground lead should be of the same gauge cable as the burning lead. It too is carrying current

Because great difficulty is usually encountered in lowering leads to a diver, some companies make up the cutting lead, with its O_2 hose, and the ground lead in one unit, taped together throughout the entire length of lead, with a 10- or 15-foot extra length of ground lead extending beyond the torch. This unit is very heavy for deck personnel to handle but it is a much simpler and more efficient system for the diver.

THE KNIFE SWITCH

A large, easily operated single-pole knife switch firmly fastened in a stable position close to the diving telephone is a very important link in the burning outfit. It must be situated so that the tender can instantly open or close the switch at the diver's command. The switch must always be in the open position except when the diver is actually burning.

Severe welding-type eye flashes can be received from the knife switch arcing and so the tender must not look directly at the flashes and he must operate the switch with firm, swift, positive strokes when opening or closing.

The handle of the switch must be well insulated and large enough to grasp without any danger of the fingers or hand becoming part of the electrical circuit. For the tender, holding a wet hose with one hand and operating the knife switch with wet gloves or in a careless manner is an excellent way to solder his teeth fillings. *The knife switch is a potential danger. Operate it with caution.*

BURNING ELECTRODES

The electrode does the work in oxy-arc burning. The arc is produced between the electrode and the material to be cut and the electrode is consumed during the burning process. Depending upon the heat setting of the welding machine, the electrodes discussed below will last between 30 and 60 seconds.

The most common commercial burning electrodes are tubular steel, $5/16$-inch O.D., 14 inches long with a bore of approximately $1/8$ inch. The electrodes are covered with a flux which produces a gas bubble at the point of contact with the steel to be cut, within which the arc burns. This bubble helps to stabilize the arc and keeps the heat from rapidly dissipating to the surrounding water. The rods are also waterproofed to protect the flux and to prevent the escape of electricity through the length of the rod.

The flux is somewhat fragile, and if the rod has to be poked through a hole to effect the cutting, or if the rod has to be jammed into the kerf for cuts in very heavy metal (for instance, a shaft), or if the rod is going to be used in any position where the flux might be chipped or damaged, then it is well worth the time to additionally cover the rods with masking tape. This will provide sufficient protection for the flux and prevent arcing out of the side of the rod. Rods should be checked for several features before being given to the diver. The flux must not be cracked or broken off. Cracked rods can be adequately repaired with spiral wrapping of masking tape. If flux is missing in large chunks, the rod should be discarded. If no other rods are available, the ones with missing flux can be salvaged by three or four wraps of tape around the area where the flux is missing. This will serve to insulate the rod, but when the rod is consumed down to this area, some difficulty will be encountered because the masking tape will not burn as fast as the rod. Rods should be additionally checked to be sure the oxygen hole is not plugged and also that the end of the rod which inserts into the torch is round, clean and not burred. One of the many frequent and unnecessary frustrations of oxy-arc burning under water is a rod stub stuck in the torch because of a rusty, deformed or burred end. A pair of pliers in a pocket of your coveralls is excellent insurance against the necessity of having the torch hauled to the surface to correct this.

It is important to have a good carrier for the rods. The oil fields are littered with thousands of rods carelessly dropped and needlessly wasted for lack of an adequate carrier. Some divers will laboriously insert dozens of rods one by one into a piece of rope. This is a fouling hazard and a very poor and time-consuming method of carrying rods to the job. Others will carry a large bundle of rods by hand, losing most of them on the descent or misplacing the bundle when they reach the bottom. Use a rod quiver. It is a simple matter to make one out of 4-inch fire or rubber hose with a wooden or brass disc for the bottom. The quiver must not be so short that the rods fall out or so long that the rods are difficult to remove; about 11 inches high is ideal. Fit it with a loop or a snap hook so it can be attached to your weight belt.

OXY-ARC CUTTING TORCHES

There are a number of oxy-arc cutting torches on the market; the most popular ones are the Craftsweld, Swindell, Desco and Arcair. They are all simple, rugged, efficient tools consisting of an insulated grip, a chucking arrangement for holding the rod, and a trigger to control the oxygen flow.

My personal preference is the Craftsweld, because it weighs less than the others and, therefore, is the easiest to hold in awkward positions and for long periods of time. What it gains in being lightweight, it must sacrifice to a degree in ruggedness and it is, therefore, a bit more delicate than the other torches mentioned.

At the other extreme is the Arcair torch, which I am sure could survive being run over by a bulldozer, but in comparison to the Craftsweld it is extremely heavy and tiring to use for extended periods.

SPARE PARTS AND ACCESSORIES

Carry an adequate supply of spare parts for your torch to cover any contingency of torch damage or malfunction. These should include spare collets or chucks, washers, springs, oxygen valves, and flash arresters. More than once, a project costing thousands of dollars an hour has been shut down because of the lack of a $2.00 torch collet.

Fig. 25. A diver burning riser clamp bolts, using a Craftsweld torch. The diver s wearing an Aquadyne mask. Photo by Brooks Institute; General Aquadyne, Inc.

An electric arc emits infrared and ultraviolet rays that can be severely damaging to the eyes. In silty or murky water this is no problem but for burning in clear water, a light No. 4 or 6 welding lens should be used. For hard-hats, an accessory is offered by several companies, including Craftsweld and Desco, which is a brass lens holder that fastens to the faceplate and operates on a hinge.

With lightweight helmets or masks, I have had good luck with wetting a standard welder's lens and pressing it to the inside of the faceplate. Surface tension of the water will hold the lens onto the faceplate, permitting you to push it around with your nose into the best position. The lens will not fall off as long as it is wet. If the air flow across the faceplate should tend to dry the lens, admit a little water from time to time and slosh it around your faceplate and then purge it. In order for this method to work, both the faceplate and the lens must be clean, free of dirt, oil, grease or grit. If the standard lens will not fit in your mask, the corners can be trimmed with a glass cutter.

A pair of heavy-duty rubber gloves is essential for oxy-arc burning. There is considerable electrical energy liberated in the vicinity of the torch, especially just before the arc is struck and when it is broken. Rubber gloves provide the proper insulation to prevent what could be a dangerous shock and what is at any time an uncomfortable and distracting shocking sensation. Many divers use Playtex gloves as manufactured for dishwashing, wearing them under a pair of cotton work gloves. Since I strongly dislike the sensation of even mild electric shock, Playtex gloves are not adequate protection for me. There is a slightly heavier type of rubber glove made for cannery workers, I find these suitable, but my real preference is for the type worn by high-tension linemen.

A 4-pound short-handled sledgehammer is an indispensable diver's tool and especially handy on burning jobs. Frequently the two pieces to be severed will be stuck together with slag. Rather than waste time and rods going over your cut, a few sharp raps with the hammer will often do the trick.

Last on my list of items required for oxy-arc burning is often one of great importance but one that even experienced divers seldom carry under water for a burning job—a scraper.

All steel immersed in fresh or salt water for just a short period will collect a coating of organic deposits, scale or rust pustules. The oxy-arc was designed to cut steel, not animal or vegetable matter. There have been instances where divers have blown their whole dives trying to burn through oystershells.

For continuous easy cutting, a clean surface and positive electrical contact is absolutely necessary. Clean all rust, scale, paint, shells and vegetation out of the line of cut, even if this requires what might seem to be a long time and a lot of effort. Generally, the easiest way to do this is with a short, stiff scraper. I have a 4-inch stiff-blade painter's broad knife made up to a snap hook, hanging from a ring on my diving belt.

TOPSIDE PREPARATION FOR UNDERWATER BURNING

Before you go in the water for a burning job, check the leads for cracks, cuts or breaks and repair as necessary. Put a rod in the torch and check the collet. Then hit the trigger to be sure you are getting O_2. Hold your finger over the hole in the rod with the trigger depressed and check the washer for leaks. After this, strike an arc to assure yourself of a complete electrical circuit. Only after all this has been done should the torch be put in the water, making sure that the knife switch is in the open position.

Check the machine heat settings, the polarity and the O_2 gauge pressure yourself, or if this is not possible, be sure it is checked for you by a responsible, knowledgeable person, such as the welding foreman.

When you are dressed out in your diving gear, be sure you have a welding lens if you will be working in clear water, a quiver full of rods hanging on your belt, your rubber gloves on or with you, and your hammer and scraper.

Rubber gloves are very susceptible to tears and punctures and are not suited for the abusive treatment received during the descent, setting up of the gear, preparing the surface to be cut, etc. For this reason, I slip over my belt a doubled rubber band, one inch wide, cut from an inner tube. I keep my burning gloves tucked under this band until ready to begin burning.

UNDERWATER PREPARATION FOR BURNING

The first order of business, after you are on the bottom and have located your cut, is to attach your ground. If there is an edge for you to grip with the ground clamp, well and good; scrape it clean and screw down the clamp tightly. If you are to cut a pipe or a bulkhead without edges, you may have to punch a hole to secure your ground.

In any burning under water, it is vital to assure yourself that you are not burning into any gas- or fuel-filled area, with the attendant possibility of explosion. For attaching the ground to pipelines, it is easiest to use the welder's pipeline ground which is a half section of pipe, the same size as that to be cut, about two inches wide, with three sharp points welded to the inside, connected to the ground cable. This is forced down over the pipe after it has been cleaned, making a secure ground.

The concrete weight coating will, of course, have to be removed first, a difficult, laborious job that might use up two or three divers. This is best accomplished with a pneumatic demolition gun and an 8- or 10-pound mall. A pair of tin snips or side-cutting pliers will also be required to cut the chicken-wire reinforcing in the concrete.

Under the concrete there will be a heavy coating of tar or Somastic. This material must be completely removed before burning is attempted, but it is very brittle when cold and usually responds to vigorous belaboring with a sledgehammer. As in many types of work (painting, for instance), the actual preparation for cutting is quite often more time-consuming and difficult than the cutting itself.

However, proper preparation is essential to good work. Part of the preparation for making a good cut on a pipeline might very well be jetting a hole under the pipeline large enough to allow you to make a continuous and unimpeded cut.

PIPELINE CUTTING TEMPLATE

It is impossible for the most skilled underwater burner to make a square cut around a pipeline and end up where he started if burning free-hand. All oxy-arc torches are awkward and difficult to operate, even in clear water. Use a burning guide or template for underwater burning whenever possible, especially for pipeline cuts.

A suitable pipeline cutting template can be made out of a 4-inch-wide strip of ⅛-inch plywood, stiff rubber or gasket material. A piece of rope tied around the pipe is frequently used as a burning guide, but this is not as satisfactory as any of the suggested materials. The burning guide or template must, of course, be fashioned out of some non-conductive, non-metallic material.

John Becksted, former Operations Manager for Sub-Sea Services, Inc., has made an excellent burning guide for pipe. It is a piece of heavy belting about 6 inches wide with a pair of grommets punched into it every few inches. Attached to one end are two pieces of Bungi cord with hooks. The belting is wrapped around the pipe and the cords are stretched and the hooks inserted into the pair of grommets that will assure a tight fit around the pipe.

MAKING THE CUT

Now that all the preparatory work is completed, and your ground is securely installed, you are ready to begin the cut. As important as any of the preceding is a comfortable working position that leaves you with both hands free. Proper burning is virtually impossible without using both hands. As you assume a comfortable position, be sure that you have not placed yourself between the ground and where you are going to burn because this will increase the possibility of shock, especially with an AC machine, and result in the rapid electrolytic corrosion of the metal parts of your diving equipment.

Put a rod in the torch and be sure it is tightly held by the chuck. A loose rod will arc against the insides of the torch and become fused to the collet or otherwise damage the torch. Press the trigger to be sure you have O_2 and then get yourself in position to start your cut.

On vertical cuts it is sometimes better to start at the bottom and work up so that the slag will be dropping free of your line of cut. This is especially true when cutting very heavy steel or shafting.

On a pipeline cut where there is the possibility of an end dropping, it is doubly important to start at the bottom so that you can be in the clear when the cut is completed. Always study your intended cut and plan your moves to keep yourself and your air hose and other gear free of falling pieces. Try to determine the probable stress in the piece you are cutting and direct your cut so that you are gradually relieving the stress as you burn, rather than have the pieces come apart with a sudden jolt at the end of your cut.

For example, if you are cutting an overhanging beam or pipe, start your cut at the top, so that as you advance your cut the piece to be severed will gradually sag. For cutting a pipeline or beam suspended between two supporting points, start your cut on the bottom and as the cut advances, the member will gradually sag downward. As a general rule, make your most difficult cuts (the ones requiring uncomfortable or contorted attitudes) first, leaving the easiest cutting (the portion that will leave you physically in the clear) for last.

As stated earlier, the oxy-arc torch is an inherently awkward and difficult tool to manipulate and it requires both hands. This is because a slight force or pressure has to be exerted in two directions simultaneously, forward along the line of cut, and inward as the rod consumes itself.

Hold the torch in your right hand and, if possible, lean your left shoulder or elbow against something solid to steady yourself. Some divers hold the torch chuck with the left hand but I find it much steadier to grasp the rod itself 3 or 4 inches up from the point of contact. After steadying my left elbow, with my left hand I form the type of fist you might use to hold a poker hand and slide the rod along the side of my thumb and over the tip of my index finger. When using this method, be careful not to burn your left glove by getting it too close to the arc. Some of the synthetic rubber gloves on the market today are highly flammable and they can suddenly flare up if they come in contact with the arc and severely burn your hand. For this reason, as well as to protect your rubber gloves from normal wear and tear, wear a pair of work gloves over them.

When you are ready to start burning, hold the trigger down, giving you a steady flow of O_2, and tell your tender to close the knife switch. The most common phrase for this is "Make it hot," to start burning and "Make it cold" to stop the current or, more simply, "Hot" or "Cold." Push the rod against the steel to be cut and you should get an instantaneous arc. You may have to forcefully tap the rod tip against your work, knocking the waterproofing off the tip before the rod starts to burn. When the arc is initiated, push it lightly into your work and guide it slowly along the line of cut.

With some rods, the inward pressure will have to be quite firm in order to crumble the flux as the steel tube burns up inside of it. In any event, the rod must be pushed into the work and hardly any of the arc should be visible to you on your side of the cut. As you proceed with the cut, if you are burning completely through the metal, the orange glow in the arc will be visible through the kerf behind the rod, but not visible at the forward edge of the rod. Move the torch forward along the line of cut as fast as you can, consistent with maintaining this condition. It is not necessary to poke your rod entirely through the plate to assure yourself that it is completely severed; doing so will be difficult because of the accumulated slag on the backside of your cut and it will slow your cutting rate considerably. As long as you can see the light of the arc through the kerf behind you and the arc has been uninterrupted, you have cut through the steel. If there is no tension in the steel you are cutting to separate the two edges of the cut detectably, you can pass the tip of your broad knife through the kerf occasionally to assure yourself of a complete cut.

When the arc has been struck, it is important to proceed with your cut just as fast as possible, consistent with complete burn-through. This is so because any delay in the movement of your arc will just result in melting your rod and accumulating slag at the point of your arc. Slag bubbling through the kerf and collecting on your side of the cut is an indication of incomplete cutting and can result from several causes. Moving the torch too fast is the most common; insufficient heat is another, as is insufficient O_2 pressure.

Situations where the steel you are burning is backed up by heavy paint, tightly packed mud, clay or sand, or worst of all, concrete, present the most difficult problems. Heavy paint on the backside of a cut can generally be overcome by a liberal increase in O_2 pressure. This is also effective when burning against mud or other overburden but in some instances it can lead to a possibly dangerous situation. The burning process creates hydrogen gas which can collect in lethal concentrations behind a cut. Oxygen mixed with this hydrogen produces an explosive mixture and it needs only the arc of your torch to detonate it. This is a fairly common occurrence during inside jacket-leg pile cut-offs, and the resulting explosions have severely injured and even killed divers. Whenever possible, establish a vent for accumulating gases, even if it means more burning.

When burning against concrete, the rod will have to be moved in a series of shallow forward-directed arcs flipping the slag out of the way ahead of you as you proceed.

Most instructors recommend that the rod be held perpendicular (at a 90° angle) to the work, but I am in agreement with this only for very heavy metal—over one inch thick. I find that holding the rod so that it points in the direction of the cut at a slight angle has several advantages. It is a much easier and more comfortable attitude to maintain and it helps in directing the cut in a straight line. It also projects the heat of the arc to the metal ahead of the cut, contributing to a faster cut. On painted or dirty surfaces it tends to burn and lift off this interfering material a fraction of an inch ahead of the arc, enabling you to maintain a steady, unbroken arc because your rod tip is continually in contact with clean conductive material. When the material to be cut is backed up by overburden, the slight forward angle results in the oxygen stream jetting out a small pocket ahead of the cut which will accommodate the slag. In metal an inch or more thick, the rod should be held perpendicular so that all the heat and the oxygen are directed over the shortest route through the material to be cut.

If, while you are burning, your rod tends to stick to the surface of the metal you are cutting, it is an indication of cracked rod flux through which current is escaping. If it is not the rod, and the machine is set up hot enough, it means a current leak through the insulation of your leads, or a poor ground connection. Suspend burning until the cause is located and corrected. If oxygen is leaking around the head or the back of the chuck, it indicates a worn or missing washer. Correct this immediately since it not only results in an unnecessary waste of oxygen and poor cutting because not enough oxygen is being directed at your cut, but the rod may arc and fuse to the brass parts inside the chuck and ruin the torch.

At the end of your cut, break your arc smartly and immediately call for "Power off." Don't attempt to remove the stub until this is confirmed by the tender. Don't burn the rod too short; this will only subject the torch head to unnecessary damage from the heat of the arc, and make removal of the stub difficult or impossible. A 2-inch stub is reasonable to handle and consistent with thrifty utilization of expensive cutting rods.

It is possible, especially with heavy rubber gloves, to change rods with the power on and some divers do so, trying to save time. This is bad practice because of the increased risk of shock and also because the rod will arc against the chuck and other internal parts of the torch, quickly wearing it out or damaging it. Call for "Power off" to change rods.

Efficient oxy-arc burning depends more on the interpretation of tactile than on visual sensations and most of your burning will probably be done with limited or no visibility. However, it is important to practice burning where you can see so that you can draw a comparison between the physical sensations of burning with the visible results. *Get the "feel" of burning.*

CARE OF TORCH AFTER BURNING

When you are through burning, the torch should be rinsed in fresh water. Be sure the rod stub is removed because this will corrode within the chuck, making it difficult to remove and possibly ruining the chuck. The moving parts of the oxygen valve and any threads of the chuck should be lubricated with a silicone or other non-flashing lubricant. The tension on the oxygen regulator diaphragm should be removed by backing off the T-handle adjusting screw when it is not in use.

CUTTING A BUCKLED PIPELINE

A common job offshore is cutting a buckled pipeline in two so that that the ends can be lifted to the surface for repair. Caution must be exercised in this situation because, although buckled, the pipe may not be perforated and, therefore, not flooded.

When you first strike your arc, water will rush into the pipe at a velocity, depending on the depth of water. If this situation is encountered, the diver must wait until the line is completely flooded before he continues his cut, a process that can take hours or days, depending on the length and size of the pipeline. With a buckled pipeline, the object is to get it severed as quickly as possible and we are not concerned with square or even cuts. Find a place where a part of the weight coating is broken away and hit the pipe a few licks with your sledgehammer to remove the dope. Then, assuming the pipe is flooded, cut a large hole or window through the top of the pipe and remove this piece. Generally, the remaining bottom section of the pipe can then be cut from the inside without removing the rest of the weight coating and dope, or jetting to get under the pipe.

When using this method for large-diameter pipeline, be sure the hole is large enough to keep your hands in the clear if the pipe is in stress. For smaller pipe, only the rod need be introduced into the inside of the pipe.

CUTTING JACKET-LEG PILES

Cutting jacket-leg piles is another frequent chore offshore. These are the piles that are driven down through the legs of the structure, and upon which the entire weight of the platform rests. When the structure is to be salvaged, either for removal or transfer to another location, these piles must be cut off below the jacket, often 10 to 20 feet below the mud line. If the pile is large enough to accommodate a diver, the cut is easiest to make from the inside.

The reason the cut-off must be made below the jacket is that after the pile is inserted into the jacket leg and driven, the space between the inside of the jacket leg and the outside of the pile is pumped full of concrete grout under high pressure.

The first requirement is a suitable ladder for getting down to the water inside the pile. The water level will quite often be 15 or 20 feet below the top of the pile and a rope ladder or Jacob's ladder is the easiest for all concerned to handle for this particular application.

Another requirement is a stable platform from which the diver can work. This can be made from 3-inch pipe welded together in a T, with the lower, horizontal part three or four inches shorter than the inside diameter of the pile. This is lowered on a line, or air tugger, and the diver can sit or stand on the horizontal bar and lean for support on the vertical bar. It is a simple matter for the diver to rotate this T-bar as he advances his cut.

Another important piece of equipment for an inside burn-off is a guide or template. This can be made from a long strip of springy green oak lath, such as is used for the framework on fish traps. A 3-inch strip of 1/4-inch plywood with saw cuts halfway through every two or three inches at right angles to the length of the strip also makes a suitable guide. Tie your guide into a circle of smaller diameter than the pile and when in position, you can cut the string allowing the guide to spring out against the inside wall of the pile. This provides an excellent edge to drag your rod along and a perfect cut can be made in zero visibility.

A guide also reduces the length of cut because it prevents you from zig-zagging and it automatically brings you back to the point where you started. You must be careful not to accidentally move your guide by excessive downward pressure with your rod tip.

As mentioned earlier, the possibility of explosion is constant and unpredictable as a result of the accumulated gases produced by your burning forming in a pocket outside the pile. The only recommendation I can make to reduce this possibility, or at least limit the size of the pocket, is to punch a series of holes about a foot above your cut, in the area where your cut begins and ends. This might allow any collected gas to vent and escape up the inside of the pile. Your best and safest course of action is to make the cut as fast as possible, thereby limiting the amount of dangerous gas produced. A 30-inch I.D. pile, 1 inch thick, can easily be cut in six to eight minutes by an experienced burner. Because the danger of explosion does exist during an inside burn-off, I also recommend that this work be done only while wearing a helmet with a Lexan® (a registered trademark, Du Pont Chemical Co.) faceplate.

CUTTING NONFERROUS METALS

Cast iron and nonferrous metals such as bronze, aluminum and stainless steel can also be cut with an oxy-arc torch. With these metals the process is actually one of melting rather than burning and the oxygen serves only to blow the molten matter out of the way. In situations where there is a great deal of this metal to be severed, compressed air can be substituted for oxygen for economy. Compressed air will not work for the proper burning of steel.

Bear in mind that the slag and molten material from burning maintains its destructive heat for several seconds under water. Avoid situations where the slag will fall on you or your equipment.

Practice burning whenever the opportunity presents itself.

THE THERMAL LANCE

A recent industrial development with genuine underwater potential is the thermal lance. This is a $3/8$-inch pipe, 10 feet, 6 inches long, packed with a number of different metal alloys, such as aluminum, magnesium, thermite and steel. High-pressure oxygen is forced through the pipe and the lance, once ignited, burns with tremendous heat. One manufacturer claims a temperature of 10,000 degrees for his product.

The 10′6″ bar lasts for about six minutes, and it is this feature that poses the greatest problem for underwater use. A number of bars can be coupled together for longer burning, but the extra length makes it extremely awkward to handle. The advantage of the burning bar is that it will burn or melt through almost anything—steel, nonferrous metals, rock, and concrete.

Its greatest use is for very heavy steel, such as shafting. The manufacturer claims that it will punch a hole through 12 inches of metal in one minute, consuming 6 to 12 inches of the bar. It could also be used to advantage for cutting a damaged pipeline quickly without the necessity of removing the weight coat and dope.

The thermal lance must be ignited by some outside source of heat (such as a burning torch), or under water, an oxy-arc torch. The only way to stop the bar from burning is to shut off the oxygen. The lance will continue to burn for several seconds after this, and for safety, and to keep from burning up the oxygen fittings and hose attached to the top of the lance, the oxygen must be shut off with at least a foot of the bar remaining. Because of its awkwardness and the high rate of oxygen consumption, the burning bar is practical only for very heavy steel or other materials that cannot easily be cut with the oxy-arc torch.

The burning bar can be purchased from Burning Bar, 13273 Ventura Boulevard, Studio City, Calif. 91604.

THE CLUCAS THERMAL-ARC SYSTEM

A decided improvement over the thermal lance for underwater use is the Clucas thermal-arc cutting equipment. The Clucas system is in principle the same as thermal-lance cutting, but instead of the rigid bar, the consumable agent is a tough, flexible plastic-covered cable which comes in 100-foot lengths. The cable, as opposed to six minutes for the bar, has an average burning time of approximately one hour. Because of its flexibility, it is less awkward and, therefore, much easier to control and use. The Clucas equipment comes complete with a surface control panel which includes an oxygen regulator, off-on flow valves for the oxygen and a knife switch and electrical connections for igniting the cable. The cable is hooked up to an oxygen hose and welding cables similar to oxy-arc equipment, but the electrical circuit is used only for igniting the thermal cable and is not used during the actual burning.

The manufacturer's operating instructions follow.

THERMAL-ARC CUTTING EQUIPMENT

Setting Up Equipment.

1. Connect three bottles of O_2 to Control Unit at position marked HIGH PRESSURE O_2 IN using the 10 ft. of high pressure hose supplied and stamped "Panel" on the connection.
2. Connect extension leads to unit at positions – NEG AMPS OUT TO CUTTING CABLE and L.P. CUTTING O_2 OUT.
3. Connect Kerie cutting cable to the extension leads and slide red insulating sleeve over the joint.
4. Connect D.C. generator or supply (max. 400 amps, 50-80 volts open circuit) or one or two 12-volt car batteries (in series) – negative side to control, + positive side to ground or metal to be cut. Use 200 or 300 amp welding cable.

Preparing to Cut. First try out equipment on the surface to familiarize yourself with it. Set up as described and use a thick piece of mild steel as practice material.

Set O_2 pressure at about 200 psi and switch O_2 on, close knife switch and touch practice metal with tip of Kerie cable. The cable should ignite immediately; lift knife switch when ignition takes place. Proceed to cut metal starting at the edge. No dark glasses are required and with care there will be no splash-back.

There are two oxygen off positions, namely CUTTING O_2 OFF and CUTTING O_2 OFF & VENT. In the first "off" position the Kerie cable will continue to burn until the O_2 in the line is used up. However, to kill the flame quickly, use the OFF & VENT position although this is rather wasteful of oxygen.

Cutting Under Water. O_2 pressure should be set at approximately 250 psi (over ambient); the diver takes down the Kerie cable and when ready to cut, orders "switch on." The attendant turns on the O_2 and closes knife switch. When the diver reports ignition (the needle on the ammeter will drop to zero), the knife switch is lifted and the diver proceeds to cut, using the same method as in oxy-arc cutting.

Further Information. It is advisable to use two 12-volt car batteries in series, although it is possible to gain ignition with only one. Ignition is difficult to obtain when ammeter reads less than 110 amps when striking arc. When using batteries, the Kerie cable should be chopped clean before cutting commences.

With DC welding generators hooked into the system to gain ignition, supplying say, 400 amps, 60 to 80 volts open circuit, ignition will take place regardless of the condition the end of the Kerie cable is in, providing the wire can make contact with the work. When switching off, the diver should rub the end of the cable against the work to prevent the molten nylon from insulating the end of the cable, thus ensuring further ignition.

Mild steel should start to cut as soon as the flame touches it, but should the pressure to the Kerie cable drop below approximately 120 psi it will cease to cut metal, even though the cable is still burning furiously. Do not waste cable trying to cut—change bottles.

Two-and-one-half-inch plate can be cut at one pass with O_2 pressure at about 280 psi. When cutting thicker metal, pressure can be raised to 300 psi.

Pressures just below 200 psi should be used when cutting wires from propellers, etc., to prevent damage to shaft or vessel. As the cutting cable is consumed and becomes shorter, O_2 pressure can be reduced; e.g., from 250 psi to 220 psi when approximately 50 feet of the cable is left. When cable is used up, always switch off in plenty of time to prevent extension leads from becoming damaged. Do make sure that all joints in cables, etc., are screwed up tightly.

Further Hints. In an emergency when no generator or batteries are available the cable can be ignited in the following manner: Chop the end of the cable clean and open up the center. Poke down a small amount of wire wool or paper (about the size of a filter tip) for about one inch, leaving a small amount protruding; turn on oxygen (5 psi) and light wire wool with a match; immediately hold the end of the cable against wood or concrete and raise the pressure to 50 psi. The burning cable can now be passed to the diver, who, when in position to cut, asks for the pressure to be raised for cutting.

The burning cable can also be used very effectively as an underwater light at O_2 pressures around 80 to 100 psi. When cutting metal covered in thick bitumen, etc. (e.g., field joints on pipelines) the ground or + lead can be made up into the diver's shot and the metal weight used as a striking plate to gain ignition.

As this system is intended for heavy duty cutting, O_2 consumption on thin metal is obviously rather high for the work done; therefore, when cutting sheet piles it may be advantageous to the operator to cut the clutches or joints only with the Kerie cable and the pans or recesses with the carbon arc system. Speed- and costwise the thermal arc system will hold its own against any other form of underwater cutting in any conditions on metal over $^3/_4$ inch thick.

Consumption—Cable. Kerie cable is supplied in 100 ft. lengths and will burn off at a rate of 1 ft. in 40 seconds during normal cutting. When cutting 2-inch thick plates each foot burned should cut 6 inches of plate.

Consumption—O_2. 3 × 240 cubic ft. O_2 bottles will cut approximately 6 feet of 2-inch plate.

Explosions. Any underwater equipment can cause explosions regardless of gas pockets if the temperature of the metal reaches a critical point, resulting in an explosion deep in the cut. As thermic cutting equipment can bring on this critical temperature quickly, care must be taken when cutting metal over three inches thick. When cutting thick metal, the cable (or lances) should be withdrawn momentarily every three or four seconds to allow water to enter the cut.

Under no circumstances should any attempt be made to cause a fire or inferno deep within thick-section steel. The cutting of thick metal, that is, propeller shafts, etc., should be done from the outside working around the circumference. If a considerable amount of cutting is to be done, it would be advisable to replace the diver's front glass with thick Perspex (particularly if it is thin laminated glass) as Perspex will repel hot slag, preventing pitting and thermal cracking.

See that the equipment is kept free from oil or grease contamination. However, the Schrader on-off valve and connections between extension leads and Kerie cable may be lightly lubricated with *pure silicone grease only.*

When work is completed, isolate panel from O_2 supply. Always use high-pressure O_2 hose supplied *next to the panel,* and if extensions are required they should be joined to this, as the original hose is specially insulated.

Exclusive United States' agent for the sale of the Clucas Thermal-Arc cutting equipment is: Taylor Diving and Salvage Company, Inc., 795 Engineers Road, Belle Chasse, La. 70037.

WELDING

Underwater welding in a wet environment remains an incomplete technology and poses seemingly insurmountable problems at the present time. Chief among them are the consistently weak and brittle welds produced by this method when compared to similar welds made in a dry atmosphere. This is because of the instantaneous and severe quenching action of the water.

United States Navy experiments indicate that welds made in the water can achieve 80 percent of the tensile strength and 50 percent of the ductility of similar welds made in the dry, but this is insufficient to meet the stringent code requirements for welding on petroleum and natural-gas transmission lines. Welding under water to code specifications is accomplished in a dry environment by use of specialized underwater weld huts or habitats and the diver's participation in this complex operation is discussed in Chapter 14.

Welding in the wet does have certain limited applications offshore and it will behoove you to learn the technique and acquire the necessary skills. The installation of the Weld + pipeline leak-repair clamp requires welding in the wet (Chapter 11). Very satisfactory tempo-

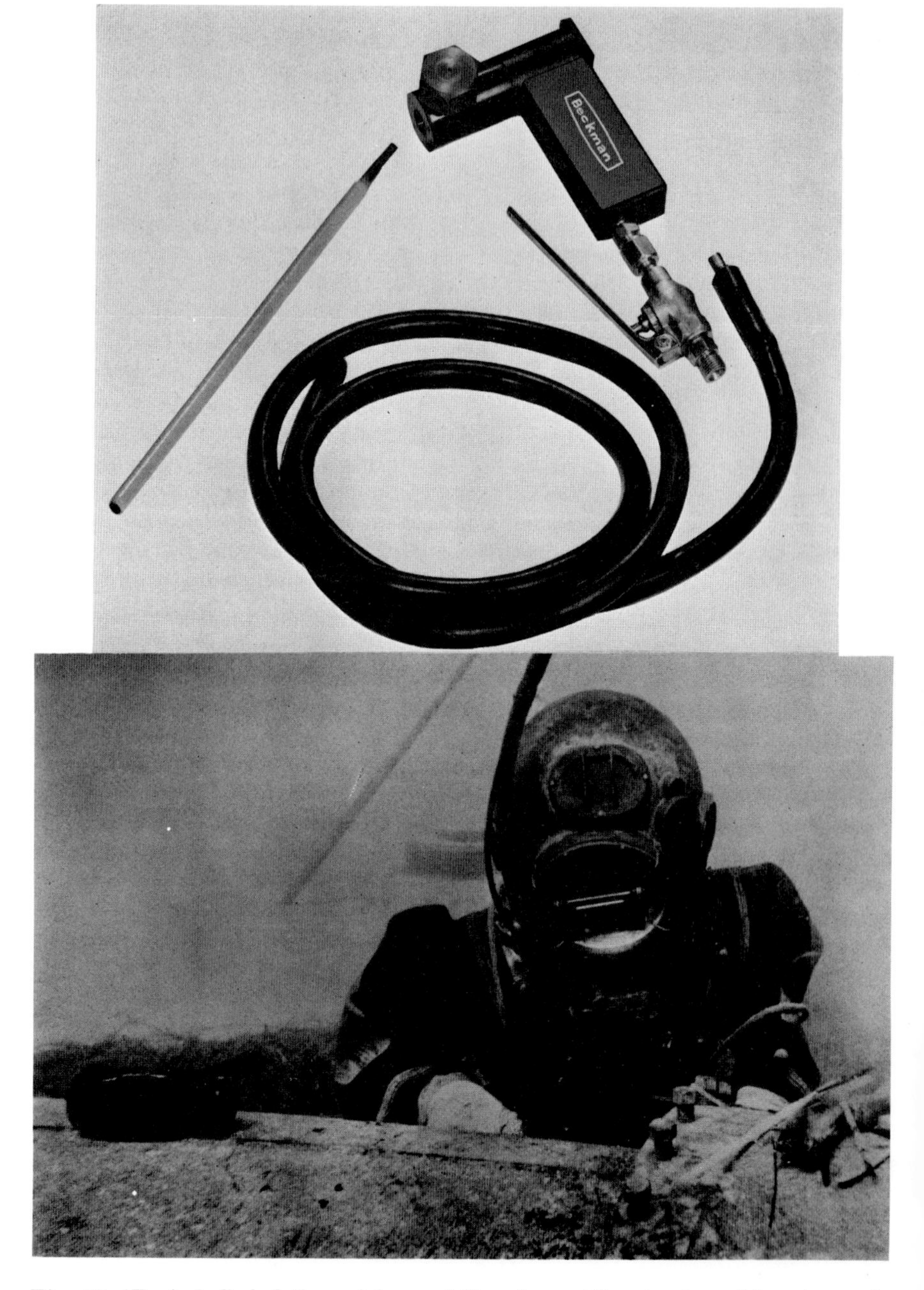

Fig. 26. (Top) A Swindell, or Advanced Oxy-Arc cutting torch and burning rod. Photo: George Swindell and Beckman Instruments, Inc. (Bottom) A diver welding on a pipeline repair sleeve. Note the eye shield on the helmet faceplate. Photo: Divcon, Inc.

rary repairs can be made by wet welding on barges, crew boats and other steel vessels or structures that develop leaky rivets or split seams. Wet welding is often an excellent method for erecting staging platforms for underwater work.

Any experience or schooling in topside welding will be a tremendous help to you in welding under water. Take a short course in welding at any of the numerous industrial schools, and when you are on a job, practice at every opportunity. Again, the welding foreman will probably be delighted to help you out.

Basically, everything required for burning under water is also necessary for welding under water, with the exception of the cutting torch, the burning rods and the oxygen paraphernalia. The Arcair torch can be supplied with special collets for welding rods, but its excessive weight makes it very difficult to use for welding. Craftsweld manufactures an excellent underwater welding stinger.

A suitable underwater welding stinger can be made by completely taping all the exposed parts of a regular topside stinger, leaving bare only that section that grips the rod.

Any straight-polarity, mild steel all-purpose rod can be used for welding under water. The rods must be waterproofed to prevent the loss of current and deterioration of the flux. Any synthetic paint or plastic spray coating can be used. Two coats, well set up between applications and before use, are generally necessary.

Welding under water without visibility is almost impossible. Just try to run a straight bead topside with your eyes closed.

If anything, a stable platform is more necessary for welding than for burning. All of the preliminary checks and precautions pertaining to oxy-arc burning also apply to welding. As with burning, welding is done with a straight polarity current.

For most purposes, $\frac{3}{16}$-inch rod is the best because of the thickness of the weld deposit; $\frac{3}{16}$-inch rod will lay down a bead almost $\frac{3}{16}$ inch thick, and a single pass should suffice for most jobs that would be attempted using the wet welding method. For sealing holes or plate splits a doubler plate or patch is the easiest to weld because it provides a guide for your rod while applying a fillet weld.

When repairing a split or a tear, it is always wise to drill or burn a small hole at the ends of the split to keep it from progressing under the patch. It is sometimes difficult to hold a patch in place long enough to get it tacked, especially on a vertical or overhead surface such as the side or bottom of a barge. A magnet with a pull of from 5 to 50 pounds works very well to clamp small patches to the work surface. For large patches, a velocity power cartridge stud driver will have to be used.

Second and third beads are almost impossible to apply because underwater welding in the wet is accomplished by the self-consuming technique. This requires a strong downward pressure on the rod and the crumbling of the flux is inconsistent and requires varying pressures causing the bead to wander unless it is contained in a U or channel.

Underwater welding will require 200 to 300 amps, depending on the length of leads and the weld position.

If you are a welder, set your machine, using the rods, ground and welding leads you will actually be using under water and make several practice welds on the surface. When you have determined the heat settings for a proper topside weld, increase the machine settings about 15 volts and 20 amps. This should be enough to compensate for the loss of heat and current under water.

If you are not an experienced welder, have the welding foreman do this for you. The requirement of a clean surface for welding is more critical than it is for burning and you will have to add a wire brush to your tool kit.

A welding lens is compulsory because you have to stare directly into the arc to direct your bead.

When all the preliminary work is done and you are ready to commence welding, hold your rod against your work at an angle of about 20 degrees with the tip of your rod pointing away from the direction of your bead, and call for "Power on." If the rod does not start to burn, you may have to tap or scratch it to remove the waterproofing.

When the rod is burning, keep a downward pressure on the tip, maintaining the angle of the rod until it is consumed. Break the arc and call for "Power off" before inserting a new rod. The end of the bead will have to be chipped and wire brushed before you start the next rod on an overlap of the bead. While the rod is burning, you may have to apply a slight pressure into the bead with the rod to form a wider bead. A 3/16-inch rod should lay down a bead almost 3/16 inch wide and from eight to ten inches long for a 14-inch rod. You do not hold an arc as in topside welding. The rod tip is held in contact with the material to be welded at all times.

The angle of the welding rod may have to be varied from 15 degrees to 45 degrees, depending on the particular application, and this will have to be determined by personal experimentation.

In general, welds made with the self-consuming technique produce uniform, good-looking welds. If your rod tends to stick to the work, it is a sign of insufficient heat. Dripping globular beads are an indication of too much heat.

You can easily build a water glass so that you can practice topside in a split 55-gallon drum.

Wear rubber gloves.

See the list of underwater burning and welding suppliers in the Appendix.

Chapter 8

USE OF EXPLOSIVES

Situations requiring the placement of explosives by divers are extremely rare in the offshore petroleum industry. However, explosives are occasionally used for the demolition and removal of storm- or fire-damaged well platform structures, for demolition and removal of wrecks and other obstructions in the pipeline right of way, and for trenching through rock or coral. There are a number of other possible applications for explosives in various phases of offshore work that are to my knowledge ignored or unexplored. This could result in reduced underwater man-hours and lower job costs. These will be discussed near the end of this chapter.

The designing of shots and the selection of the types of explosives to be used is the job of the professional blaster or explosives engineer. Presented here is a rudimentary introduction to explosives, their potential for useful work and what you, as a diver, should know concerning their safe and effective use.

To begin with, explosives are designed and manufactured to do what the name implies, *explode*. A common-sense rule then would be, treat *all* explosives at all times as though they might *go off*. Your job is to see to it that they *don't* go off until conditions of safety and desired effect are ideal, and then that they *do* go off at the instant intended. Emphatic and peremptory "do's" and "dont's'" form an important body of rules for safe explosive handling. A list of "Do's and Don'ts" appears at the end of this chapter. Study them and abide by them in all explosives' applications.

An explosive is a chemical or combination of chemicals which, when initiated (or set off) by heat or shock, convert instantaneously into a very large volume of gas which creates intense heat and pressure.

Low explosives, such as gunpowder and black powder, convert into the gaseous state relatively slowly and exert a heaving or pushing effect rather than a shattering one. Low explosives are almost never used offshore.

DYNAMITE

High explosives detonate or convert into gas almost instantaneously creating a powerful shock and pressure wave. High explosives are what we are concerned with here, and those most frequently used for commercial purposes offshore are dynamite and various combinations of chemicals such as ammonium nitrate, called blasting agents.

Dynamite is almost always made with varying percentages of nitroglycerine and an inert filler or desensitizer such as sawdust. Straight dynamite is commonly available in percentages of from 40 to 60 percent. The percentage designation indicates the amount of nitroglycerine by weight of the cartridge, the remainder being the filler. Dynamite generally comes in waxed Manila paper wrappers, in

Fig. 27. An oil well structure, destroyed by a hurricane, freed from the ocean bottom and recovered by use of explosives. Photo: Westinghouse, Inc.

cartridge sizes 1¼ inch in diameter by 8 inches long, and numbering about 105 sticks to the 50-pound case. Dynamite can also be supplied in a wide range of other cartridge sizes, up to six inches or more in diameter and numbering only two to the 50-pound case.

Straight dynamites are suitable for underwater use if they are detonated shortly after placing. If straight dynamite remains in the water over 24 hours, its explosive efficiency becomes questionable.

GELATIN DYNAMITES

Because of the extreme dependence upon weather of all open-sea operations, it makes economic sense to use gelatin dynamite for difficult-to-set multiple charges. If the operation should have to be suspended for several days, because of bad weather, the charges already set, if gelatin dynamite was used, will still be good. Commercial gelatin dynamites or submarine blasting gelatins come in a range of strengths from 20 to 100 percent. They are more expensive than straight dynamites or blasting agents and their use is economically justifiable only for small charges or specialized applications.

BLASTING AGENTS

Blasting agents come in a wide variety of types and packaging. They are generally safer than dynamites because they are more stable and will not ordinarily detonate with a blasting cap. They require a booster of high explosive or a stick of dynamite to set them off. They are the least expensive of commercial explosives and, for this reason, more suited to very large charges such as would be required for wreck dispersal. Almost all blasting agents have poor water-resisting qualities and depend entirely on the packaging to keep them effective for underwater use.

Blasting agents can be supplied in metal canisters, waterproofed cardboard cylinders, or 50-pound polyethylene sacks. When loading or placing sacks under water, you must, of course, be careful not to puncture or rip them. A stick of dynamite in the center of the charge is the cheapest "booster" or primer for detonating blasting agents.

ELECTRIC BLASTING CAPS

Electric blasting caps are the only dependable type of primer or initiator for underwater use. The cap is a copper alloy or aluminum tube about 1/4 inch in diameter and 1 1/8 inches long or longer, depending on the strength designation, with two insulated electric wires protruding from one end. A No. 6 cap is the minimum size that should ever be used for underwater work and a No. 8 is preferable because of its higher impact and the greater certainty of detonating the main charge. Caps come with lead wires from 4 inches to 300 feet long, but those with the shorter wires are much less expensive. Inside the electric blasting cap is a high-resistance wire called the bridge wire, which heats up when electricity is applied. The hot bridge wire sets off an ignition mix which detonates a primer charge; this in turn detonates the base charge, all within the cap. The cap base charge detonates the dynamite or other explosive charge. Although electric blasting caps are reasonably safe, they should never be abused by throwing, dropping or any other impact-producing activity. The leg wires must never be jerked, or pulled out of the cap.

The greatest danger present in the handling of electric blasting caps is from exposure to accidental or extraneous sources of electricity. As delivered, the ends of the wires of electric blasting caps are shunted, or short-circuited by a strip of aluminum called the shunt, which is crimped over the bare wire ends. This shunt must never be removed until the last second before hooking up the wires to the galvanometer for testing, or to the blasting wire. The opposite ends of the blasting wire should be kept shunted or twisted together until the last moment before they are hooked up to the blasting machine for firing the charge.

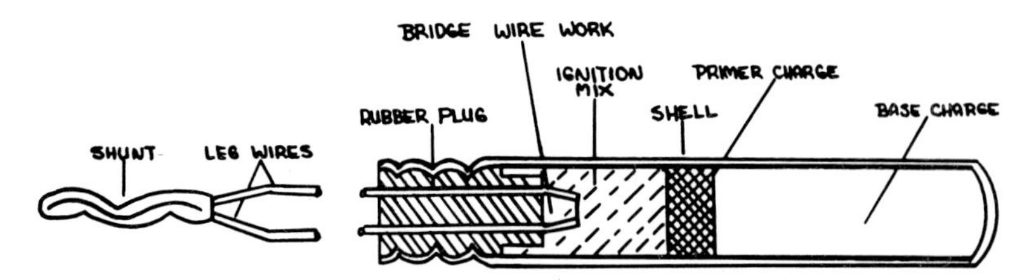

Fig. 28. Electric blasting cap. Drawing by Art Herman.

Electric blasting caps must be kept and stored away from heat, moisture (because of corrosion), any source of electricity, any other explosives, or any place where they might be knocked about or have things dropped on them.

PRIMACORD

Primacord,® a trademark of the Ensign-Bickford Co., is a high-velocity detonating cord, used to detonate other explosives. It has an outside diameter of about ¼ inch and consists of an outer covering of waxed cotton braid over a plastic jacket. The inside core is of high-velocity high explosive. Primacord comes in spools 500 or 1000 feet long and must be initiated with a blasting cap. It has a velocity of detonation of 21,000 feet per second, which means that if a length of Primacord four miles long were laid out with a blasting cap on one end and a stick of dynamite on the other, the dynamite would explode one second after the cap was detonated. Primacord is indisputably the best and safest method to use for detonating explosive charges and this is especially true for underwater work. Many charges can be set up under water and linked together with Primacord. When all the charges are set, a final Primacord leader can be run from the surface and tied into the Primacord on the bottom. The diver need never handle a blasting cap as this is applied to the Primacord at the surface just prior to shooting the charge. Standard reinforced Primacord is suitable for use under water if it is to be detonated the same day it is set. If the charges are complex and require several days to set, then the more expensive plastic-reinforced Primacord should be used. Primacord, although very stable, is a high explosive and it should be handled with all the caution due any high explosive.

The electric wire used for underwater blasting should be copper for lower resistance, and it should be very well insulated as well as flexible. For underwater work requiring long blasting wire leads, the wire should be kept on a spool or drum with a quick winding capability. The longer the leads, the larger-gauge wire required to carry the current. The blasting-machine ends of the wire should be shunted or twisted together at all times, except when being hooked up to the blasting machine. Immediately after the shot is fired, remove the wires and twist them together. If the lower or cap end of the wire was close to the blast, several feet will have to be cut off because of possible damage to the insulation and to the copper wire.

THE GALVANOMETER

A galvanometer is a very sensitive electrical measuring instrument used to test the electrical continuity of the blasting caps and the firing circuit. The galvanometer is an extremely delicate instrument and must not be abused or tossed about. It should never be tampered with, and if it is not working properly, it should be returned to the manufacturer. It uses a special low-current battery and only the battery specified by the manufacturer should be used. Any other battery might detonate the blasting cap.

THE BLASTING MACHINE

The usual kind of blasting machine found offshore will be of the impulse-generator type which gets its power from a sharp twist or downward thrust of the handle of the machine. The harder you twist or push the handle, the more power you will generate.

Activate it vigorously when it is time to shoot the charge. The blasting machine should be kept dry and stored in a dry place.

BLASTING A SUNKEN WRECK

It is always easier to illustrate a procedure with a "fer-instance." In this case we have a seismic survey boat sunk in about 120 feet of water with one ton of explosives aboard. The boat has been surveyed and written off as a total loss, not worth salvaging. It is not a menace to navigation, but the Coast Guard is uneasy about all that explosive lying around, and the decision has been made to try to dispose of it by sympathetic detonation. The explosives are in cans, stored in the cargo hold, and they are accessible through an open deck hatch. The wreck is buoyed, and you will be working from a crew boat anchored on a two-point moor directly over the wreck.

A down line is attached which will land the diver on deck beside the cargo hatch. A 50-pound case of 90-percent gelatin dynamite has been selected to do the job because of its high brisance or shattering effect.

PREPARATION FOR BLASTING

The case of dynamite is brought out and placed in the center of the crewboat deck. A jackknife can be used to slit the tape holding the cover on the box, but be careful not to strike the metal staples on the box. A fixed-blade knife is always preferable for use with explosives because of the possibility of a jackknife snapping shut and producing impact.

Only those explosives to be immediately used should be brought out on deck. Never prepare more than one charge at a time. If this one case of dynamite is all you have aboard, remove one stick and put it away in a safe place. I will explain its purpose presently. Remove another stick of dynamite and with a 3/8-inch wooden dowel, or a sharpened stick, poke a hole down into the center of one end of the cartridge, 3 or 4 inches deep. Steel or iron tools should never be used for this purpose because of the danger of producing static electricity. The hole should not be jabbed, but rather pressed with a slow, even pressure. Set this stick of dynamite aside and break out the roll of Primacord. Open the case, observing the same precautions as for dynamite, and remove the spool. Using a knife against a wooden surface, cut off about three inches of Primacord and throw it overboard. Don't saw Primacord; cut it with firm strokes in one direction only. Use the Black Magic from your diving kit, or any other quick-drying cement, and swab the cut end of Primacord. This is to waterproof the end, preventing water from seeping into the explosive core and desensitizing it. If you have no Black Magic or other cement, use tape, but do a good job. Cut a 6-foot piece off the end of the coil and treat both newly-cut ends as above. Always take time to waterproof the ends of Primacord as this will greatly reduce the probability of misfire. Push one end of the 6-foot length of Primacord down into the hole in the stick of dynamite. Knead the end of the stick so that the dynamite is pressed firmly around the cord inside the stick. Bend the cord in a shallow loop and run it back, up alongside the stick of dynamite, and tape it securely at both ends and in the middle. Be sure that the Primacord is in intimate contact with the stick of dynamite with no tape between them. Replace the stick in the box, preferably in the center under several other sticks, and poke a hole through the box cover. Pass the Primacord through this hole and replace the cover. The case is cardboard, and it will crush under pressure and rapidly disintegrate under water. For this reason, the case must be securely bound with 1/4-inch rope in at least three places to keep it a cohesive unit under water. With small line or marline secure the Primacord to one of the case bindings with a constrictor knot or a rolling hitch. This is to prevent the Primacord from being accidentally pulled out of the dynamite. This charge is now ready, so set it aside while we check out and prepare the rest of our gear.

Fig. 29. (l. to r.) A hand-held ten-cap electric blasting machine; a "Hell Box" or electric blasting machine; a properly made-up primer, with Primacord® securely taped to a stick of dynamite. Photos: E.I.DuPont De Nemours & Co., Inc.

Put a broom handle or other wooden dowel through the center of the 1,000-foot spool of Primacord in preparation for fast unwinding. Cut off a 3-foot piece and again waterproof the exposed ends. Set this piece aside. Inspect your blasting wire for cut or damaged insulation. Be sure that all the wire ends are clean and shiny. The blasting wire should be checked for electrical continuity by twisting the wires of one end together and attaching the other ends to the galvanometer. The galvanometer should be on a stable platform for testing and not held by hand. If the needle on the galvanometer moves, you have a continuous circuit. If the needle does not move, you have a break in the blasting wire, and it must be located and repaired before the shot is attempted. If the wire checks out, remove the ends from the galvanometer and twist them together.

Next, take the blasting machine and activate it vigorously four or five times to be sure it is working mechanically. The machine can be tested by hooking one end of the blasting wires up to it and jumping an arc across the other ends by twisting or pushing the handle. The brass terminals for attaching the blasting wire to the blasting machine should be wire-brushed until shiny.

TESTING THE BLASTING CAP

The next step is the most critical and most liable to accident—testing the blasting cap. Only one cap should be brought from the cap storage area at a time and this cap should be taken to a place aboard as remote as possible from all other caps and explosives. If possible, insert the cap into the end of a piece of heavy iron pipe. If there is no pipe, hang the cap over the side of the boat in the water, being sure there are no explosives stored inside the boat in that vicinity. In the absence of sophisticated testing equipment to locate stray electrical currents, all electrical equipment on board should be shut off. This includes generators, motors, radio transmitters, radar sets, etc. Emissions from radar and radio transmitters can set off electric blasting caps. Do not handle electric blasting caps while this equipment is running. Another possible source of dangerous stray electricity is the flywheel and drive belt of your air compressor. Keep caps as far away as possible from all running machinery. Shut down all machinery when testing caps or preparing for a blast. Dangerous static electricity can be produced by your clothing. Do not wear nylon jackets or heavy wool shirts when handling blasting caps. The most severe danger of accidental detonation of electric blasting caps comes from lightening. Under no circumstance conduct underwater blasting operations or handle electric blasting caps when an electrical storm is detectable, even on the distant horizon. The great danger of stray electricity entering the blasting circuit is the reason why all electric wires to be used in blasting operations must be thoroughly insulated, and why all circuits must be shunted until the moment before firing the charge. Place your galvanometer in from the edge of the boat several feet. Remove the shunt from the blasting-cap wires and attach them to the galvanometer. If the needle moves, the cap is good. If the needle does not move, the cap is defective and should not be used. Don't just throw it overboard because it could be washed up on the beach eventually where a child might pick it up, or it could be swallowed by a fish and end up exploding under a fish-cutter's knife. Put the cap in a remote, safe place and after the job is completed it can be taped to a good cap, lowered to the bottom on a weight and detonated. Assuming the cap is good, remove the wires from the galvanometer and immediately twist them together.

"Static electricity can be generated by the movement of particles, especially under dry conditions, whether the particles are freely suspended or are imbedded in a moving insulator material, such as a motor-driven belt." (From the *Blaster's Handbook,* by DuPont.)

In underwater blasting operations, the blasting wire cannot be stretched out before the cap is attached as is possible in blasting operations on land. The diver might have to carry the cap and blasting wire under water with him (I strongly oppose doing this) or the blasting wire with cap attached will have to be slacked off as the boat or barge moves away from its position over the charge. Because of the possibility of induced static electricity as pointed out in the DuPont statement, if the blasting wire is stored on a spool or in a coil, I recommend that it be unwound and flaked out in long lengths free for running, before the cap is attached.

Again, be sure that the machine end of the blasting wire is shunted, and then connect your blasting cap to it. The wires must be clean and shiny before they are twisted firmly together. The most frequent cause of misfires that I have experienced has been a poor connection between the blasting-machine wires and the blasting-cap leg wires, and poor insulation of this connection. Because of the great difficulty in dealing with a misfire during submarine blasting operations, this connection deserves all the care you can give it. Hold a bared end of one blasting wire and a bare cap leg wire together, pointing in the same direction, and double them over in a tight loop. Hold the doubled-over ends of these wires firmly together between your left thumb and forefinger and twist the eye of the loop tightly, five or six times. Repeat the operation with the other two wires, then paint them liberally with Black Magic or other quick-drying cement. Allow the cement to dry between coats and then carefully tape the splice with electrical or rubber tape. The completed circuit of blasting wire and blasting cap should then be tested with the galvanometer as above. Shunt the machine end wires when you have finished.

The lead wires of electric blasting caps are not very strong, and tension should never be put on them. I find it best to refold the cap wires the way they are folded when they come out of the box. Then I double these folded wires back onto the blasting cable and securely tape them there, leaving the cap extending about two inches beyond the end of the blasting wire. The next step is to lay the previously prepared 3-foot length of Primacord alongside the blasting wire with the cap overlapping the Primacord about one foot. Tape the cap firmly to the Primacord with five or six wraps of tape, taking care that no tape comes between the cap shell and the Primacord. The cap must always be pointing in the direction of the charge when taping it to the Primacord or when you detonate it, it might just shoot off the short end of the Primacord without initiating the main charge. If caps smaller than No. 6 are used, two of them should be taped together as above on opposite sides of the Primacord, hooked up in parallel.

Leave a bight in the two free inches of the cap leg wires and securely tape six or eight inches of the blasting wire to the Primacord. Now, if any strain is applied, it will fall on the blasting wire and the

Primacord and not on the blasting-cap wires. The tensile strength of the blasting wire will depend on the kind you use, but it will probably be stronger than the Primacord, which is only 150 pounds for the reinforced type.

Leave the cap with Primacord and blasting wires attached in a safe and remote spot, possibly the bow, and prepare for the diving operation.

PLACING THE CHARGE

A 50-pound case of dynamite is rather heavy and awkward for a diver to carry with one hand during a descent, so a ½-inch line with a snap hook on one end should be made up for lowering the charge. There should be a 3- or 4-foot running line secured two or three feet above the snap hook with a constrictor knot. The other end of this line is secured to a shackle for sliding up and down the down line. It is especially important to make up such a line if there is more than one charge to set. The diver can then pick up his charges at the base of the down line with a minimum of lost time. All the slack must be taken out of the down line when it is used for a trolley line.

For multiple charges (e.g., if we were going to try to disperse an entire wreck), several cases of explosives would be set in various compartments. One case for each compartment would be made up as already described, but with longer leads of Primacord. Each primed case would be put under or alongside in contact with the other cases to assure total detonation, and the Primacord would be led up on deck of the wreck for attachment to a Primacord main line. The main line connects all the charges in the various compartments. The best knot for attaching these trunk lines to the main line is a clove hitch. The hitch must be drawn up as tightly as possible with a 4- or 5-inch tail. When laying out multiple charges using Primacord, great care must be taken that none of it gets pulled around sharp corners, forming angles in the Primacord. If this happens, the Primacord might detonate in a straight line shooting off into the water at an angle, without detonating around the corner. Another reason to avoid sharp corners is to prevent the Primacord from cutting or chafing. If it must be run around sharp corners, tie it back with marline or small line to hold it in a gentle arc out of contact with any metal. With multiple charges, the last job for the diver is to bring the main Primacord line down from the surface and tie it into the main trunk line. This is best done with a square knot, and again, it must be drawn up as tightly as possible. The knot should have 5- or 6-inch tails in order to draw it up properly.

In our present job, it will be just as easy to hook the main line Primacord to the charge before the diver takes it down. The square knot can be additionally secured with several wraps of tape around the knot tails. As the diver descends, the Primacord must be slacked off and tended from a point as far away from the down line as possible, to keep clear of his hose and the down line. The diver must be sure the Primacord stays clear on the way down. On the wreck, unhook the lowering line and have it picked up out of the way. Make your way into the cargo hold and place the charge on top of the explosives there. Be sure your charge is in contact with main bulk of the explosives and in a stable situation where it will not slide or topple. Come out of the hatch and check your Primacord lead. If it is leading across the hatch coaming, tie it back to an opposing rail or deck fitting. If all looks well, ascend, checking the Primacord on the way up. If the Primacord should cross or be wrapped around the down line, the down line will be severed when the Primacord is detonated. Be careful that you or your hose do not become entangled in the Primacord, inadvertently pulling the charge out of position. This is another good reason for tying the Primacord off with small line to some part of the wreck.

MOVING AWAY FROM THE WRECK

The easiest way to move the boat away from over the charge is to slip your moorings and slowly drift downwind or downcurrent. In the absence of wind or current, the engine will have to be started and the clutch engaged intermittently, just enough to give you headway. It is important to leave some kind of buoy on the wreck so that you can check your relative position. Without a buoy, it is conceivable that you could drift back over the top of the wreck without knowing it. Pay off the Primacord from the spool as you move away, taking care not to put any strain on it. As you near the end of the spool, unreel the last 40 or 50 feet long before it will go overboard. Tie your Primacord pigtail (which has been made up to the cap and the blasting wire) into the end of the main line Primacord with a tightly drawn square knot. Taping the knot tails onto the standing parts insures good contact. Pay your blasting cable overboard as required and allow no strain to come on it. If things are moving too fast, the engine might have to be used to check the boat's movement. Because of the obvious danger of fouling the Primacord or the blasting wire in the propeller, they must be tended with extreme care. If possible, keep the bow of the boat pointed toward the charge and tend these lines from the bow. If there are any other explosives aboard, in this case a stick of dynamite and some blasting caps, they should not be in contact with the skin of the boat. They should be stored toward the center of the boat on some kind of cushioning such as a mattress. This is to keep them insulated from shock waves, which travel long distances through water.

FIRING THE SHOT

On this particular project, a 1000-foot length of Primacord and an additional 500 feet of blasting cable between you and the charge should put you in the clear. Rely on the recommendations of an explosive engineer for safe distances over the water for large charges. The shallower the water and the larger the charge, the farther away you will have to move the boat. Keep a lookout posted, and if any boats come into the area, the shot will have to be aborted. When about 100 feet of blasting wire is left on board, hook it up to the blasting machine. When about 30 feet of cable remains, check again for all clear, pass the word FIRE IN THE HOLE, and twist or push the handle of the blasting machine as hard as possible. The electricity is released at the end of the stroke, and the faster the stroke, the more electricity will be generated.

You should hear the sound of the blast quickly followed by a geyser of water, indicating a successful shot. If you hear or see nothing, it means a misfire. If you hear a slight report like a .22 rifle discharged in the distance, it probably means that the cap detonated, but not the Primacord. If you hear a louder report, like that of a .38 being fired, followed by small bubbles and wispy smoke, it probably means that a portion of the Primacord fired, but not the main charge. In the case of a misfire, you have nothing to lose by trying again. Pull up the handle of the blasting machine and ram it home. I have saved an apparent misfire by removing the wires from the blasting machine and attaching them to a fully charged storage battery. This points up another important reason why the insulation on the blasting wire and the insulation of the cap splice must be perfect—to prevent current loss when it is time to fire.

If the shot has fired successfully, head the boat back to the wreck slowly, retrieving the blasting cable as you go. Allow enough time for the cloud of explosive gases to disperse, because they are toxic.

HOW TO HANDLE A MISFIRE

After a charge of this size has been fired, unless a strong current is running, you will have to wait several hours for the water to clear before making an inspection dive. Any underwater explosion will kill fish in the immediate vicinity. If sharks are indigenous to the area, they will be there in numbers right after the blast. This is another good reason for waiting several hours before making an inspection dive.

In this situation, a misfire is a touchy problem. It is recommended that you wait at least 30 minutes before investigating a misfire, but this is difficult to do in the open sea when your boat is not moored.

Remove the wires from the blasting machine and check the circuit again with the galvanometer. If the test shows a closed circuit, try once again to detonate the charge. If the circuit is open, remove the wires from the galvanometer and twist them together. Slowly move the boat in the direction of the charge and retrieve the cable without pulling too hard on it. When you come to the cap, cut the Primacord below the knot to isolate the cap from the charge. Hang the cap in the water, or put it into a piece of pipe and cut off about 10 feet of the blasting wire. Shunt this piece with the cap attached immediately. Make up a new pigtail of Primacord with another cap. Tie it into the main Primacord line and back off once again. Repeat the firing procedure and, hopefully, the charge will fire this time.

If the cap fired, but not the Primacord, or only a portion of the Primacord fired, you will have to remoor over the wreck and repeat the whole job. This is why we saved a stick of dynamite. It will have to be primed and inserted into the charge on the wreck, and the entire foregoing procedure will have to be repeated. Hopefully, you took sufficient care during every step of the operation, and the shot fired the first time.

CUTTING STEEL WITH DYNAMITE

Dynamite is sometimes used to cut or shear steel under water, as in the dismantling of a damaged oil-well structure. Dynamite cannot compete economically with the oxy-arc torch or the thermal lance for cutting steel, and it should only be used in special circumstances where burning would endanger the diver. Dynamite must be hard against the steel to be cut. If the steel is covered with barnacles or other marine growth, a preliminary light charge will have to be fired to clean the steel. For greatest effect, the shock waves of dynamite must be contained. If the charge is to be laid on a flat surface, cover it with sandbags to confine and direct the shot at the steel. For flat vertical or horizontal surfaces, the charges will work better if they are made up in half-sections of pipe, split longitudinally, or in angle iron, and tied off as securely as possible to the object to be cut. Double charges made up as above, on either side of the plate or member to be cut, staggered just above and below the line of cut, are much more effective than single charges. Whenever dynamite has to be used to cut steel, take every step possible to firmly anchor and back up your charge. For hollow members, such as pipe braces, etc., the charge will be much more effective if placed on the inside. For cutting steel, the highest strength dynamite available should be used because of the greater shearing force.

CUTTING STEEL WITH SHAPED CHARGES

Far better than dynamite for cutting steel, but much more expensive, are shaped charges. Shaped charges operate because of what is called the lined cavity or Munroe effect, so-named for the man who

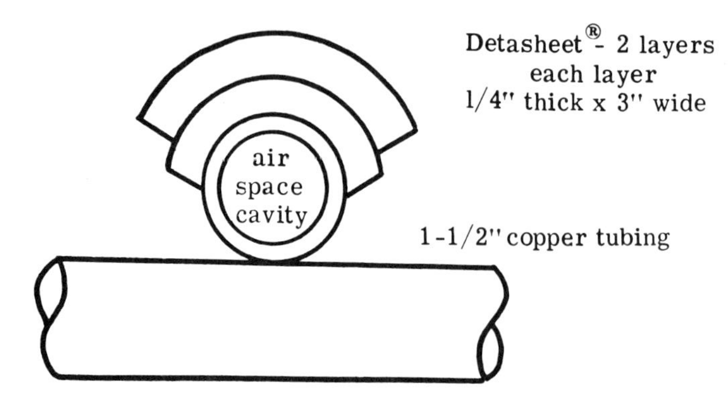

By closing ends of tube to prevent flooding of cavity, the assembly can be used under water.

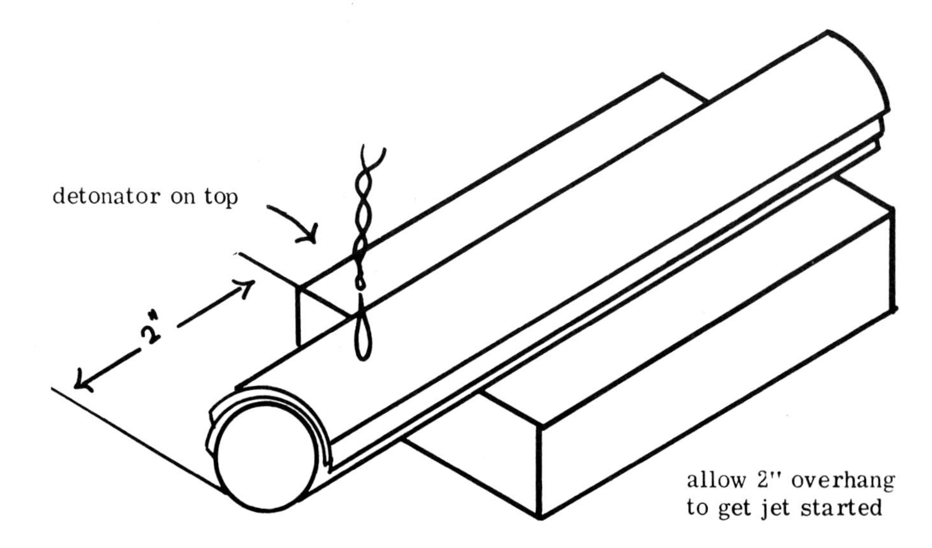

Fig. 30. Linear shaped charge cutter.

first described it. It was found that explosives were much more effective for cutting or piercing when they were molded around a cone with the point aimed away from the object to be cut or pierced and with the base of the cone facing the object. The inside of the cone, and the standoff, must remain hollow. It was further discovered that if the inside of the cone was supported or lined with thin metal, the

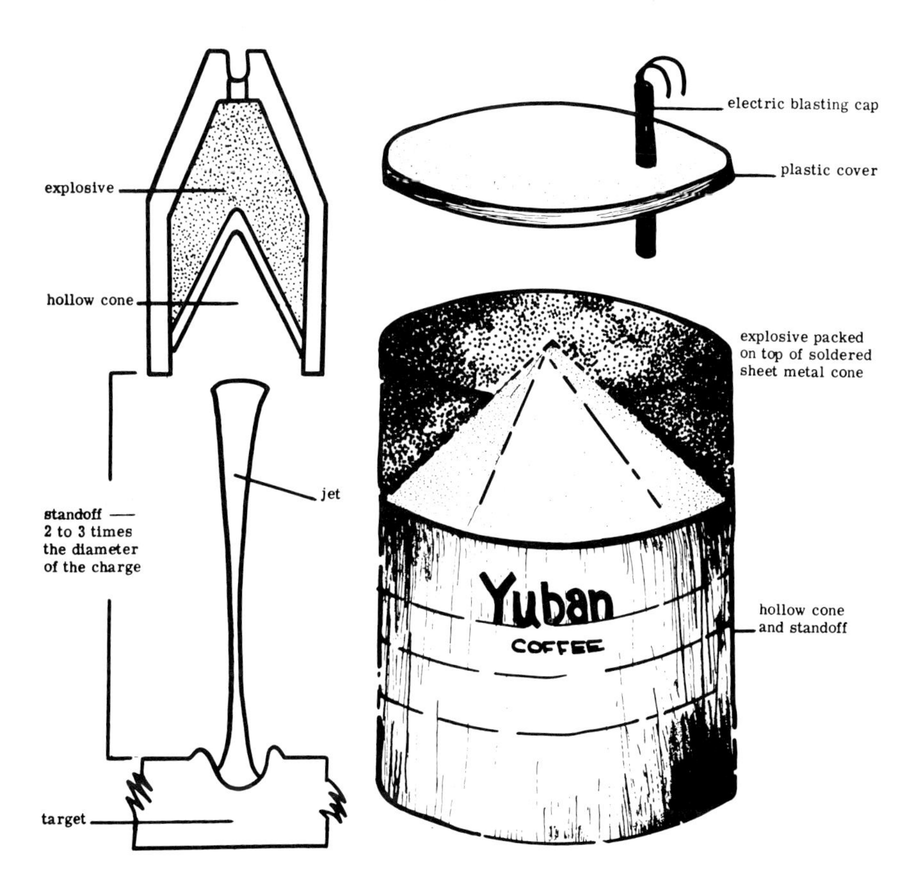

Fig. 31. A shaped charge (l.). A shaped hole-punching charge (r.) made from a coffee can or 4- to 6-inch pipe, should be weighted on top. Diver can evacuate water from can by inserting pneumohose under it when placing. Drawings by Sue Zinkowski.

effectiveness of the charge was greatly increased, because part of the metal vaporized upon detonation, adding to the heat, and also that particles of the metal lining were driven against the target, contributing to the jet effect. Through much experimentation, it was found that the shape of the cavity and the distance of standoff were critical to the end effect of shaped charges and the best shapes and distances have been tabulated. The lined cavity effect is easily adapted to line cutting by molding the explosive over angle bar or tubular configurations. Figure 30 shows a homemade linear shaped charge cutter. Figure 31 shows a homemade hole-punching shaped charge.

STANDARD SHAPED-CHARGE DEVICES

If water is allowed to enter the standoff or cavity, the effect is all but nullified. Several companies recognized the potential of shaped charges for use under water and they manufacture a wide range of standard shaped-charge devices. They will custom-build any type of shaped charge to meet a customer's particular needs. They manufacture O-ring cutters to cut pipes and braces of many standard dimensions. The O-ring cutters are tubular shaped, bent to fit precisely

Fig. 32. Diving superintendent Dennis Webb and diver "Frenchy" Collins inspecting an explosive shaped-charge ring cutter, prior to installation on the jacket leg of a damaged well platform structure. Photo: Taylor Diving & Salvage Co., Inc.

around a pipe of a given size. They are made in two exact semicircular sections, held together at one point by a hinge to allow the cutter to be placed around the pipe, and with a mechanical latch to firmly fasten the charge. Both the ring cutters and the straight-line cutters are provided with Bungi cord to help in holding the charge to the object to be cut. These specially made cutters are completely waterproof and the booster for the explosive charge is an integral part of the unit. They are initiated with Primacord which attaches mechanically to the booster with a special fitting. Much the same as with dynamite, these cutters must be in intimate contact with the steel to be cut. Barnacles or other interfering substances must be removed before the cutter is attached. If the ring cutter does not fit tightly on a plane at right angles to the pipe to be cut, shooting would be a waste of time; the pipe will not be completely severed. When tight-fitting and properly applied, the cuts made by these devices resemble a precision torch cut. Straight-line cutters for flat steel can be manufactured in any practical lengths. Primacord for use with these devices must be used in accordance with the previous instructions.

DEMOLITION OF CONDUCTOR PILES

Manufactured ring-shaped charge cutters, although very expensive, have an ideal application, effecting a saving of many diver man-hours in the demolition and removal of conductor piles when a well is being abandoned. The conductor pile can be 36 or more inches in diameter with 10 or more casings of diminishing size within it. The space between the casings or annulus rings is always filled with cement. When conductor piles are demolished in the conventional way, the first steel shell is cut away with an oxy-arc torch. The cement must then be laboriously chipped away with a pneumatic demolition gun, a very time-consuming and expensive process, especially in deep water. When the next casing is reached, it too is burned, and then again a layer of cement must be chipped away. This process is repeated until the conductor pile is weakened sufficiently to be pulled over with a tugboat, or until it is completely severed and lifted with a derrick. With the shaped-charge ring cutters, two are placed around the conductor pile, one at the mud line and the other six or more feet above it, depending on how many casings there are, and how much room will be needed to get into the center of the pile. The two ring cutters are joined to two straight-line cutters, attached vertically between them on opposite sides of the conductor pile. All four charges are fired simultaneously and when the diver returns to the bottom, he should find two split sections of shell lying there. The shock from these charges will shatter the annulus cement, and a few taps with a sledgehammer will be all that is necessary to clean up the next casing for the same procedure to be repeated. The exact size of the casings must be known beforehand, so that proper-size cutters can be fabricated for each one. As indicated, the vertical distance between the ring cutters for each layer of casing will depend upon the number of casings to be dealt with.

BLASTING ROCK

Occasionally, pipelines must be run through rock or coral, and this material must be blasted. In extremely hard rock, a series of closely spaced line holes will have to be drilled to break it up for removal. The best way to drill these holes, especially in calm, shallow water, is

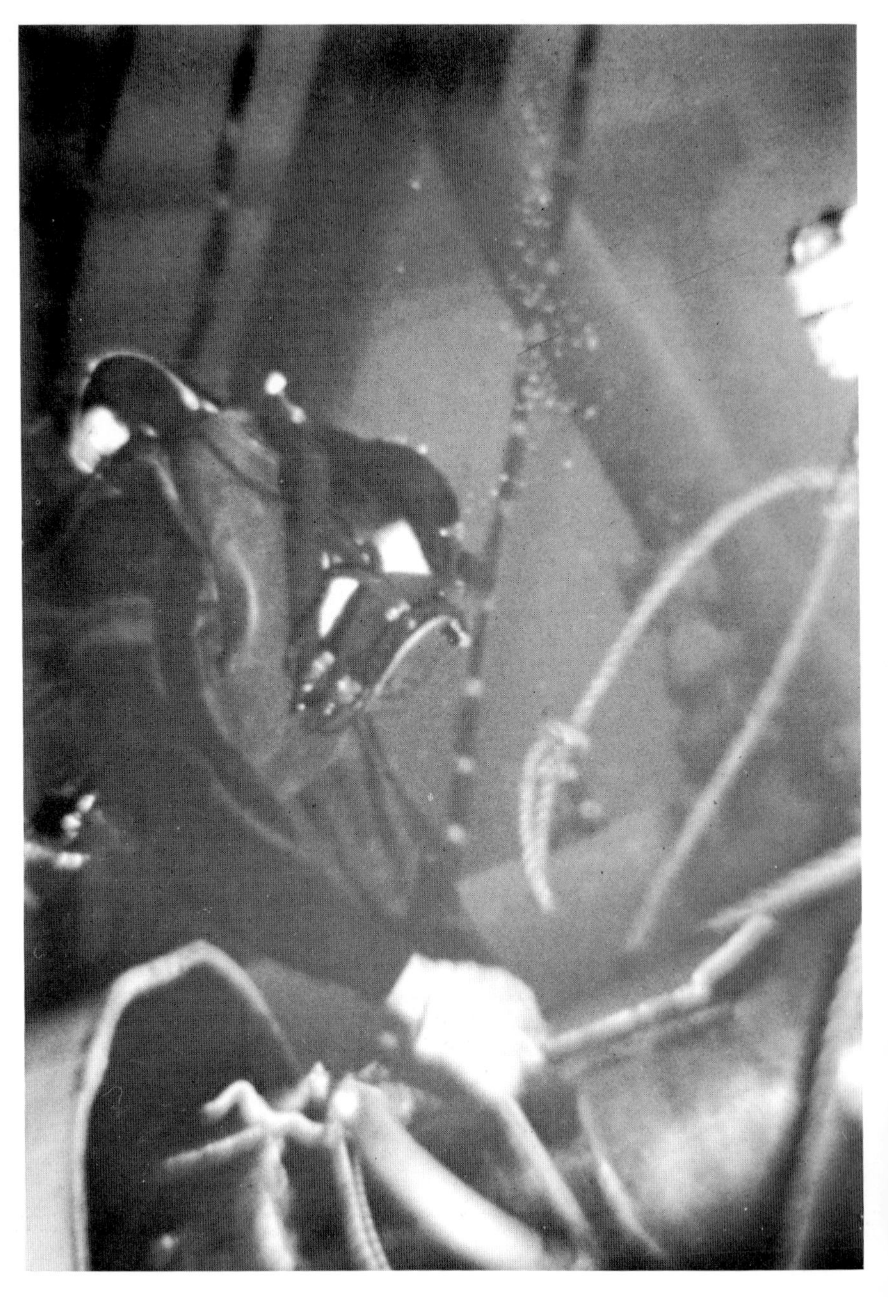

Fig. 33. Diver installing shaped-charge ring cutter on a well casing. Photo by Ted Cox; courtesy of Westinghouse, Inc.

with pneumatic wagon drills, mounted along the side of a barge. The diver will be used to place the drill tip in a crack or depression in the rock to start the hole. After each hole has been drilled, the diver might have to insert a rod in it, to keep its location marked until all the holes in the series have been drilled. This is especially necessary in a current where mud or sand is swept over the drill holes. In deep water or areas of rough surface conditions, the wagon drill might have to be put directly on the bottom, and moved between successive holes by a crane with the diver placing it each time.

A jackhammer, such as is used to drill holes on the surface, is almost useless under water, because of the difficulty in applying sufficient pressure to it, but with great perseverance and hard work, holes can be drilled with it. Jobs requiring ten or more holes should never be attempted with a jackhammer; use a wagon drill. After the holes are all drilled, the diver will remove the marking rods, one at a time, and load the holes. The charges should be made up with Primacord. If more than one stick is required per hole, the individual hole shots should be made up on deck so that all the diver has to do is slip the charge into the bore hole. Charges made up with Primacord pigtails in sections of plastic pipe slightly smaller in diameter than the bore hole work very well. If the bore should be filled with sand or mud, a water jet attached to a six-foot length of ¾-inch pipe will clean it out very quickly. After the diver has slipped the charge into the bore hole, he should gather up whatever overburden is handy and pack it on top of the charge. If there is no overburden, sacks of sand should be provided for this purpose. The charge should never be forced into a bore hole. If it will not fit in easily, abandon the hole. After all the holes are loaded, the diver will hook all the trunk-line Primacords into a main line with clove hitches. The barge is then moved away, and the shot is fired. Misfires should never be dug out of a bore hole; jet the sand out of the hole and put another primed stick on top of the charge.

In coral, shale, or soft or fractured rock, charges of dynamite secured to the underside of 100-pound sacks of sand are easy to make up and place, and sometimes very effective. Rows of shaped charges of the canister type are very effective for trenching in coral. This type of charge has often blasted holes ten or more feet deep in coral.

BLASTING CLAY OR CEMENTED SAND

For hard clay or cemented sand it is often possible to bore holes with a high-pressure water jet. This type of material should be shot with a slow or low percentage dynamite to exert a lifting force rather than a shattering force.

Small charges of explosive could very effectively be used to break the weight coating away from a damaged pipeline in preparation for torch cutting. If the line is buckled but not holed, a ring-shaped charge cutter would be ideal for severing the line, without the necessity of waiting for the line to flood.

Primacord has many excellent uses by itself, without the addition of other explosives. Five or six tight wraps of Primacord around a wooden piling will destroy enough of the wood fiber so that the piling can be knocked over or picked off the stump with a crane. A length of Primacord laid along the line of cut is a good way to clean steel of marine growth prior to oxy-arc burning. One or two turns of Primacord around a frozen nut, such as a propeller shaft nut, is an excellent way to free it. Three or four wraps of Primacord around a shaft, forward of and hard against the hub, works well to start a propeller off its taper.

CLEANING STEEL AND CONCRETE PILINGS FOR VISUAL INSPECTIONS

A frequent job offshore is cleaning jacket legs or structure piles in the splash zone area (the area between wind and water that is repeatedly wetted and dried by wave action) in preparation for applying Splash Zone or other epoxy resin coatings. The surface must be shiny clean before Splash Zone is applied, and this is usually accomplished by sandblasting. This is a lengthy and difficult job, complicated by wave action at or near the surface. Albert Teller, of Explosives Engineering, told me of using a process similar to the following for cleaning steel and concrete pilings for visual inspections in Hawaii:

Cut a roll of 90-pound roofing felt into 4-inch sections, using a long stiff knife lubricated with fuel oil. With Bungi cord to anchor one end of the strip of felt, unwind the roll in an overlapping spiral around the jacket leg, keeping the felt tight. Secure the top end the same as the bottom. Anchor your Primacord under the bottom Bungi cord, and wrap the Primacord over the roofing felt, working around the jacket in a spiral, with the Primacord wraps on approximately 2-inch centers. Secure the top end. Tie in a Primacord pigtail with cap attached as described earlier, and shoot. The area will be completely cleaned of marine growth, rust and scale, and at most, will need a lick or two with a wire brush to make it ready for the application of Splash Zone. This same system can be used on any section of the structure for visual inspection of welds, etc.

DETASHEET

Detasheet® is a new, flexible high explosive developed by DuPont that has an excellent potential for underwater use. You will note that it is the explosive used on the homemade linear-shaped charge shown in Fig 30. It is composed of an integral mixture of the high explosive PETN, and an elastomeric binder. Detasheet is flexible over a wide range of temperatures. It is easy to cut and handle, yet retains all of the explosive properties of PETN alone. In addition, it is safe to use, waterproof, and available in sheets, cords and a variety of extruded shapes. Detasheet is colored olive drab and is available in sheets of various thicknesses. It is flexible, tough, durable, and has a uniform velocity of detonation. One of the safest of high explosives, it is not affected by repeated flexings, can be molded to any shape, is completely waterproof, extremely resistant to water erosion and remains relatively unchanged by extreme hydrostatic pressures. It is highly insensitive to impact and cannot be detonated by a .30-caliber bullet being fired into it from a distance of 40 feet. It may be cut to any desired shape with a common fixed-blade knife.

Detasheet can be quickly and efficiently attached to any item by simply cutting, forming to shape and applying with an adhesive. The recommended adhesive, DuPont 4684, is brushed on the surface to receive the explosive charge and also onto the contact surface of the Detasheet. This insures complete contact between the explosive and the item to be cut. While the adhesive on each surface is still tacky, the Detasheet is pressed firmly in place by hand. The charge is then ready for initiation. For consistent initiation of Detasheet, a pair of No. 6 or stronger caps should be used. Masking tape or a similar material should be used to secure intimate contact between the detonator and the explosive. The entire butt end of the detonator must be in contact with the explosive. With very thin sections of Detasheet, a small additional thickness should be bonded to the larger section at the point of detonator contact to insure delivery of the full force of the detonator. Because different requirements call for varying amounts of explosive force, layers of Detasheet can be built up to increase the effective explosive charge. Subjecting small particles of the explosive to a grinding or shearing action must be avoided. Detasheet, therefore, should never be cut with scissors. I know of no suitable adhesive for bonding Detasheet to metal under water, and until one is developed, its uses will be limited to applications where the explosive device can be assembled on the surface, as in Fig. 30.

I have not given any instructions for handling electric blasting caps under water. I feel very strongly that they should never be carried under water by the diver because of the possibility of current leakage from the diver's telephone, and the danger of accidental detonation from rough handling inherent in all underwater work. Always use Primacord for detonating underwater explosives and attach the electric blasting cap to the Primacord on the surface.

RULES FOR HANDLING EXPLOSIVES

Following is a list of "Do's and Dont's" for handling explosives. Most of them have been excerpted from a list adopted by the Institute of Makers of Explosives. Abide by these rules.

1. Don't purchase, possess, store, transport, handle, or use explosives, except in strict accordance with organizational, local, state, and federal regulations.
2. Don't leave explosives lying around untended.
3. Don't store explosives anywhere except in a location which is clean, dry, well-ventilated, substantially constructed and securely locked.
4. Don't drop, throw or slide packages of explosives or handle them roughly in any manner.
5. Don't permit metal to contact cases of explosives. Metal, flammable or corrosive substances should not be transported or stored with explosives.
6. Don't store blasting caps or electric blasting caps in the same box, container, or magazine with other explosives.
7. Don't store explosives in a wet or damp place or near oil, gasoline, cleaning solution or solvents or near radiators, steam pipes, exhaust pipes, stoves or other sources of heat.
8. Don't store any sparking metal or sparking metal tools with explosives.
9. Don't smoke or have matches or any source of fire or flame in or near explosives.
10. Don't shoot into explosives or allow discharge of firearms near explosives.
11. Don't open kegs or wooden cases of explosives with metallic tools. Use a wooden wedge and wooden, rubber, or fiber mallet. A fixed-blade knife may be used for opening fiberboard cases, provided that it does not come in contact with the metal fasteners of the case.
12. Don't accumulate any cases, paper products or other materials used for packing explosives. At sea, throw them overboard, being sure that they are empty.
13. Don't place explosives where they may be exposed to flame, excessive heat, sparks or impact.
14. Do replace or close the cover of explosive cases or packages after using.
15. Don't carry explosives in the pockets of your clothing or elsewhere on your person.
16. Don't attempt to reclaim or use fuse, Primacord, blasting caps or any other explosives that have been water-soaked, even if they have dried out.

17. Don't use explosives that are obviously deteriorated.
18. Don't make up a primer or explosives charge in a magazine or near other explosives.
19. Don't strike, tamper with or attempt to remove and investigate the contents of an electric blasting cap or try to pull the wires out of a blasting cap.
20. Don't store electric blasting caps in the same box, container or compartment with other explosives.
21. Don't leave electric blasting caps exposed to the direct rays of the sun.
22. Don't handle, use or remain near electric blasting caps or explosives during the approach or progress of any electrical storm.
23. Don't attempt to fire electric blasting caps except with an adequate source of current.
24. Don't use in the same circuit electric caps made by more than one manufacturer. Resistance in the bridge wires may differ.
25. Don't spare force or energy in operating blasting machines.
26. Don't tamper with or change the circuit of a blasting machine.
27. Don't store explosives so that cartridges stand on end.
28. Don't force cartridges or any explosives into a bore hole or past any obstructions in a bore hole.
29. Don't spring or shoot a bore hole near another hole loaded with explosives unless it is a part of and connected to the shot.
30. Don't tamp explosives with metallic bars or tools; use only wooden sticks with no exposed metal parts.
31. Don't use combustible material for stemming.
32. Don't tamp dynamite that has been removed from the cartridge; avoid violent tamping. Never tamp the primer.
33. Do confine the explosives in the bore hole with sand, earth, clay or other suitable incombustible stemming material.
34. Don't kink or damage Primacord or electric blasting-cap wires when tamping.
35. Do avoid placing parts of the body over the bore hole when loading.
36. Don't uncoil the wires or use electric blasting caps during dust storms or near any other source of large charges of static electricity.
37. Don't uncoil the wires or use electric blasting caps in the vicinity of radio-frequency transmitters or radar antennas.
38. Do keep the firing circuit completely insulated from the ground or other conductors, such as the steel deck, bare wires, pipes or other paths of stray currents.
39. Don't have electric wires or cables of any kind near electric blasting caps or other explosives, except at the time and for the purpose of firing the blast.

40. Do test all electric blasting caps, either singly or when connected in a series circuit, using only a blasting galvanometer specifically designed for the purpose.
41. Don't test electric blasting caps unless they are put in a pipe or behind a barrier.
42. Do keep the leg wires of electric blasting caps shunted out at all times until ready for firing a shot.
43. Do immediately remove wires from the blasting machine after a shot and twist them together.
44. Don't carry under water or require another diver to carry under water electric blasting caps. Do use Primacord instead.
45. Do be sure that all wire ends of electric blasting caps and blasting wire and the terminals of the blasting machine are bright and clean.
46. Don't allow near the blast area or where charges are being made up or tested, any persons not essential to the blasting operation.
47. Don't fire a blast until all surplus explosives are in a safe place, and all persons and boats or vessels are at a safe distance.
48. Don't return to the blast area until the smoke and fumes from the blast have been dissipated.
49. Don't use electric blasting caps in any wet work unless they have adequate water resistance and have suitably insulated wires.
50. Don't use any means other than a blasting galvanometer containing a silver chloride cell for testing electric blasting caps.
51. Don't use damaged leading or connecting wires in a blasting circuit.
52. Don't attempt to investigate a misfire too soon. If possible, wait as long as one hour.
53. Don't drill, bore or pick out a charge of explosives that has misfired. If possible, try to detonate with another charge, placed near or on the misfire.
54. Don't abandon or throw away any explosives. Dispose of or destroy them in strict accordance with regulations.
55. Don't make up more charges than needed to be loaded and fired in one shot.
56. Don't stack boxes of explosives so that the roll or shift of the boat will cause them to fall.
57. Don't store explosives against the skin or hull of boats and barges.
58. Do cushion surplus explosives aboard before firing charges.
59. Do waterproof the cut ends of Primacord.
60. Do tie all knots in Primacord tightly.
61. Do anchor or tie off the Primacord main line with small line to prevent it from being pulled out of the charge.
62. Don't kink Primacord.
63. Don't run Primacord around sharp corners or over sharp edges.

64. Do point the cap in the direction of the charge when taping it to Primacord.
65. Don't handle raw dynamite or explosives bare-handed.
66. Do wash thoroughly if skin comes in contact with raw explosives.
67. Do move the boat or barge a safe distance away before firing any charges. Too far is always better than too near.

And finally, consult your explosives' manufacturer for any questions or problems related to explosives. It is in their best interest to cooperate fully in promoting the safe and effective use of their products.

Explosives are dangerous. Use them with caution and intelligence.

Chapter 9

DIVING FROM A PIPE-LAY BARGE

A lay barge is a floating factory that remains in continuous operation 24 hours a day, in all but the severest weather. It moves ahead along the pipeline route, pulled by its anchor winches in smoothly flowing increments of 40 feet, the average length of a joint of pipe.

Lay barges are built in many different sizes, but average dimensions would probably be 350 feet long by 80 to 100 feet wide. The frontispiece is the Brown and Root lay barge *L. B. Meaders,* one of the world's largest.

The size of the work force depends on the size of the barge and the size of the pipeline being laid. It can range from 100 to 200 men. Some of these men are ship's company, involved with the maintenance and internal operation of the barge, and include the captain, the timekeeper, cooks and room stewards, mechanics, and electricians. The rest of the men aboard are concerned only with the actual mechanics of laying the pipeline. These are the superintendent, the shift foremen, welders and helpers, crane operators, X-ray technicians, riggers, dope hands, and laborers. Almost all lay barges today carry at least two permanently assigned divers and their tenders. Everybody aboard works a daily 12-hour shift, with the exception of the captain, the superintendent and the divers. These men are on call 24 hours of the day.

A lay barge uses enormous quantities of pipe, welding rod, sand, cement, tar, fuel, food, and so on. These materials are delivered by tugboats and material barges in a constant shuttle between the lay barge and shore. Fresh groceries, mail, and crew replacements are brought out by a daily crew boat. Most lay barges also have a helicopter landing platform to facilitate the transportation of company dignitaries and for the evacuation of injured workmen.

There are generally at least two deck cranes aboard a lay barge—one at the bow and one at the stern. These cranes are in constant operation, handling the anchors and off-loading the many materials used in pipelining.

The bow crane unloads the 40-foot joints of pipe from the supply barges (Fig. 34), and loads them in a rack at the bow. Also at the bow is a conveyor system that holds 10 or 15 joints of pipe and advances them one at a time to the line-up station.

Fig. 34. (Top) A modern lay barge. Note the helicopter (center), the pipe supply barge (right center) and the pipeline going overboard on the stinger (bottom center). Photo: J. Ray McDermott & Co., Inc. (Bottom) Unloading pipe from a supply barge. Photo: Bethlehem Steel Corp.

PRELIMINARY ASSEMBLY AND WELDING OF PIPELINE

The pipeline ramp, where the work of assembly and welding of the pipeline takes place, is almost always on the starboard side of the barge.

Work on a pipeline will commence even before the barge is on location. The workers will assemble and weld together as much pipe as will fit on the ramp, so that they can begin pushing pipe as soon as the anchors are run. This first work is called filling the shoes, because the pipe, as it moves from bow to stern down the ramp, is supported every 30 feet or so by pedestal-mounted clusters of rubber-covered rollers. These rollers, or "shoes," are raised and lowered by hydraulic rams, and they are adjusted to support the pipe in a gentle arc down the length of the pipe ramp.

The back end of the first joint of pipe is sealed with a welded dome-shaped cap. This is to keep water from entering the pipeline while it is being laid. There will be a heavy welded pad eye on this cap on the beginning end of the pipeline, and its purpose will be explained presently.

After the cap is welded, the first joint is pulled down the ramp about 40 feet, and the second joint of pipe is rolled into place by the conveyor.

Each joint of pipe is delivered with the outside covered with a corrosion-resisting dielectric coating of somastic or other substance, called the dope coat, and encased in a thick shell of concrete. This shell of concrete, called the weight coating, neutralizes the buoyancy of large-diameter pipe, and provides valuable mechanical protection for the dope coat and the pipeline itself. About a foot at each end of each joint of pipe is left bare and the pipe ends are beveled in preparation for welding.

These beveled ends are inspected for nicks or dents while the pipe is still in the rack, and are cleaned of rust with a power wire brush. When the conveyor rolls a joint in the line-up station, it is ready to be welded.

The shoes at the line-up station are set up to control each joint of pipe through all directions—up, down, forward, back, and from side to side. Each joint is then perfectly butted and mated to the section already in the ramp, and the first bead of weld is laid on, joining the pipes.

Two welders, working one on either side of the pipe, start at the top and weld the two pipes together. This first pass is called the stringer bead, or the root pass, and the larger the diameter of the pipe, the longer it takes to complete the weld. When this first pass is completed, the pipe is pulled aft down the ramp 40 feet, and another joint is rolled into place, lined up and welded on. Spaced at intervals of 40 feet down the ramp are additional work stations. The next two are also welding stations, where the filler and capping beads are applied. After the final welding pass is completed, the next stop on the pipeline's progress down the ramp will be the X-ray station, where the entire weld is X-rayed and inspected for flaws.

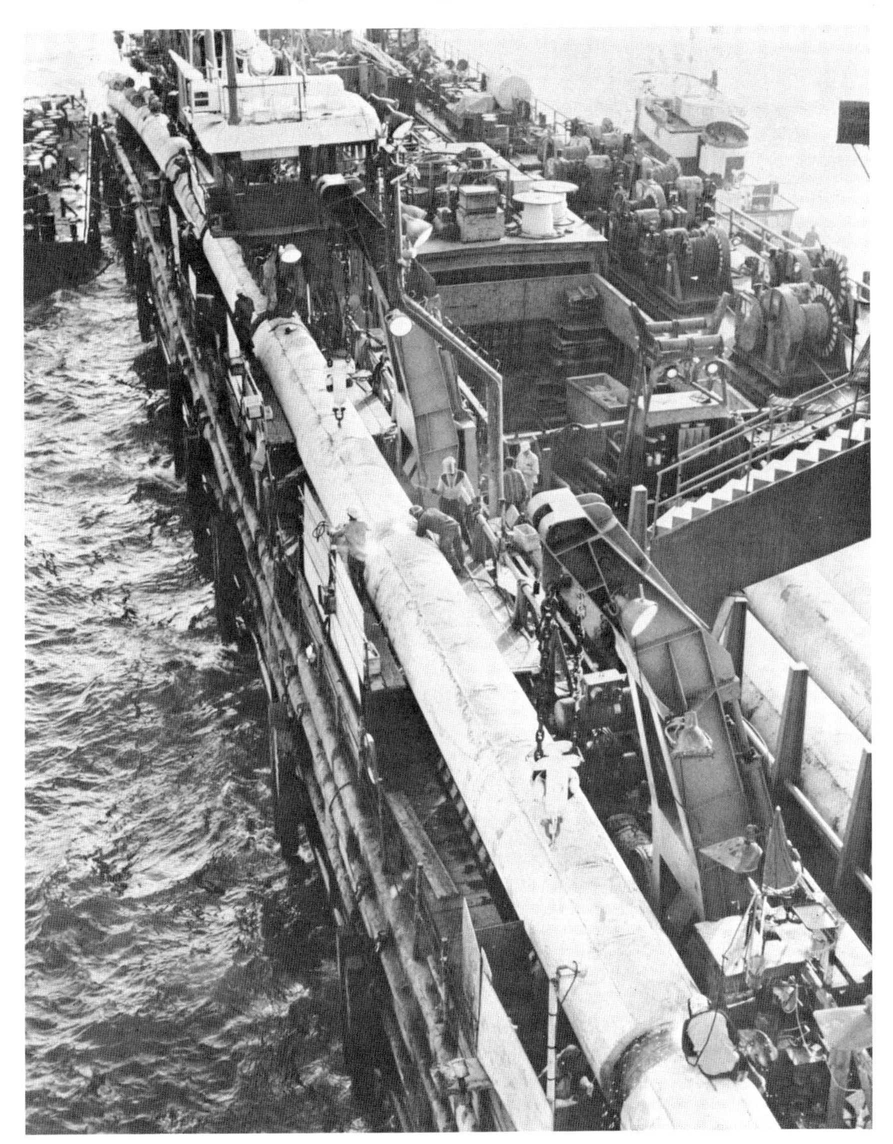

Fig. 35. The pipe ramp on a modern lay barge. Note the welders working at three different positions along the pipeline, the davits (lower right-hand corner and center), and the anchor winches (upper right-hand corner).
Photo: Brown & Root, Inc.

Fig. 36. (Top) The dope station on a lay barge. Last stop on the barge on the pipeline's journey to the bottom of the sea. Note the direction of the flap on the field joint tin. Photo: Bethlehem Steel Corp. (Bottom) The massive shaft that connects the pontoon to the barge hitch. Photo: John Violette.

THE DOPE STATION

If any flaws are discovered, they will be ground or burned out, and the joint rewelded. After the weld passes this inspection, the joint is moved down the ramp to the final work station at the stern of the barge. This is the dope station. All the exposed metal between the two joints of pipe is here wire-brushed clean and painted with hot somastic. A piece of sheet metal is wrapped around the joint and secured to the weight coating on either side with steel bands (Fig. 36). A little flap is left in the sheet metal, and through this, the void between the joints is filled with either concrete or hot liquid tar or asphalt. When the void is filled, the flap is closed and secured with a steel band. This joint, called the field joint, is now completed, and this section of pipe is ready to be pushed into the water. In this way, step by step and 40 feet at a time, pipelines 100 miles long or more are constructed.

FIELD JOINTS

The field joints on a pipeline are very helpful to a diver. Because they occur every 40 feet or so, counting them is an excellent way to measure distance traveled when working on or inspecting a pipeline on the bottom.

When the dope crew assembles the sheet metal over a field joint, they work to a system, and the exposed edge of the field-joint tin, usually near the top of the pipe, is invariably pointing in the same direction on every field joint on the entire pipeline. Whenever you are working on a pipe-laying operation, check the direction in which the field-joint tins are pointing.

If, as you are standing on deck facing forward, the edges of the field-joint tins are pointing to starboard, this will mean that as you move along the pipeline on the bottom, the edges of the field-joint tins are pointing to your right, you will be traveling towards the bow, and vice-versa. Whenever you become disoriented while working on a pipeline on the bottom, you have only to locate a field joint to set yourself straight.

THE CONTROL TOWER

High above the ramp, about halfway down the barge is the control tower. The eight or twelve anchors that hold the barge firmly in position while it is slowly pulled along the pipeline route, 40 feet at a time, are controlled from this tower. There is a red light and a green one, with operating buttons, at each of the work stations along the ramp. When the cycle of work on each joint is completed at each station, the light is changed from red to green. These lights are connected to a corresponding panel in the control tower, and when the anchor foreman has a green light from each work station, he will sound a warning hoot on the horn, and pull the barge ahead another 40 feet.

DAVITS

Spaced about every 60 feet along the ramp are the davits, four to six in number, depending upon the length of the barge. These are the side booms, used to put the pipe overboard, and to pick pipelines up off the bottom for a variety of operations. Specific uses of the davits will be treated later in this chapter.

TENSIONING SHOES

Halfway down the ramp, aft of the welding stations, there will be one or two massive machines called tensioning shoes. These will consist of two rows of double rubber tires, or two rubber-covered caterpillar treads, one above and the other below the pipeline. These rows of tires or treads are controlled by hydraulic rams, which squeeze the pipeline between them, and they can be set for a predetermined tension on the pipeline. The tires or treads can be hydraulically rotated, to push the pipe down the ramp, or to pull it back aboard. As the pipeline is laid, that section between where it leaves the back of the barge and where it rests firmly on the sea bottom, if unsupported, will sag, because of the weight of the pipe. This sag produces dangerous bending stresses in the pipe, and if severe enough, could buckle the pipe. The tensioning shoe, by holding back on the pipeline as the barge is pulled ahead on its anchors, relieves some of this sag, permitting pipelines to be laid in moderately deep water without a pontoon, or with a shorter pontoon than would be necessary if the pipeline were to be laid without tension.

Holding tension also helps to hold the pipeline in a straight line over the pipeline route. In deep water, the sag in the pipeline has to be supported by a pontoon.

PONTOONS

A pontoon is a controlled, variable buoyancy device, used to support the pipeline between the barge and the ocean bottom. Pontoons are generally constructed of two large pipes, 20 inches or more in diameter, joined together along the bottom side by a series of interconnecting U-shaped pipe braces. These braces, fitted with rollers, form a trough or a channel between the main pipes of the pontoon, through which the pipeline runs. The pontoon is divided into a series of watertight compartments about ten feet long, each with its individual flood valve and blow valve. A pontoon can be from 100 to 1000 feet long. One end of the pontoon attaches to a hitch at the base of the pipe ramp at the stern of the barge, and the other end either drags along the bottom or floats a few feet above it.

Laying a submarine pipeline is a complicated procedure, and the many intricacies of equipment and technique can be more easily explained and understood if we simulate an actual job.

You have been selected as lead diver for J. Houston-McRoot lay barge No. 3. It is the beginning of the season and the first job will be 30 miles of 18-inch pipeline to be laid between two platforms, complete with risers at each end. The job will begin in 40 feet of water and end in 150 feet. Because the job is to begin at the shallow end, only one diver is initially assigned to the barge, and so it is the responsibility of you and your tender to load aboard and properly set up all the diving equipment. This is a rare opportunity to set the barge up so that you and other divers working from it can do your jobs safely and efficiently.

SETTING UP THE DIVING EQUIPMENT

The two largest pieces of equipment, the recompression chamber and the compressor, will require the most work and care in setting up. The chamber should be located in a place that is shaded, remote from the vibrations of deck winches, and protected from the weather. It must also be accessible, with the entrance hatch clear of obstructions. A small permanent grating or threshold should be installed at the entrance, with sacking or a mat for the diver to wipe his feet on. If the structures of the barge do not provide shade, then a pipe frame and a canvas awning should be built. A few company cigarette lighters and fountain pens slipped to the welding foreman at this beginning stage will assure his cooperation. The initial setting up of your equipment requires quite a bit of welding. If you can't do it yourself, take the trouble to cultivate the welding foreman. Build a rack close to the chamber to hold four or five bottles of oxygen. Check out the oxygen masks for the chamber, the communications and the internal lighting. The light bulbs must never be larger than 25 watts. You should have spare O-rings or gaskets for the chamber hatches in your kit.

The same considerations of location apply to your compressor, with the additional requirement of setting it up away from any possible source of contamination from the exhaust of other machinery on the barge. It is often quite simple to extend the air-intake filter on a length of hose or flex pipe and mount it in the fresh air, high above the deck.

Be sure you have ample lubricating oil for the engine and the compressor, as well as filter elements for fuel, lube oil and air, as well as other basic spares.

The air-distribution system from the compressor to the chamber and the various diving stations should be hard piping. If you take the trouble to do this at the start, you can spend the season free of concern for crimped or oil-soaked hoses. Run your pipe over the most protected route possible, and have the welder anchor it down at intervals with U-clips tacked to the deck. You will need a permanent diving station at the bow, at the stern, and one or more, spaced along the pipe ramp, depending upon the length of the barge.

Each diving station should have a small volume tank, and a manifold of valves. In addition to your diving air, the barge's internal air system should be hooked into each diving station for emergency use. The manifold should have valves to control the incoming air from the compressor and from the barge supply, so that either system can be shut off and isolated, and the other one turned on, in case of a failure. Each manifold should also have valves for two or three diving rigs and a drain cock to periodically drain off moisture. There should be two gauges, one to indicate the primary air-supply delivery pressure, and the other for the emergency air. I find it a distinct advantage to hook the pneumohose for the diving rigs into the emergency supply system. In this way, the diver carries with him two independent air supplies at all times.

In the event of a failure in the primary system, or even in the communications, the procedure on deck would be to open wide the pneumo air-supply valve. Thus, while the emergency air is being cut into the manfold, the diver will have air coming out of the end of his pneumohose. He can insert the end of the hose under his helmet or mask, and start his ascent with assurance of a couple of lungfuls on his way up.

At each diving station, build a rack to hold your diving hoses up off the deck. This will save you the work of coiling and tying up hoses between dives, or, as often occurs, save the hose from being trampled on and abused by the barge workers. When the dive is finished, your tender can coil the hose on the rack, and it will be ready to go for the next dive.

It is important at the outset to build at least two good ladders. They should be wide enough to climb comfortably, extend into the water at least 5 feet, and have the rungs spaced no more than 10 inches apart. One ladder can be kept at the bow and the other at the stern, and shifted around as needed. While working on a lay barge, you will quite often be called upon to dive from either the crew boat or the tugboat, so make up a light ladder, about 10 or 12 feet long, for this.

There will be a little room designated aboard the barge as the diving locker. Check off and stow the rest of your equipment there, and be sure the room can be securely locked. The diving equipment will be checked out to you personally. Most of it disappears if not properly guarded.

You will have two or three diving telephones. Check them out and be sure you have spare batteries and fuses. Have spare batteries and bulbs for your diving lights. Have a full supply of fittings and adaptors for all the different-sized hoses, as well as tape, teflon tape for pipe fittings, spare phone speakers for your masks and helmets, spare non-return valves, silicone grease for wet-suit zippers and chamber gaskets, wet-suit glue and spare wet-suit material, neat's-foot oil for leather belts and harness, work gloves, diving knives, and any other gear you can think of to keep you completely independent of the shore. You must have a full tool kit with wrenches, pliers, screwdrivers, etc., and a small soldering outfit is indispensable for repairs to the telephones and communications wires.

Check out your cutting gear according to the recommendations in Chapter 7. If you are going to use scuba gear, be sure you have a bottle gauge, spare tank O-rings, a wrist depth gauge and compass, a life vest with spare CO_2 cartridges, masks, flippers, and so on.

When you have all your gear hooked up and stored, fire up everything to be sure it is operating, and that you are ready to dive. A lay barge can cost $2000 an hour or more to operate. There is no surer way to endear yourself to the barge superintendent than to spend 30 or 40 minutes scrambling around looking for a ¼-inch pipe-to-oxygen-hose fitting when he is waiting for a diver.

THE HITCH AND PONTOON

At the beginning of a pipeline job, virtually all of your diving will be done from the stern of the barge. On the starboard side, at the end of the pipe ramp, is the pontoon hitch, a massive fabricated steel block, weighing from 30 to 50 tons. This hitch locks the end of the pontoon to the barge. It also acts as a giant hinge upon which the pontoon can swing in a vertical arc. The hitch is equipped with a number of hydraulic rams, some to actuate locking fingers which grip the end of the pontoon, and others to raise and lower the hitch vertically on T-beam tracks. The vertical movement is necessary to adjust the end of the pontoon under the pipeline as it leaves the barge, to maintain it in a gentle supported arc on its way to the bottom. The hitch is controlled from a small hydraulic console mounted on deck just above it at the starboard stern corner of the barge. The hitch will be tested to be sure that it moves up and down freely and that the fingers open and close. Quite often, pieces of tar or concrete or other debris will get jammed into the fingers, and the diver will have to clear them. The hydraulic-control hoses often leak, and will have to be tightened or replaced by a diver. When the hitch has been checked operating, the pontoon will be maneuvered to the stern of the barge by a tugboat.

There are two pad eyes at the barge end of the pontoon, and a diver will have to shackle a pair of slings connected to the stern crane into these. In calm weather, this job can be easily done wearing just a pair of coveralls, because the pad eyes are above the surface. In rough weather, it will be much easier if you wear your diving mask. You will have the added safety of your hose attached to you if you should get swept off the pontoon, and you will not have to hold your breath when the waves roll over you. You will also be able to give directions for the crane operator to lower or pick up on the slings by telephone rather than by frantically waving your arms and shouting. In rough weather, the hitch and the lower end of the pipe ramp is by far the most dangerous area to work on the entire lay barge. Because it slopes down beneath the water, there is constant wave action funneling up the ramp and rushing out with terrific force, even in moderate seas. Stay away from the hitch and the base of the ramp in rough weather. The force of the seas sucking in and out of the ramp is tremendous and no man is able to hold on against it. If the weather is very rough, insist that the barge be warped around bow into the seas, to give you a lee while working on the pontoon and hitch.

When the slings are hooked up, the tugboat will hold the outer end of the pontoon in alignment with the barge, and the crane will lift the barge end of the pontoon out of the water a few feet and drop it into the hitch. The connecting point of the pontoon is a massive horizontal steel shaft that fits into hooklike receptacles on the hitch. The steel fingers are hydraulically closed over the shaft and mechanical dogs are then set to lock the fingers in place. On some hitches, the diver must set these dogs, tapping them in place with a sledgehammer.

When the pontoon is firmly attached to the barge, the diver must hook a number of control hoses to it. The actual number depends upon the design of the pontoon, but there will be at least five. These hoses will carry air for blowing water out of the pontoon, and air or hydraulic fluid for actuating various flood valves. There will also be at least one pneumohose for depth readings at the lower end of the pontoon. These hoses run from the control console, down through a pipe called the J-tube, and they come out underwater near the hitch. Usually a plumber's fish, with one end tied to the hoses, is run down the J-tube. The diver will pull the hoses through with the fish, and then connect them to a panel mounted on the back of the pontoon. The fittings are generally of the spring-loaded quick-disconnect type, and require only a minute or two to hook up. It is wise to daub these fittings heavily with grease before you hook them up. This will keep them lubricated and they will be easy to remove when the job is finished. After they are connected, all slack in the hoses must be pulled back aboard the barge, to keep them from getting fouled on the hitch.

TYPES OF PONTOONS

Pontoons are built to a number of different designs, but the essential features and the principle of operation will be the same for all of them.

Running down the entire length of the pontoon, usually under the port side, will be one or two pipes, 2 or 3 inches in diameter. These pipes are called headers, and they carry the air supply for blowing the individual compartments on the pontoon. Branching off the headers, at the barge end (or upper end) of each compartment, will be the valve controlling the flow of air into the compartment. On some pontoons, the headers are also used to vent the air from the compartments when the pontoon is being flooded. Some pontoons have individual vent valves at the top of each compartment. The U-shaped pipes connecting the two members of the pontoon are called crossovers, and the valves for flooding each individual compartment will be found at the bottom of every second crossover, at the back or lower end of

Fig. 37. An excellent view of an extremely long pipe-laying pontoon. Note the rollers. Photo: Paul Stafford and J. Ray McDermott & Co., Inc.

each compartment. All of these valves, the air, or blow valves, the vent valves and the flood valves are operated by the diver, according to a flooding schedule, or on instructions from the superintendent. On most pontoons, there are one or two large compartments that can be flooded and blown from the control console at the stern of the barge.

Fig. 38. An articulated pipeline pontoon. The many control hoses are shown in the bottom right-hand corner. On many pontoons these hoses must be connected by a diver, five or six feet under water. Photo: Paul Stafford and J. Ray McDermott & Co., Inc.

The air and vent valves on these compartments are operated by hydraulic or pneumatic rams, and they are called the automatic valves. At the very end of the pontoon, there will be a double-pontoon section, mounted beneath the main pontoon. This section, called the slide section, because it sometimes slides along the bottom, can be 30 to 40 feet long. Its purpose is to give the very end of the pontoon additional buoyancy, and it is almost always controlled by automatic valves.

Some pontoons are constructed of continuous pipe sections for their entire length, while others are articulated. An articulated pontoon will be made up of hinged sections, 60 to 100 feet long.

The headers are bridged across these hinged sections with rubber hoses. Some articulated pontoons have the sections connected by screw jacks, so that the pontoon can be pulled into an arc, or bow, to conform to the arc of the pipe as it is laid.

At the beginning of the season, the pontoon for your particular barge probably can be inspected on the dock at the company yard. Take the opportunity to make this inspection, as pontoons vary, and you will have a better understanding of the mechanics of operation and your particular duties after such an inspection. If an inspection is not possible, seek out your pontoon blueprints and study them carefully. The blueprints should be carried aboard the barge.

TESTING THE CONTROLS

After the pontoon hoses have been hooked up, the various controls will be tested out. The valve for the pneumohose will be opened and if air bubbles appear at the extreme end of the pontoon you can assume that it is operating. The automatic valves will be opened one at a time, and air from the blowing system admitted into that particular compartment. If a large eruption of air bubbles shows on the surface, the valve is open. If the bubbles cease as soon as the valve is closed, the valve is operating properly.

FLOODING PONTOON COMPARTMENTS

The next step will be for the diver to flood certain compartments of the pontoon according to an engineered flooding schedule. After these specified compartments have been flooded, the pontoon, with pipe in it, should be controllable by flooding or blowing the compartments having automatic valves. The compartments on the pontoon will be identified by numbers welded to them, with No. 1 compartment closest to the barge. Because the pontoon will be floating, and the valves are located only a few feet under water, some divers do this preliminary flooding wearing a scuba face mask and breath-holding. I prefer to to use the hose and mask for a number of reasons. When you have completed your work, your tender can drag you back to the barge. The hose is a safety line if you should get swept off of the pontoon. You need not worry about holding your breath as the waves roll over you in rough water and you will be in telephone communications with the barge, in case the flooding schedule is changed because of unexpected reaction of the pontoon during flooding. This last is the best reason. Pontoons have been known to sink during preliminary flooding and having something to breathe and your tender's strong arms and back on the end of the hose are reassuring when this happens.

A 300-foot pontoon is the longest that can be worked on using surface-supplied hose gear from the stern of the barge, and this can be done only if there is no current. Longer pontoons will have to be inspected and worked on with scuba gear, or working from a live-boat (live-boating is covered in Chapter 11).

Whenever you go out on a pontoon, either on a routine inspection or to flood or blow compartments, always carry a crescent wrench to operate the valves. Carry the wrench on a lanyard, or on a snap hook on your belt. Pulling yourself 300 feet out on a pontoon is hard work, and to have to do it twice because you forgot or lost your wrench is a bit embarrassing.

Most pontoons are very slick, with little to grab onto for handholds. Always wear flippers when working on a pontoon, even with hose gear. Choose the open-heel type, easy to remove and slip over your arm when you are ready to climb back up the ladder. During this preliminary period, before the actual work has started, it is an excellent idea to take the time to rig a guideline along the entire length of the pontoon. Take the end of a coil of ½-inch line out to the extreme end of the pontoon and tie it off there. Then, with short pieces of ¼-inch line, secure the ½-inch line to every other crossover, all the way back to the stern of the barge. You now have a convenient hand line to use in pulling yourself back and forth along the entire length of the pontoon. It will make your dives on the pontoon shorter, safer and a lot less exhausting.

After the hand line is rigged, you can get on with the flooding. Let us say that you have to flood compartments 4,6,8,10,16,22 and 24. Go overboard with your crescent wrench and pull yourself along the hand line until you reach the number "4" welded on the side of the pontoon. If the pontoon is the type that vents through the header, you will have to open the header valve at the barge end (forward end) of the compartment. If the pontoon has vent valves, these will be on top of the pontoon, and you will have to open the ones on compartment No. 4. The flood valve will be mounted at the bottom of the next crossover aft of the vent valve. When you open the flood valve, water will flow into the compartment and air will flow out through the vent valve or the header valve. Treat all open valves under water with caution. In deep water, the flood valve on a pontoon is capable of sucking the flesh off of your body, or trapping you to the valve. Stay away from the open end of all valves under water, and whenever possible, approach them and work on them from behind.

When not flooded the compartments will be filled with air, sometimes to a pressure of 150 pounds or more. This is another good reason to stay away from the front of valves. When the valves are first opened, this high-pressure air will escape in a powerful rush. When two compartments to be flooded are near each other, such as Nos. 4 and 6, it is okay to flood them simultaneously. If they are widely separated, it will take longer to flood the pontoon, but it will be easier to flood them individually as you work your way down the pontoon. When air is no longer escaping from the vent valve, the compartment is probably flooded. However, at the surface, the pressure differential is so slight that the compartments may not flood entirely until pipe is in the pontoon. The weight of the pipe will then force the pontoon down far enough to flood the compartments completely.

After each compartment has been flooded, the flood and vent valves will have to be closed. Whenever you are given a flooding schedule by the superintendent, write it down in your logbook. In this way you will know which compartments are flooded without having to rely on memory. This will save you a lot of guesswork at the end of the job when the pontoon has to be blown dry. When you finish the flooding schedule, have your tender pull you back to the ladder.

When the flooding schedule is completed, the pontoon will be awash. The weight of the pipe together with more water in the compartments controlled by the automatic valves will be required to sink it.

PUSHING PIPELINE INTO THE PONTOON

The next step in the operation is to push pipe until the pontoon is filled, and about half a joint extends beyond the end of the pontoon.

This brings us back to the dome cap and the pad eye welded to the beginning end of the pipeline. Just before this cap is pushed into the pontoon, the riggers will shackle a 30- or 40-foot-long sling into the pad eye, lay it back along the top of the pipe, and lash it into place with soft line. The purpose of this sling is to secure the end of the pipeline to an anchor or a structure leg, to hold it in place as the barge pulls away laying pipe. Be sure that the soft-line lashings are secure before the pipe goes overboard, because if they come undone before the sling passes out of the end of the pontoon, the sling will foul on the pontoon rollers, requiring an unnecessary dive to free it. Be sure also that there is a shackle in the end eye of the sling and that it is safetied. It is irritating to get on the bottom 300 or 400 feet from the barge and find that you cannot complete your job for want of a shackle. Remember, in a situation like this, the onus is not on the rigger who forgot the shackle, but on the diver who could not complete his job.

As the pipeline is pushed out into the pontoon, frequently the end gets caught in a roller, stopping the pipe. The stern crane will then be boomed down as far as possible, over the pontoon; the diver will have to go overboard to hook a sling between the crane and the pipe-

Fig. 39. Two lay barges working away from the same platform. Note the stinger floating on the surface, behind the upper barge. Photo: Brown & Root, Inc

line to lift it clear. This might have to be done several times before the end of the pipe clears the end of the pontoon.

When half a joint of pipe or so is protruding from the end of the pontoon, the barge will be winched around and backed up to the structure with the end of the pipe a few feet away from the jacket leg to which it is going to be attached.

SHACKLING THE SLING AROUND THE JACKET LEG

Your next job will be to shackle the sling at the end of the pipeline around the jacket leg, at the bottom. With a 300-foot pontoon, this point can be as much as 400 feet from the stern of the barge—much too far to use hose gear from the stern. You will also have to give directions for moving the barge as close as possible to the jacket leg, and this will require telephone communications, so scuba gear is ruled out.

Scuba gear should never be used for a working dive for several reasons. Prime among them is the rapid consumption of air during heavy work; dragging a cumbersome sling and shackle around on the bottom is certainly heavy work, so do not attempt this job using scuba gear. You will have to load your compressor and gear aboard the crew boat or tugboat and go over to the platform. Some of the newer and larger tugboats have an air supply adequate for diving. Before you use this air, however, check out the location of the air intake. Take along a pair of watermelon volume tanks with a good filter, and be sure you have adequate pipefittings with you to tap into the system. If you are the least bit dubious about the tugboat's air, take your own compressor. Be sure to take a ladder, and whenever you are going to work from a boat remote from the barge, check and double-check to be sure you have brought all the equipment you will need, before the boat leaves the barge.

Have the boat tie up to the platform with the stern close to the jacket leg you are going to work on. Over the boat radio, direct the barge to back up until the end of the pipe is about 10 feet away from the jacket leg.

If the water is dirty, or it is nighttime, it is a good idea to tie the end of a ½-inch line into the end of the sling on the pipeline, before the pontoon is sunk. As the pontoon sinks, pay off this line. When the pontoon is on the bottom, you can make your descent on the jacket leg, holding on to the bight of your line. On the bottom, tie the line off to the jacket leg, and you will have a positive guide to help you locate the end of the pipeline. Ideally, it will be no more than 8 to 10 feet away from the jacket leg. The ½-inch line will be a big help to you in pulling the sling over to the jacket leg. Pass the sling around the jacket leg, secure the shackle with the pin in the eye of the sling, and safety the pin. The barge is now ready to lay pipe. As the barge pulls ahead on its anchors while each joint is welded on, the cable will hold the end of the pipeline in place. Incidentally, the pipeline route, or right of way, will have been previously laid out by a survey boat, and marked with weighted buoys.

INSPECTING THE JOB

After you return to the barge and unload all your equipment, the foreman will want you to make a check of the pontoon. From this point on, if the job runs smoothly, these routine inspections will be about all that will be required of you. The superintendent will want the pontoon inspected at least twice a day, and it is best to schedule these inspections for daylight hours, say 7 A.M. and 7 P.M. In areas of strong current, inspections should be made during the periods of slack water.

Rig a guideline to save you a lot of time and make your twice-daily pontoon inspections easier. Take a 30-foot length of ¼- or ⅜-inch line and tie a 4- or 5-pound shackle into one end. Take this out to the extreme end of the pontoon and tie it off there, allowing the shackle to drag along the bottom behind the pontoon.

Your inspection procedure will be as follows. Pull yourself over to the barge end of the pontoon and check out the control hoses to be sure there is not too much slack in them. Check the hitch to be sure that all the fingers are closed and that the dogs are in place. Constant motion of the pontoon, as in heavy weather, can force the fingers open if the dogs are not properly engaged. This could result in the total calamity of the pontoon coming unhitched and buckling the pipeline. Bearing in mind the previous warnings about the area of the hitch and the end of the pipe ramp, check the hitch fingers on every pontoon-inspection trip. If the hitch and hoses are satisfactory, pull yourself down the pontoon along the guideline, checking the pipe at each set of rollers as you go. Stay below the pontoon and make your checks visually by looking up at each set of rollers.

Stay outside of the pontoon. The pontoon and the pipeline are very flexible, and with even a moderate sea running, they will be springing apart and coming together. An arm or leg caught between the pipe and a roller or the side of the pontoon will shorten your diving season drastically. You must also guard against your hose being swept into the pontoon and caught in one of the same positions. Because of this possibility, and because 300 feet horizontally is a long way to be from the barge in case of air-supply failure, never make the trip down the pontoon without a Nemrod or Bouey-Fenzy life vest, or a CO_2 vest and bailout bottle.

If the current is running from starboard to port, you will have no trouble with your hose slack, as it will be swept away from the pontoon. Current running in the opposite direction, from port to starboard, could cause a problem. In this case, have your tender move as far to the port side of the barge as he can, to give your hose lead some angle and reduce the possibility of the bight being swept into the pontoon. Keep your hose as tight as possible and try to keep any existing slack hanging in a bight below the pontoon. If, as you work your way down the pontoon, the pipeline is not bearing on the rollers at the barge end, but gets closer to them as you progress, and finally bears on them near the outboard end, it will mean that the end of the pontoon is too buoyant and is trying to float the pipeline. However, if the pipeline is bearing hard on the rollers at the barge end, but the pontoon is sloping downward away from under it towards the far end, it will mean that the pontoon is not buoyant enough. Both are bad situations. In the first one, the pontoon could develop enough upward force to float the pipeline, buckling it. In the second situation, the pontoon might sink out from under the pipeline entirely, again causing a buckled pipeline. Check the rollers carefully as you proceed down the pontoon, to be sure the pipeline is bearing firmly on each set. If not, report the situation. You will have to flood or blow various compartments to correct either of these situations. Because of this possibility, always carry a crescent wrench with you on each inspection trip. At the end of the pontoon, check to see that the pipe is not bearing hard on one side or the other as it leaves the pontoon, If it is, it will mean that the anchor foreman is not pulling the barge along in a straight line. In this situation, the field-joint tins could be scraped off as they pass out of the end of the pontoon, and if the side pressure is strong enough, it could damage the weight coating on the pipeline.

At the outer end of the pontoon, have your tender take a pneumoreading on you, and check it against the pontoon pneumogauge. Then, drop down on your weighted line and take a pneumoreading on the bottom. These two readings will indicate how far off the bottom the end of the pontoon is riding. This dimension will vary from 0—30 feet, depending upon the size of the pipe, the length of the pontoon, and the depth of the water. After you take the bottom pneumoreading, return to the end of the pontoon. In dirty water, in a strong current, or when the barge pulls ahead between joints of pipe, you could lose the end of the pontoon. Holding on to your weighted line will enable you to find your way back. Your next chore will be to run down the pipeline and check how many joints are in suspension between the end of the pontoon and where the pipeline touches the bottom. This could vary between two and five joints, again depending on many variables. If your hose is not long enough (with a 300-foot pontoon, it probably will not be), you will have to make your pontoon inspections with scuba gear which will take less than ten minutes each on a 300-foot pontoon. Do not use scuba gear unless you have been trained in its use. Do not make decompression dives using scuba gear. Do not use scuba gear at night, or in rough weather, or in strong currents. Always wear an inflatable life vest when using scuba gear. Never make a dive with scuba gear unless your tank or tanks are fully charged.

On pontoons longer than 300 feet, inspections should be made from a live-boat (Chapter 11 covers live-boating procedures). There will be a buoy connected to the end of the pontoon, and you will make your descent on this, and make your inspection working up the pontoon toward the barge. Starting the dive from a buoy at the end of the pontoon and working back to the barge, you can inspect pontoons longer than 300 feet using scuba gear.

When making an inspection, stay outside of the pontoon. If you are using hose gear, every time your tender hears the horn signalling that the barge is going to pull ahead another joint, he must warn you to stay clear and hold on.

In rough weather, the pontoon can be a very dangerous place. The roll or heave of the barge will be transmitted to the pontoon, and because it is flexible, it can develop a severe whipping action making it impossible to hold on. A tap on the head, 300 feet out on the end of a wildly whipping pontoon, could spell finis to a promising diving career. Use discretion. If holding onto the pontoon requires all your strength, back off and abort the dive.

At the end of your inspection with hose gear, have your tender pull you back aboard so that you will not develop hose slack. You can help him out by pulling yourself along with the hand line, but do not get ahead of your slack.

MAKING YOUR REPORT

When you return to the barge, make a report of your inspection, and if you found anything unusual, mark it in your log.

Brief mention should be made of the occasional troublesome problem of oil-field ethics. You are working for the contractor and your loyalty is to him. The contractor's representative is the superintendent, and he is the one who must make the moral judgments between truth or the withholding of it. Make all reports to him, accurately and honestly, and to no one else. On every job there will be a customer's representative or inspector, and naturally he will want to know everything concerning the progress and condition of the pipeline. His information should come from the superintendent. Advise him that it is company policy (and it is) for the diver to make all his reports directly to the superintendent. By following this procedure, you can remain clear of all contractor-client arguments. Be sure, however, to record any damage or unusual conditions in your logbook.

INSTALLING RISER CLAMPS

Three or four days into the job, if all has been going well and the pontoon has presented no problems, the superintendent will want you to install riser clamps on each of the two platforms. As you know the riser is that vertical section of pipeline which extends from the bottom up to the platform deck above water. Risers are permanently attached to the structure to prevent their being moved by wave action or other causes. This is done with mechanical clamps which bolt around the riser and around a jacket leg or structure bracing. There are many types of riser clamps. The clamps are attached to th structure at predetermined locations and depths, specified in the construction drawings. Sometimes, if the lay barge has a heavy schedule, only the top and bottom clamps will be attached. The others will be installed later from a workboat or a smaller barge. Installing riser clamps is hard work, but with proper setting up and rigging you should be able to put on any clamp in 20 or 30 minutes.

SETTING UP AND RIGGING

The superintendent will arrange for a workboat for your use (perhaps one of the barge supply boats), and he will detail laborers and riggers to help you. Try to have at least five helpers, for there is a lot of topside work and rigging to be done. You will need all your diving gear, and if you are working in decompression depths, you will need a chamber and a couple of other divers. You will need a large-volume air compressor to power the pneumatic impact wrench and at least two air tuggers. Take ample tool hose to reach from the compressor to the air tuggers and enough to get the impact wrench to the bottom clamp. Take a coil of ½-inch rope, a coil of ¼-inch rope, and an assortment of slings, chains, and shackles. You will also need five small-cable snatch blocks. Be sure the sockets for your impact wrench are the correct size and take a few spares; have back-up wrenches of the proper size.

If possible, make a preliminary check of the layout of the platform from which you will be working, and have U-bolts or J-bolts to fasten down the air tuggers. Check at this time for the number and length of the slings you will need to hang the snatch blocks. Determine how you will moor the workboat to the structure in order to weld an air tugger to the boat deck to give you a fair lead for hoisting the other air tuggers to the structure platform.

PREPARING JACKET-LEG CLAMPS

Before the clamps are loaded aboard the workboat, some important preliminary work must be done to them. Each bolt should be removed and the threads cleaned with a thread-runner or die and then oiled. This is very important, every nut and bolt must be capable of being fully run up by hand.

Jacket-leg clamps will have to be measured and identifying numbers painted on them according to their locations as shown on the blueprint. The bottom clamp is usually designated No. 1. The differences in clamp sizes will be due to variations of the jacket-leg diameter, occurring because of double thicknesses of steel in the vicinity of braces. Check the blueprints carefully. Make a list of the clamp numbers, with their exact locations in terms of angle to the structure and elevation from the bottom. Paint the numbers on the correct clamps and on the corresponding faceplates or caps. because parts of different clamps will not be interchangeable. Paint a vertical stripe on the inside of each clamp where the exact center line of the riser will fit. Paint another vertical stripe down the center of the back side of each jacket-leg faceplate; the purpose of these stripes will be explained presently.

Recheck the blueprint to be sure that you have the proper jacket leg noted. Frequently riser clamps are installed on the wrong jacket leg, because the blueprints were not read carefully. A good job in the wrong place is useless. Read the blueprints for every job carefully. Clamps are always designed by a desk-bound engineer who has no feeling for nor understanding of the many problems a diver must overcome to install them. The most troublesome design deficiency is that the clamps are seldom provided with an accurately balanced pickup point. This one detail means the difference between installing a clamp in 15 or 20 minutes, and two hours or more. Strip the clamp of all bolts and remove the faceplates. Try to judge the point of proper balance and have the welder tack a pad eye in that spot. Have the crane pick up the clamp and if it hangs properly at the attitude and angle that it must have to fit around the jacket leg, weld up the pad eye. If it does not hang properly, keep moving the pad eye until you find the right location for it. Do this with each clamp. Some types of clamp, notably the telescoping kind, are impossible to hang from a single point; two or three pickup points will have to be welded to these. A come-along on one sling leg on this type of clamp provides a means for the diver to adjust the clamp angle in the water. The faceplates are usually hung on three rope slings, one each through the top bolt-hole on each side and one around the middle. Installing the faceplates is greatly simplified if you weld pad eyes to them also for proper hanging. Do all this work aboard the barge where you have the help of welders and riggers.

When all this is done, load all of your equipment and supplies aboard the workboat. Try to place the clamps in positions corresponding to the order of their installation, so that they will not have to be moved around unnecessarily. Clamps weigh from 300 to 1500 pounds, and you will not have a crane to handle them on the workboat. A little sensible preplanning at this stage will save a lot of work.

RIGGING SNATCH BLOCKS

When you reach the platform, the first chore after the workboat is moored is to rig a snatch block high up on the structure. Using the air tugger welded to the deck of the boat, land the other two air tuggers on the structure and secure them. Placing air tuggers on the structure to lower the clamps and faceplates is essential, even when working directly from the lay barge. This completely eliminates the annoying and dangerous movement of the clamps due to movement of the barge or workboat in the waves. Snatch blocks will have to be rigged in the upper works of the structure so that the clamps and faceplates will hang in the most advantageous positions prior to bolting them up. Most jacket legs are on a batter or angle, so that the deeper the water, the farther out from the top of the structure will be the bottom of the jacket leg. With this in mind, the snatch block for the bottom clamp must be hung as far out over the water as possible in a vertical plane over the jacket leg. For the upper or shallower clamps, the snatch block may have to be moved in closer to the center of the structure.

Clamp No. 1 is picked up by an air tugger and put in the water. To avoid confusion, the air tugger for the clamps should be designated No. 1, and the air tugger for the faceplates, No. 2. Quite often other tuggers will be needed to pull the clamps around into proper position, and these should be designated No. 3, 4, etc. The faceplate for clamp No. 1 is also picked up and put in the water on the back side of the jacket leg, and the clamp and faceplate are then lowered simultaneously.

HOW TO HANDLE NUTS AND BOLTS

The nuts and bolts to fasten the clamp and faceplate around the jacket leg can be handled in a number of ways. They can be put in a tool bucket and lowered on a ½-inch line, tied onto a line secured to the clamp, or put through the bolt holes of the faceplate and secured with two or three turns of the nuts.

Here is an excellent way to handle nuts and bolts: Take a 3-foot piece of ¼-inch line and securely tape each end. Put two or three tight wraps of tape around the rope every four inches until you have six positions taped, the usual number of bolts per faceplate. Slide a nut up the rope above the top tape position, and then insert a welding rod through the rope to keep the nut from slipping off. Do this with all the nuts and then tie the rope to the clamp or the faceplate. When you need a nut, all you will have to do is pull out a welding rod and slide the nut off the bottom of the rope. The remaining nuts will be secured until you need them. Cut a number of square pieces (about 2 by 2 inches) out of some gasket rubber for the bolts. Make a hole in the center of each square of rubber with a punch which is ¼-inch less in diameter than the bolt. Put the bolts through the holes in the faceplates and force one of the rubber washers over each bolt. This will hold the bolts firmly in place, but you will still be able to twist them or move them in and out or back and forth to line them up with the clamp holes. This system can save 10 or 15 minutes diving time on each clamp.

LOWERING THE CLAMP INTO POSITION

Several methods have been developed to eliminate the need for a back-up wrench when installing riser clamps. One method uses a specially made nut, 3 or 4 inches square, which cannot turn when it is in position behind the clamp. Another way to achieve the same end is to weld a standard nut to a 3- or 4-inch square washer.

Two pieces of ½-inch line, at least 6 feet long, should be attached, one to each of the top two bolt holes in the clamp, before it is put in the water. Take a 4-pound hammer and a scraper with you and hit the water. Flippers are useful for swimming around the jacket leg and working on those clamps not located near bracing to stand on. Stirrups of ½-inch rope, long enough to pass around the jacket leg, with a spliced eye to stand in, are helpful for this type of work. You will have to guide the clamp and the faceplate as they are lowered, to prevent them from hanging up or from dropping on the wrong side of the structure braces. Your tender will be watching your pneumogauge as you descend, and when you come within 5 feet of the clamp location, he will stop you and the tuggers. Hold the end of your pneumohose on the jacket leg and, with your tender taking readings, find the exact location for the clamp. In the Gulf of Mexico, where the tide range is only about a foot and a half, a pneumoreading is accurate enough to establish this location. In areas where there is a great tidal difference, the pneumoreading will have to be correlated with the time of day and stage of tide, or the clamp positions will have to be determined by actual measurement. The area of the jacket leg where the clamp is to go must be cleaned of all marine growth. The scraper will work fine in most cases, but the hammer will be needed for oysters or large shells. This cleaning is very important. The clamp is designed for a close fit and that part of it in contact with the jacket leg is generally coated with rubber, making it very difficult to move or slide if there is any marine growth on the jacket leg.

After cleaning, lower the clamp to its exact elevation and push it around to the high side of the jacket leg. Because of the batter of the leg, and because the platform deck rarely extends far enough over the water for you to rig the snatch block from a point directly above the clamp location, the weight of the clamp hanging from the tugger at an angle will hold it in place around the high side of the leg. Lower the faceplate or cap until it will mate with the clamp, and thread the small lines from the clamp through the corresponding holes in the faceplate. Pull the faceplate into the jacket leg as far as it will go with these lines, and tie them off. Insert the two center bolts and screw them up by hand. You should then be able to put in the two bottom bolts, cast off the two lines, and insert the top bolts. Now all you have to do is slide the clamp around to its specified angle and tighten up the bolts. In rigging the cap or faceplate, I recommend that pad eyes be welded to it, rather than use the top bolt holes for the slings. This is why. In order to install the top two bolts, the rope slings will have to be removed from the bolt holes. This leaves the entire weight of the cap hanging from the lower bolts, binding the cap to the jacket leg and making it difficult or impossible to rotate the clamp to its final position. With pad eyes, the weight of the cap will remain suspended from the tugger while the clamp is located and secured.

The final location of the clamp on its axis around the jacket leg is very important, and each clamp must be installed at exactly the same angle, so that the riser will fit into all of the clamps. This angle is variable, and it can range from 45 to 135 degrees from the point of intersection at the center line of the jacket leg of any two faces of the structure. The vertical lines painted in the center of the riser section of the clamps and down the center of the back side of the jacket-leg caps are to help you line up the clamps on the jacket leg at the proper angle.

Down the back side of each jacket leg, usually centered exactly between the intersection of the horizontal braces, there will be a grout pipe. This 2- to 3-inch pipe runs all the way from the surface to the mud line. If the clamps must be located at an angle of 135 degrees from either platform face, the white line on the back of the clamp caps can be lined up with the grout pipe. If the clamps are to be positioned at an angle of 45 degrees from either platform face, then either joint between the clamp and the cap can be lined up on the grout pipe. In most cases, the grout pipe will be welded to the jacket leg with stand-offs, leaving enough room between the jacket leg and the grout pipe for the cap. If there is not enough room, the grout pipe will have to be burned away before the clamps are installed. Rather than burn the entire grout pipe free, just burn out pieces where the clamps are to go. The grout pipe remaining in place will provide convenient handholds for you while you are working. If large sections of the grout pipe must be burned free, be sure they fall all the way to the bottom, and do not hang up on the bracing where they might come loose and fall on you later. After the bottom clamp has been installed in its proper location, a ¼-inch polypropylene line can be run down from the surface, over the top of the center line, painted inside the clamp, and secured around the jacket leg. A good strain is taken on this line, and it is secured at the surface in a position corresponding to the center-line location of the back of the riser to be installed. As each successive clamp is installed, it is adjusted for axial location so that the polypropylene line lies over the painted center line.

BOLTING THE CLAMP

In making up the clamps with the faceplate caps, sometimes it might be necessary to start out with two long all-thread bolts through the center holes, drawing the clamp and the faceplate together with these bolts and the impact wrench until they are close enough to insert the proper bolts. Another consistent design error for riser clamps is that the bolts supplied are often barely long enough when the clamp and the cap are perfectly made up. They should always be at least four inches longer than this to give the diver some leeway when assembling the clamps. The bolts should never be wrench-tightened until the clamp is located perfectly, vertically and axially. If you cannot push the clamp around into position by hand, you might have to rig a come-along from a structure brace to give you a lateral pull, or you may have to rig a snatch block and pull the clamp around with an air tugger. When the clamp is in position, tighten the middle bolts a few threads first on one side and then on the other. If all the threads are run up on one side only, the clamp may be cocked out of position by as much as several inches. Draw up both sides evenly. Do not tighten all the bolts completely at this time; the two middle bolts will be enough to hold the clamp in place. Even if the clamps are in their exact called-for position, when the riser is put overboard the pipeline measurement might be a few inches too long or too short and the clamps will then have to be rotated one way or another to meet the riser. You can tighten all the bolts completely after the riser has been installed.

When using an air or hydraulic impact wrench, familiarize yourself with the reversing switch or lever. This is usually tiny and difficult to find under water, especially with gloves on and no visibility. Before you put your impact wrench on the bolt head, be sure it is rotating in the proper direction to avoid undoing all your hard work in a fraction of a second. Test the rotation of the impact wrench before you put it on the bolt. Impact wrenches have tremendous force; keep your hands and fingers away from the socket and the nut. If you are using a back-up wrench, tie it off with a lanyard, so that if it slips off the nut, it will not be lost. Keep your hands clear of the handle of the back-up wrench. It is quite easy to get yourself jammed between the wrench handle and the clamp. If a pneumatic impact wrench is not functioning properly, send it to the surface and have them send you another one. The most common causes of a malfunctioning pneumatic impact wrench are low air pressure or volume and lack of lubrication. All pneumatic tools should have an automatic lubricator installed in the air-supply line. If you do not have one, have your helper occasionally take apart one of the hose connections and pour a pint of diesel oil or light lubricating oil into the hose. It is helpful to rig your pneumatic wrench with a short lanyard tied between the hose and the forward part of the wrench so that the wrench is hanging at the proper angle to be pushed over the bolt heads. When you complete the clamp, unhook the tuggers and move the tools up to the next clamp location above you, usually about 20 feet, and repeat the procedure. When all the clamps have been installed on both structures, you should have nothing more to do than your twice-daily pontoon inspections and an occasional wheel job. You and your tender will have plenty of idle time. Use it to keep your gear checked out and properly maintained.

GETTING RID OF THE PONTOON

When the pipeline has been laid up to the structure, a cap will be welded over the end of it, and the forward crane and all the davits will be hooked up to it with cable or nylon slings. A diver will have to go overboard to hook a sling from the stern crane to the pipeline 30 or 40 feet down the pontoon. The end of the pipeline will then be picked up by both cranes and all the davits. The pontoon will be sunk from under the pipeline, the cranes will swing, and the davits will boom down until the pipeline is overboard and out of the pontoon.

A dive will have to be made to be sure that the pipeline is clear of the pontoon, and then the pipeline will be lowered slowly to the bottom.

Fig. 40. (Left) A tube turn for a large diameter pipeline. The lower end will be welded to the pipeline and the vertical riser will be bolted to the flange on top. Photo: Brown & Root, Inc. (Right) Swinging the riser into place to weld it onto the pipeline. Photo: Paul Stafford and J. Ray McDermott & Co., Inc.

With very long pontoons, this system will not be possible. If there is room at the structure, the barge will pull ahead on its anchors, and the pipe will be allowed to slide off the back of the barge and out of the end of the pontoon. Where it is not possible to pull the barge out from under the end of the pipe, another system must be used to free it from a long pontoon. The bow crane and the davits are hooked up to the pipeline, but the stern crane is hooked up to the pad eyes on the pontoon. The pontoon hoses are disconnected, and the pipeline, with the entire pontoon, is lowered to the bottom. All the slings from the cranes and the davits are unhooked, and the barge backed down until it is centered over the section of the pipeline still in the pontoon.

All the slings, from both cranes and the davits, are hooked up on the pipeline which will be lifted free of the pontoon and laid on the bottom. Next, the slings are hooked up to the pontoon to lift it slowly to the surface. If the pontoon is very long, the compartments at both ends may be blown dry before it is lifted. At the surface, the entire pontoon is blown dry and towed away after the slings are removed. The barge will then pull ahead over the end of the pipeline, and again all the slings are hooked up to it.

The diving operations are then conducted from the bow of the barge.

SETTING THE RISER

Before the pipeline is put overboard, a ¼-inch line is tied to the cap at the end. If the barge had to back up to get rid of a long pontoon, this line was buoyed. In any event, when the barge is back in position, pulled up to the structure, the diver makes his descent on the jacket leg to which the riser is to be clamped, letting the bight of this line slide through his hands as he descends. The line is to help the diver quickly locate the end of the pipeline on the bottom. Once on the bottom, the diver can cut this line and tie his end around a structure brace, close to the jacket leg. Using this line, the diver measures the approximate distance between the jacket leg and the end of the pipeline. If the distance is more than 20 feet, the pipeline should be lifted back to the surface, and the necessary length to bring it close to the jacket leg added on. This final measurement, between the end of the pipeline and the jacket leg, is the most critical on the whole job, for it determines whether or not the riser will fit into the clamps. The closer the end of the pipeline is to the jacket leg, the more accurate this measurment will be. If this distance is 15 feet or less, a fairly accurate measurement can be made. The most consistent error divers make when taking this measurement is failure to line up the pipe laterally, so that the pipeline is straight and heading exactly into the clamps.

It is not likely that the pipeline was laid on the bottom in a precise line to the jacket leg. The diver should have the end of the pipeline picked up and the barge moved from one side to the other so that this lineup is perfect. If the distance is 15 feet or less, a fairly accurate measurement can be made with a piece of polypropylene line. Polypropylene line is used because new Manila will shrink when it is first put in the water, and nylon will stretch.

Tie one end of the line into the pad eye on the end of the pipe with a bowline and stretch it taut to a point on the jacket leg exactly under the center of the bottom clamp, and tie an overhand knot in the line. The length of the clamp can be subtracted from the measurement later. To further refine this measurement, however, if the bottom clamp is less than 6 or 8 feet from the bottom, take down with you a 10- or 12-foot rod, or length of small-diameter pipe. Stand this pipe vertically and accurately against the painted line at the inside center of the clamp, and push the pipe into the bottom. Now you can pull your measuring line taut and knot it where it crosses the vertical rod. At the other end, cut the measuring line at the center of the loop of the bowline, leaving the knot tied. You must be sure that this measuring line extends along the line of the pipe accurately, and that it is stretched parallel to the bottom. On deck, the two ends of the cut bowline loop can be held together, the line stretched taut, and measured with a tape.

If the pipeline is too long and extends beyond the jacket leg, it must be moved so that it is hard up against the jacket leg. A line can then be squared with a rod or a piece of angle iron, from the face of the jacket leg across the top of the pipe, and the weight coat marked at that point with a grease pencil or a hammer and chisel.

Before any measurement is taken, all the cranes and davits must be completely slacked off so that the pipeline is lying firmly on the bottom, or the measurement will come out too long. This measurement is critical; take every precaution possible to guarantee that it is accurate. If it is not, the riser will not fit into the clamps and the line will have to be picked up, the riser cut off, and a piece welded into the pipeline. Doing this will take about 12 hours and cost a lot of money.

After taking the measurement, you should have a few hours off. The line will have to be raised to the surface and the riser welded on before you return to work. Use this time to check out the riser-clamp caps. Be sure they are numbered and properly rigged. Check all the bolts to be sure the nuts can be run fully by hand.

When welding of the riser to the pipeline is almost complete, the foreman will send for you in plenty of time to get into your gear. After the pipeline and riser are lowered slowly and evenly until about halfway to the bottom, you descend on the riser to the tube turn. The tube turn is the radius bend in the pipeline where it changes from the horizontal plane to the vertical plane of the riser.

Under the best conditions, in daylight, and with clear water, you will be able to see the jacket leg from the riser. If it is nighttime, have the barge electricians hang floodlights over the side to light up the work area. If the water is deep and surface lights will not reach down far enough to help you, have a diving light secured to the riser about ten feet above the tube turn. If the water is very dirty with no visibility, you may have to tie a line to the tube turn and descend on the jacket leg. Then use this line to guide the riser into the first clamp. If you have visibility, sit on the tube turn and guide the riser as it is lowered. It is very important to stay on the barge side of the riser at all times during the lowering operation, and to have your tender keep a good strain on your hose. This is to prevent you or your hose from getting jammed between the riser and the jacket leg or the clamps. As the pipeline and riser are lowered and you approach the bottom (your tender can tell you by watching your pneumo), be sure the riser is lined up laterally. If not, stop the lowering and move the barge to one side or the other for perfect line-up before too much of the pipeline is resting on the bottom. Remember, move the barge in short increments, giving it plenty of time to settle down between moves. When the pipeline is lined up, aiming into the bottom clamp, continue lowering. Remember that the jacket leg is probably on a batter, and depending on what angle the pipeline is laid into the jacket leg, the top of the riser will have to be swung over one way or the other to conform to the batter of the jacket leg, and to line the riser up with all the clamps.

ACCURATE MEASUREMENT IS IMPORTANT

If your pipeline measurement was accurate, the riser should be hard into the bottom clamp when the pipeline touches the bottom. If the riser jams into the bottom clamp before the pipeline is on the bottom, it means that the pipeline is too long. With pipelines less than 18 inches in diameter, this is not too serious. By taking a strain on the back two davits, a slight overbend will be pulled into the pipeline, thereby shortening it. After the riser clamps are bolted up, and the davits slacked off, this overbend, if it is a large one, can be laid over on its side. If it is not too extreme, it can be left as it is, to be buried later by the jet barge.

The weight of large diameter pipe will jam the riser into the jacket leg as any overbend is slacked off. This could bend or break the clamps. For a good riser installation, there is no substitute for an accurate measurement.

If the riser does not meet the clamps when the pipeline is resting on the bottom, the pipeline is too short. In this situation, perhaps the pipeline and riser can be moved in closer to the center of the platform, and the clamps rotated inward to meet the riser. If this does not work, the pipeline and riser must be lifted back to the surface, the riser cut off, a piece added to lengthen the pipeline, and the riser then rewelded to the pipeline. This is a time-consuming and expensive process, but it will have to be done if the line comes out too short. No amount of pulling, tugging, and heaving with cranes,

Fig. 41. (Top) Riser clamp and cap. Photo: Author. (Bottom) The riser is in place. After the clamp caps have been bolted on, the diver will have to unhook the heavy slings leading to the crane block. Photo: Paul Stafford and J. Ray McDermott & Co., Inc.

deck winches, and snatch blocks will stretch a too-short pipeline, but every foreman afloat will spend five or six hours trying to do it. The fear of ending up with a too-short line compels most foremen to add from 2 to 5 feet to the diver's measurement on the theory that too long is better than too short. This theory is correct with small pipe, but nothing will take the place of a careful, accurate measurement.

Some types of riser clamps operate on the telescoping principle, so that if the pipe is short a foot or so, the clamp can be extended to meet the riser. Once the riser is in the bottom clamp, it is important to get the faceplate on and bolted and the crane and davits slacked off as soon as possible, especially in rough weather. This is to eliminate movement of the pipeline and riser due to movement of the barge in surface waves. When the bottom clamp is secured and the davits and crane are slacked off, the riggers at the surface can move the top end of the riser to your directions using come-alongs and tuggers on the platform. If the clamps were carefully lined up when they were installed, the rest of the riser should fall into place after the bottom clamp is secured. If the clamps do not line up, each one will have to be laboriously swung into line. You might have to pull the riser laterally for a few inches, to swing it into a clamp. This can be done with a sling around a structure brace, and a snatch block and tugger. This can be done on small pipe with a come-along, perhaps. If a clamp must be swung to meet the riser, remember that you will have to suspend the clamp before loosening the bolts, or the clamp will slide down the jacket leg out of vertical position.

TIGHTENING RISER-CLAMP FACEPLATES

When tightening up the riser-clamp faceplates, again be careful to tighten the bolts evenly on each side. With an impact wrench, all the hard work is in shifting the wrench from one bolt to another, not in pulling the trigger. Many divers have the tendency to put the wrench on one bolt, and then hammer away until the threads are stripped, or at best, until flanges of the faceplate and clamp on one side are drawn together. Remember, a good inspection diver will not accept a lopsided faceplate. Draw the bolts up evenly on both sides of the cap. This work will be much easier for you if you take the time to rig your wrench so that it is hanging at the proper angle, then, all you have to do is push it over the bolt head. When all the bolts are hammered up, be sure that by tightening the last one, you did not loosen one of the others. An inspection diver's greatest delight is finding and reporting loose bolts. Be sure that *all* the bolts are tight before you leave the clamp, including the ones holding the clamp to the jacket leg.

INSTALLING RISER-CLAMP CAPS

Riser-clamp caps are always installed by starting with the bottom one and working each successive clamp on the way up. Long risers, even of large diameter pipe, are quite flexible, and a sag or a bight can easily develop. If you start by securing the upper caps, there is no place for this bight in the riser to go, and the lower or middle caps will be impossible to secure.

If the riser does not fit tightly into the clamp, you may have to start your caps with two long all-thread bolts through the center holes, drawing these up with the impact wrench until the standard bolts reach.

Beware of fastening your diving hose between the cap and the riser; the consequences can be more than embarrassing. When all the bolts on all the clamps have been tightened, make one more dive to check them all over again. Quite often, when drawing the riser into the upper clamps, one or more of the lower caps is loosened. Before leaving the water, be sure that all your tools have been sent up and that all the tugger and winch wires have been cast off and are free. When all the work on the riser has been completed, the davit slings will have to be unhooked from the pipeline. On an average barge (i.e., one not too long), if this is to be done by one diver, it is best to work from the amidships diving station. Have your tender drag sufficient hose and your mask or helmet up to davit No. 1, and enter the water there. You will need an extra man on the telephone, as your tender will be too far away to hear it.

When you are on the davit chain ready to descend, have your tender move aft to davit No. 2 and tend you from there. This will give a good lead to your hose and help to keep you from spiraling around the davit chain during your descent. It will also indicate the direction of travel when it is time for you to move aft. When you are on the bottom, unhook one eye of the sling, and wait there while the davit chain is raised, to be sure the sling pulls clear from under the pipeline. Be sure your hose is clear of the sling and the chain before they are raised to the surface. When ready to move to davit No. 2, have your tender precede you down the deck to No. 3, and be sure he keeps a good strain on your hose as you advance. As you unhook each sling and move on down the pipeline, be sure to inspect the pipeline carefully for damage, breaks in the weight coat, buckles, or overbends. Report any unusual findings immediately.

As you move aft from davit to davit, you must be patient with your tender. Your hose could be hanging up on each of the barge fenders as you move along, and your tender will need time to clear it. It may be necessary for him to go in the water to do so. If, as you move along, your hose gets fouled on one of the davit chains, calmly go back to the point of entanglement, and clear your hose before moving on down the pipeline. If the stern crane is still hooked up to the pipeline, this will be at a point well aft of the stern of the barge. In this case, don't unhook the last davit but pass it by until you have unhooked the stern crane. You can then make your way back to this last davit, and have something solid to make your ascent on after you have unhooked it.

Be sure your tender has a ladder placed for you near the last davit so you can get back aboard the barge.

With the davits unhooked, your work should be over for a while. The barge will have to pick up anchors and tow back to the platform where the pipeline began. The procedure for installing this second riser will be exactly the same as that just outlined, except that at the beginning the diver will have to hook up the davits to the pipeline on the bottom, and he will also have to cut the cable anchoring the pipeline to the structure.

FREEING THE CABLE TO INSTALL A SECOND RISER

As you recall, your first chore at the start of this job was to secure the beginning end of the pipeline to a jacket leg with a cable. Now, this cable must be freed so that the pipeline end can be picked out of the water and a riser welded on. As the barge pulled away, there was so much strain put upon this cable that it is highly improbable that either of the shackles can be loosened. It will be much easier to burn a shackle than fight through the separate strands of the cable. The shackle around the jacket leg will be up out of the mud, so cut that one. In this particular case, a ground attached to the structure above water will probably suffice. Though it should not take more than half a rod to cut the shackle, have several on hand.

HOOKING UP SLINGS AND DAVITS

After the cable is freed from the structure leg, the slings from the davits and from the bow and stern cranes will have to be hooked up to the pipeline. On most barges, when setting a riser, the bow will be closest to the structure. Hook the bow crane into the sling in the end of the pipeline first. As always, when working with the davits, if only one diver will be doing all the hookups, dive from the diving station halfway down the pipe ramp. There will be a soft-line guide rope connecting all the davits and the cranes so that the diver can find his way easily from one to the other as he moves down the pipeline. Most davit cables or chains will terminate in double hooks. One eye of the sling will be attached to the hook and safetied, and the other eye will be hanging free. There is generally some sort of hinged gate over the hook, to lock the eye in place, and these gates are safetied with a welding rod poked through them and bent over. Before the slings go overboard, be sure that the one end is properly safetied, and before you go overboard to hook up davits, be sure you are carrying enough welding rods to safety the hooks after you pass the sling eyes.

After the bow crane is hooked up, you will move aft to davit No. 1, holding onto the guide line and following along the pipe. If the guide line does not follow along on top of the pipe, but veers off to one side or the other, you will have to move the barge to line it up properly. Remember that you will probably be moving aft, and therefore your barge directions will be reversed. If, as you move down the pipe from the bow crane to davit No. 1, the guide line veers sharply to your right, the stern of the barge will have to be moved to starboard, to line the davits up over the pipeline. Be sure to move the barge in short increments of 5 or 6 feet, allowing plenty of time for the barge to settle down between movements. If the davit chain is laying in a pile on the bottom, have the chain picked up so that it is hanging free before you give the orders to move the barge. Never try to drag the davit chains or slings across the bottom, to hook them up to the pipeline. They are much too heavy for this, and the barge should be lined up perfectly over the pipeline before it is picked up. When davit No. 1 is plumb over the pipeline, pick up on the bow crane. This will lift the end of the pipeline out of the mud so that you can pass the No. 1 sling under the pipeline without digging. Pass the sling, place the eye in the hook, and safety the eye. Pick up on davit No. 1 until all the slack is out of it and then stop. Be sure not to pass the sling over your hose when making these hookups. When number one is hooked up, tell your tender you are moving, so he can move ahead of you on down to No. 2.

To get under the pipeline at davit No. 2, you will have to pick up on the bow crane and davit No. 1, simultaneously. As you reach each davit, you must pick up on all preceding davits and the bow crane at the same time in order to pick up the pipeline evenly, with an equal strain on all slings. After each hookup, while the slack is being taken out of the sling, stand back out of the way as the strain comes on. One of the most awesome sounds in the world is that of 300 feet of davit chain piling up on the bottom several feet away from you after a link has parted on the surface. Do not hook any of the davit slings directly over a field joint, but move forward or aft at least five feet. Never get under the pipeline when it is suspended in the davits. If the stern crane must also be hooked up, complete this and return to the last davit for your ascent. On large lay barges, with five or six davits and a bow and stern crane, it is much more prudent to split the task of hooking up slings between two divers, one at the bow and one at the stern. In strong or adverse currents, the chances of a diver getting fouled are excellent if he has to pass more than one davit chain. A diver should never have to make a lateral traverse of more than 150 feet from the point at which his hose enters the water. Most davit chains or cables will be liberally smeared with a protective coating of grease. Try to avoid getting this grease over your suit, but at all times be prepared to have your tender strip off your suit before you enter a chamber for decompression.

After the davits have been hooked up, the procedure for measuring the pipeline and installing the riser will be as previously detailed.

TESTING WITH A SIZING PIG

When the actual construction work on the pipeline and the two risers has been completed, the line will be tested. The first and simplest test will be to run a sizing pig through the line, pushed through with compressed air, to ascertain that the internal diameter of the pipeline is as specified in the contract. After the sizing pig has been run, the line must be hydrostatically tested to the pressure called for in the specifications. Frequently, both of these tests are conducted from the lay barge, and the longer the pipeline, the more time they will take.

The time required for a sizing pig to run through a specified length of pipeline with a known amount of air pressure behind it is easily calculated. When that time has elapsed, and the pig has not appeared at the other end of the pipeline, assemble your diving gear and prepare for a trip on the crew boat or the tugboat. If the pig does not come out at the other end of the line, but the air pressure behind it continues to build, it is a sign that the pipeline is buckled or bent, or that a drooping bead of weld has dropped into the pipeline, obstructing the pig. If the pig does not show up, but the air pressure behind it is dropping, it is probably due to a buckle and hole in the pipeline. In this case, a slow ride over the pipeline route with the crew boat should locate the air bubbles escaping from the leak. Some pigs are fitted with sonic pingers, and if the pig does not appear, it is subsequently located with a pinger receiver operated from the crew boat or tugboat. If the pig does not have a pinger, the trouble may have to be located by divers walking over and inspecting every foot of the pipeline.

If the sizing pig comes through, but the line does not hold pressure, then again, divers may have to walk every foot of the line to locate the trouble. Sometimes, in leaking lines, dye will be introduced and this escaping dye will show up on the surface, indicating the general area of the leak. In any event, the trouble will have to be located and buoyed by a diver. (See Chapter 11 for pipeline walking procedure.)

FLANGES

Alternate methods of repairing pipeline leaks are covered in Chapter 11. This particular leak will be repaired by cutting out the damaged section of pipeline, and connecting the two ends with flanges.

The lay barge anchors longitudinally over the pipeline, with davit No. 1 close to the damaged section. The first job is to burn the pipeline in two so that both ends can be lifted to the surface. Chapter 7 covers this operation in detail, but it is well to repeat a few points.

Be sure the pipeline is flooded before you attempt to burn into it. If you do not have a hole under the pipeline that will enable you to pass completely under it with your cut, blow one out with a hand jet before you start. Be sure that all the dope coating is removed from your line of cut before you start burning. Always use a rope or other burning guide to control your line of cut. Be sure the pipe is not in tension or suspension at the point of cut.

After the line is cut, the davits must be hooked up to the line on one side of the cut, with No. 1 about 10 feet back from the end. You may have to dig under the line to pass your first sling, and a foxhole shovel will make this job much easier. If the bottom is hard clay or sand, don't burn up a dive trying to dig this first hole. Use the hand jet.

After davit No. 1 is hooked up, it can be used to pick up the end of the pipeline sufficiently to pass your other davit slings. On large-diameter pipelines, especially in deep water, the weight of the water in a flooded line imposes critical stresses upon it while the line is being lifted to the surface. To reduce these stresses, the line must be blown dry before it is lifted. This is done with a pig, introduced into the line at the other end, and pushed through the line with compressed air, forcing the water out ahead of it.

To keep the pigs from escaping out of the severed ends of the pipeline, which would result in the line reflooding, some form of retainer, or trap, must be installed. If the line is badly buckled or crimped at the point where it is cut in two, perhaps pig traps will not be necessary. However, pigs have a phenomenal ability to squeeze through improbably small holes, so it is always prudent to provide some sort of barrier. The simplest method is to burn four holes at each severed end, two on top and two on the bottom. "Tee-handled" steel rods inserted through these holes make a simple and effective pig trap at the cut ends of the line.

When the pig is started through the pipeline, the diver will be asked to make periodic inspections to see if water is being forced out of the end of the line. Always keep your body clear of the ends of any pipelines, holes, or valves under water. Check the end of the pipe from the back side, with a finger passed cautiously over the edge of the cut. When air begins to escape out of the end of the pipe, it will be lifted to the surface.

The damaged section must be cut off and a new section of pipe with a flange welded on, based upon the diver's measurements. This new section of pipe must be long enough to overlap the damaged section of pipeline still on the bottom. When the welding is completed, the pipeline is lowered to the bottom and allowed to reflood.

The length of time required for reflooding depends on the length of the pipeline and other factors. Never trust anyone's assertions concerning whether or not a line is flooded. Always stay away from the open end of a pipeline. Never work around a pipeline where there is a detectable flow into the pipeline.

The section of pipeline with the flange welded to it must be laid on the bottom as close alongside the still-damaged section as possible. With a straightedge, the diver squares across the face of the flange, and marks the other section of pipe with a grease pencil or a hammer and chisel. Then the davits are unhooked, the barge shifted over this other section of line, and the davits rehooked. Following the same procedure, this other section of pipeline will be lifted to the surface and a flange welded on at the diver's mark.

When this second flange is welded on, the line is lowered to the bottom, and the diver's work really begins. Working on flanges in any but the calmest weather can be the toughest and most dangerous work that a diver faces offshore. Any movement of the barge is transmitted through the lifting slings to the pipeline, causing the flanges to open and close at unpredictable moments. For this reason, it is imperative to get the flange mated and stabbed with drift pins, and the load slacked off just as soon as possible. Preplanning and preparatory work are just as important for flanges as for riser clamps.

Be sure that all the threads on the flange bolts are clean, and that the nuts can be run by hand. Be sure there are spare bolts and nuts. Provide yourself in advance with an adequate tool bucket to hold the nuts, bolts, wrenches, etc. Be sure you have at least three drift pins for stabbing the flanges. The drift pins should be $\frac{1}{16}$ inch less in diameter than the bolt holes, with a 6-inch taper, and at least six inches longer than the thickness of the two mated flanges.

LINING UP FLANGES

The flanges used for underwater work are precision devices. There is a circular groove machined in each flange face, and a steel O-ring fits into these grooves. The O-ring is compressed between the flange faces, effecting the seal. To create a proper seal, the flange faces have to be drawn up evenly around the O-ring with very little tolerance. Often a spacing tool is provided to precisely control the distance between the flange faces, when they are in their full, properly drawn-up position. Underwater flanges have many bolts, spaced closely together so that they can be drawn up evenly with a minimum of distortion.

With small pipelines, 12 inches or less in diameter, where rotating the pipe slightly to line up the flange holes is not difficult, both flanges are welded solidly to the pipe. On large-diameter pipe, one of the flanges is usually of a special design, allowing it to be easily rotated so that the bolt holes can be lined up with those on the solid flange. Generally, one end of the pipeline (the end with the solid flange) is left lying on the bottom. The davit slings are hooked onto the end of the pipeline with the rotating flange, and this end is brought into alignment with the dead-end flange. Before attempting to mate the flanges, be sure that there is a large hole jetted in the bottom in the vicinity of the flanges, so that you will not be working down in the mud. Line up the live end (the end suspended in the davits) in as straight a line as possible with the dead end, by moving the barge on its anchors, or booming up and down on the davits. Then be sure that both O-ring grooves are clean and free of mud or grit.

If the flanges overlap (they frequently do), you can pull them apart by lifting up on the davits farthest away from the flange. This will pull an overbend into the pipeline, shortening it. By booming up or down on the davit closest to the flange, bring the two flanges into alignment and drive a drift pin through the holes at 3 o'clock. If the holes at 9 o'clock are aligned, drive in another drift pin and slack off on the davits, bringing the flange faces together. If the holes do not line up for stabbing the second drift pin, and you are working with solid flanges, hook a davit or crane to the dead-end pipeline to impart a rolling movement. The pipe will pivot on the one drift pin, and when the second set of holes line up, drive in the other drift pin. Use a chain or a nylon sling for this rolling hookup, as a steel cable will have a tendency to slip around the pipe. Pass the eye of the sling over the top of the pipeline in the direction you want the pipe to rotate. Take two turns around the pipe and shackle the eye into the standing part at the bottom of the pipe.

With one rotating flange, try to stab a drift pin through the top hole in the solid flange and the first hole you can reach in the rotating flange. As the pipe end with the rotating flange is lowered, the weight will cause the flange to pivot on the drift pin, and when the load is slacked off, other holes should be aligned.

When all the lifting slings are slacked off, and if the flange faces are square (i.e., touching each other all around their circumference, or separated by an equal distance all around their circumference), then you can insert all the bolts below the drift pins at 3 o'clock and 9 o'clock. These bolts should not have their nuts screwed up for more than half of their threads on each side. This is to allow the flanges to be separated sufficiently for the insertion of the O-ring. Care must be taken to square the flange faces. This is the most important aspect of the entire flanging operation.

INSERTING THE BOLTS

Any attempt to insert bolts and complete the flange at this point without properly aligning will only create difficulties which will compound themselves as you proceed. Driving the bolts through with the flanges cocked will damage the threads on the bolts, and later they will have to be driven out and replaced at great cost of time and effort. If, by a stroke of luck, the bolt threads are not damaged, and they are tightened up with the flange faces cocked, you will not get a proper seal, and the flange will leak. It is imperative that the flange faces be properly squared, and this is the time to do it. If the flanges are opened at the top and closed at the bottom, insert the two bolts directly under the drift pins and screw the nuts up all the way by hand. Then, take a lift on the pipeline at a point 30 feet or so behind the flange. The bolts will keep the flanges from being pulled apart and when the faces are square, you can shim up under the pipeline at the lifting point with sandbags.

If the flanges are closed at the top and open at the bottom, put two bolts in just above the drift pins and pick up the line from a point a few feet behind the flange. When the faces are squared you can shim this point up with sandbags also. If the weather is calm, these adjustments can be made with the pipe hanging in the davit slings. If the weather is rough, saddle the pipeline with sandbags so that all lifting points can be slacked off. If the flanges are open on one side or the other, where the flanges meet the pipeline will have to be moved over the bottom laterally until the flange faces are square. Be sure to have at least two bolts in the flange before moving the pipeline, to prevent pulling the flanges apart. When the flange faces are properly squared and mated and all the bolts are in below the drift pins, knock out the drift pins. The next step is to insert the O-ring,

INSERTING THE O-RING

You must *never* put your fingers or hands between two flange faces even if the pipeline is lying firm on the bottom. There are several ways to rig the O-ring so that it can be inserted without endangering your fingers. The O-ring can be hung from several fine threads of nylon, or a strip of teflon tape. When the ring is compressed this material will liquefy and be forced out of the grooves.

A better way, but a method that some pipeline inspectors object to, is to braze a 12-inch length of brazing rod or welding rod to the top dead center of the O-ring, giving you a convenient handle. The O-ring could very well come from the manufacturer with a suitable handle attached. This is just one of many instances where devices for use under water are designed and manufactured without consulting the man who will be using them—the diver.

When the O-ring has been secured in one of the above-mentioned ways, poise it over the top of the flanges. By picking up easy on the davit farthest away from the flange, the flanges should separate. Drop the O-ring between the flanges and line it up with one of the grooves. Slack off on the davit, and the flanges should come together, holding the O-ring in place. With a wire handle on the ring, you can easily feel out the groove. With the ring hanging from nylon or teflon, you may have to use a welding rod in your other hand to hold the ring in the groove.

USING THE IMPACT WRENCH

When the O-ring is in place, tighten the bolts at 4 o'clock and 8 o'clock with the impact wrench, to be sure it doesn't drop out of the groove. Then, install the rest of the bolts. As in the case of bolting up riser caps, the hardest work is in shifting the wrench from nut to nut, not in pulling the trigger of the wrench. All the bolts on a flange must be drawn up evenly, a few threads at a time, and this is as critical as torquing up the head bolts on your Offenhauser.

On large pipelines the flange bolts can be massive—2 or more inches in diameter. The impact wrenches used to draw up the nuts on these bolts are huge, too, and sometimes virtually impossible for one man to handle, especially if the flange is up off the bottom.

A simple way to handle a heavy impact wrench is to build from 2- or 3-inch pipe a small davit than can be clamped around the pipe-

Fig. 42. A diver socking up on the flange bolts of a tanker loading hose. This job is usually done with an air impact wrench. Photo: Divcon, Inc.

line over the flange. A little handy-billy, or two-part rope falls can then be hung from the davit to support the wrench. A small cleat on the davit to secure the hauling part of the falls completes the rig. After the diver adjusts the wrench socket over a nut and secures the falls to the cleat, he can operate the wrench most conveniently without having to support its weight.

Impact wrenches, air or hydraulic, are almost never held or applied under water in the positions used for these tools above water. When working on risers, they are usually suspended from their hoses, and therefore are in an inverted or upside-down position. Underwater wrenches should be built with the hose or hoses coming out of the top so that the wrench will hang horizontally.

The trigger should be on the right-hand side with the torque bar on the left, both on a horizontal plane. If the wrench is too heavy to be supported by its hose, then a balanced pickup point for a suspending line should be built into the wrench. Both the trigger and the shifting lever for forward and reverse should be large, positive, and unequivocal controls. Pneumatic wrenches should be built with the trigger on the exhaust so that the wrench will always be filled with air, rather than water, when it is not actually in use. This would greatly reduce internal corrosion and wrench malfunction. All pneumatic tools should be fitted with automatic lubricators.

TEAMWORK REQUIRED

During this entire discussion, I have given instructions as though one diver would be doing all the work. This was to acquaint you with every possible step of the operation. In reality, especially in deep water, these jobs will be a team effort. Many divers will be used in relays, each one doing a small part of the work. Before descending, have the diver you are following brief you on the exact condition and progress of the job, as well as the location of tools and materials. When your dive is over, gather up all the tools and leave them in an accessible location. Give an accurate report of your work and any recommendations you think will help the diver after you. Don't fib about the amount of work you finished; this makes the job tougher for everyone, and you are sure to be caught sooner or later. Keep your reports factual and accurate. If you screw up a dive (and everyone does from time to time) don't try to hide it. If you dropped all the bolts and they are lost in the mud, don't neglect to tell the next diver about it. This may result in groans and a sour face from the foreman, but it is certainly better to report it and scratch only one dive, than to allow the next diver to go down and blow his dive looking for bolts and tools that are not there.

Work procedures will vary from barge to barge, from foreman to foreman, and project to project. The foregoing is intended to illustrate only a basic approach to several job problems. Use your head and your imagination. Every existing method for doing a job can be improved.

Chapter 10

DIVING FROM A PIPELINE DREDGE BARGE, OR JET BARGE

After a pipeline has been laid, it must be buried anywhere from three feet to more than ten feet below the sand or mud of the ocean bottom. The reasons for this very costly operation are to protect the pipeline from movement by storm waves, to shield it from the anchors of ships, and to keep it clear of the trawls and nets of fishing vessels.

For river crossings, or for short runs of pipe, the pipeline trench is generally excavated first by clamshell bucket or by a standard dredge, and the pipe is then either dragged or laid into the trench. For offshore oil and gas exploitation, where pipelines 60 or 80 miles long are not unusual, this method is too slow and too expensive. In a surprisingly short time, the unique and highly specialized pipeline trenching barge was evolved.

These barges are generally called "jet barges," because they accomplish their work by blasting the pipeline ditch into the ocean bottom by means of multiple high-pressure water jets. The jets are closely spaced and arrayed in two vertical rows, which, incorporated into a device called the claw, straddle the pipeline. The jet nozzles in each row are staggered so that some point straight ahead, some at a downward angle, and others in toward the center of the pipeline. Also straddling the pipeline, and mounted just behind each row of jets, are two suction pipes. They suck the mud or sand out of the ditch after it has been blasted loose by the jets. The claw, consisting of the jets and suction pipes, is mounted on two steel pontoons which also straddle the pipe and slide over the ocean bottom. This complete device is called, appropriately enough, the sled. The attachment of the claw within the pontoons is vertically adjustable, so that it can be set to cut a ditch of predetermined depth.

OPERATION OF THE CLAW

The configuration of the claw is that of a downward pointing "U." Two horizontal rollers covered with heavy rubber are mounted at the top of the claw, one at the front and one at the back. They roll along the top of the pipe when the claw is set in its deepest cutting position. Two or three sets of vertical rollers, extending for the entire depth of the claw, are mounted on each side and also straddle the pipe, rolling along and guiding the sled over the pipeline. These rollers also prevent the jet nozzles from coming into contact with the pipeline. A jet barge, like a lay barge, is pulled along the pipeline route by its anchor system. The forward travel of the barge is transmitted to the jet sled through a cable that runs from a winch, over the bow and down, under the entire length of the barge, to the sled at the stern. This pulling cable is attached to a short bridle which in turn is attached to the front end of each pontoon; it is fitted with a Tensiometer, reading out in the control tower where the anchor foreman can constantly monitor the strain upon it. All of the rollers on the claw are fitted with load cells whose gauges are also mounted in the control tower. From the readings of these roller load cells, the anchor foreman can determine the position of the sled on the pipeline and the pipeline route over the bottom. For example, if the port forward roller and the starboard stern roller each registered hard contact with the pipeline, and the opposing rollers registered no contact, then the anchor foreman would know that the sled was cocked.

He would heave in a little on his forward port anchor while slacking off a little on the starboard stern to straighten it.

JET SLEDS

Jet sleds vary considerably in size and design, as do the jet barges that service them. Some sleds use the air-lift principle on the suction pipes, sucking the mud and sand up just a short way and spitting it out to either side of the ditch. Other dredges are equipped with huge dredge pumps and suck the material all the way to the surface, discharging it over the side through a large pipe. The diameter of the suction pipes on the sled can vary from 8 to 18 inches and the delivery pressure of water at the jet nozzles can range from 300 to well over 1000 pounds.

THE WISHBONE

The high-pressure jet pumps and the suction pumps, located in the bowels of the barge, are connected to the sled by a network of steel pipes and flexible hoses. The fluid train for both the high-pressure jet system and the suction system leaves the stern of the barge through two large-diameter pipes, connected at the top end and shaped like a wishbone. These two pipes (called the wishbone, or sometimes, the stinger) can be 20 inches or more in diameter and 60 to 80 feet long. They are attached to the stern of the barge by ball joints or swivels, and the wishbone is thus capable of being rotated through a vertical arc of perhaps 140 degrees. The raising and lowering of the wishbone is controlled by a winch operating through an A-frame derrick.

When dredging in shallow water, the wishbone will be standing almost vertical, suspending the excess slack of the hoses. In deep water, the wishbone will be lowered completely under water, with its tip hanging almost vertically below the stern of the barge. Its purpose is to serve as a conduit for the various streams of water, and as a boom to control the slack of the hoses in the varying depths encountered during a jetting operation. Attached to and extending from the tip of the wishbone are the various flexible rubber hoses, which connect the barge to the jet sled. These hoses are attached to the wishbone and to the jet sled by ball joints or knuckles, to provide flexibility between the jet sled and the barge. The sled itself is raised and lowered

Fig. 43. (Top) A close-up of the jet sled, showing the catwalk around the claw. Photo: Author. (Bottom) A modern pipeline dredge barge, or jet barge. Note the jet sled hanging from the A-frame at the stern of the barge, and the large suction hoses drooping down behind the jet sled. Photo: Brown & Root, Inc.

by winch through the A-frame, which is mounted permanently at the stern of the barge. The lifting tackle for the sled is generally four parts of wire, with the sled hanging from the lower two-sheave block on four long slings, attached to pad eyes on the pontoons. Figure 43 is a picture of a modern jet barge, showing the A-frame, the wishbone and hoses, and the jet sled. Figure 44 is a picture of a jet sled, showing the claw and the jets.

Fig. 44. (Top) The "claw." This is the section of the jet sled that fits down over the pipeline; note the rollers and the two rows of jet nozzles. Photo: Brown & Root, Inc. (Bottom) A factory-made hand-jetting nozzle. Photo: Galeazzi.

How much pipeline a jet barge can bury in a day depends on many variables: the power of the pumps, the depth of ditch required, the type of bottom, the depth of water, and importantly, the skill of the tugboat captain and crew tending and shifting the barge anchors.

One jet barge has buried as much as eight miles of pipeline in a 24-hour period. Conversely, when working in Beaumont clay, or sugar sand, this same barge might have to make three or four passes over a pipeline to get it down to grade; in such a case it might average less than a mile of buried pipe in a 24-hour period.

Pipelines are seldom buried in water depths over 200 feet because of the tremendous difficulties encountered in managing surface-oriented equipment at great depths.

SETTING THE SLED

Since learning by doing is the best way, let us catch the crew boat out to Houston-McRoot Jet Barge No. 10.

Fortunately, the barge is still setting anchors when we arrive. We will have plenty of time to carefully inspect the sled and the claw before it is put into the water. Diving from a jet barge can be hazardous, and it is of extreme importance that the diver be thoroughly familiar with all aspects of the equipment he will be using. Should you pull a job on an unfamiliar jet barge and the sled is in the water when you arrive, it is imperative that you seek out the lead diver and have him carefully explain the layout of the sled and the claw. You will seldom see the sled when it is on the bottom, because of the mud and silt constantly stirred into the water during the jetting operation. If the sled is on deck, take the time to inspect it carefully.

The sled on Barge No. 10 is sitting halfway on the after deck, suspended by the A-frame and snugged into place by air tuggers. It is only halfway on deck because the claw extends about 6 feet below the pontoons and it is hanging over the stern.

This particular sled has a steel catwalk welded completely around it; an excellent feature, although not all sleds have one. Generally speaking, if your hose is in the clear, you will come to no harm from the jets or the suctions while you are on the catwalk.

Jump up on the catwalk and have a careful look around. First, check out the lifting slings, their points of attachment on the sled, and their length. When the sled is on the bottom, each of these slings is a potential hazard. As the barge lifts and rolls to even moderate waves, these slings will be coming into tension and slacking off like giant bowstrings, and contact with one of them could be lethal. The length of the slings is important, because they regulate the distance above the sled where the main lifting block will be hanging. Due to the pitching and rolling of the barge, this block can be whirling around as dangerously as an executioner's axe. There is the danger also of having your hose or your fingers drawn into the sheaves of this block. Never put your hands on the cables above the block. Keep yourself clear of the lifting block and the four lifting slings at all times. To lessen the danger of your hose being drawn into the sheaves of the lifting block, *always* pass on the down-current side of the sled when moving from front to back.

Notice the way the hoses attach to the sled. You can see the articulated ball joints mentioned earlier. These are another potential hazard. When the hoses move up and down in the waves, the flanges on these knuckles will be separating and coming together—erratic steel jaws, waiting for your hands or your hose. Keep clear of these hose knuckles. Always work on the down-current side of the sled, and always work with a tight diving hose. In the center of the sled, immediately under the back and front catwalks, are the horizontal top rollers. Check them out. Eyeball the vertical rollers, front and back, and the location of the jets and the suctions. Check the pad eyes at the front of each pontoon where the towing bridle is attached. You will not be able to see where the bridle attaches to the pulling cable, because this will be under water and under the barge. The last important thing to check out is where the divers' descending line is attached to the sled. This is an arbitrary point and varies from barge to barge. On some barges the divers attach the down line to the framework at the back of the sled, on whichever side of the barge the diving station happens to be. I prefer to attach the down line to the apex of the towing bridle. In this way, my hose remains clear of all the hoses and rigging during the descent, and I can easily determine which way the current is running on my way down. When I reach the apex of the bridle, I pull myself along the down-current leg of the bridle until I reach the sled, and go to work.

JETTING A 12-INCH PIPELINE

The anchors are set, the barge is backed up to a well platform, and we are ready to begin. We are going to jet down a 12-inch pipeline, starting at the riser on the structure. The water depth is 60 feet, there is no current at this time, the weather is calm and the visibility is good. The first job is for a diver to go overboard with an impact wrench and loosen up all the bolts on the riser-clamp caps, so that the riser will slide down through the clamps as the pipeline is jetted. The bolts must be loose enough so that the caps hang slack. The next order of business will be a quick inspection of the bottom in the vicinity of the riser. Invariably, the workers on a platform will discard enormous amounts of rubbish overboard right at the structure, not realizing or not caring that this debris will interfere with future underwater work, as well as overload the structure's corrosion-protection system. If material such as coils of cable, pieces of steel plate, short sections of leftover pipe, etc., interferes with the setting of the jet sled, it will have to be laboriously and expensively cleared away by divers loading it into cargo nets or slinging individual pieces. If (or when) the bottom is clear, the jet sled is lowered to within 15 or 20 feet of the bottom. Most sleds are equipped with a pneumofathometer, so the foreman will know when to stop it. As the sled is lowered, the descending line must be slacked off to prevent it from parting. When the barge is in position, and the sled has been lowered away, a diver will be used to actually guide the sled over the pipeline and set it.

Before you go overboard, be sure the descending line is tight. As you slide down the line, keep yourself and your hose on the down-current side of the descending line. In many locations, the current at the surface will run one way, and the current on the bottom in another. Make your descent carefully to avoid fouling your hose around the descending line.

Opinions differ as to whether it is better to set the sled from the front or back end. There are advantages and disadvantages to both ways and it is largely a matter of the particular circumstances and the individual diver's preference. In this case the sled must be set as close as possible to the riser without hanging the back end of the sled pontoons up on any structure bracing, so it is advisable to set the sled from the back end.

When you reach the sled, check again that your hose is clear, and that you are on the down-current side of the sled, and then move to the back end. Whenever you have to make a dive to set the sled, always carry a 50-foot coil of ¼-inch line with you. Tie one end of this line off at the center of the back end of the sled and drop down to the bottom. Lower the sled until the bottom of the claw is within 3 or 4 feet of the bottom. If the weather is rough, you will have to keep the sled high enough so that there is no possibility of its dropping onto the pipe at the bottom of a swell. Do not get under the sled while it is being lowered. If you lack visibility, use your small line to judge how much it has been lowered and stop the sled before you move in to inspect it. When the sled is close to the bottom, use your small line and locate the pipeline. When it has been located, give the appropriate orders for the movement of the barge to line the sled up directly over it. Once again, remember, when moving a barge to place a crane block, or a davit chain, or in this case a jet sled, the point of suspension aboard the barge of any one of these objects will reach a point along a line of movement sooner than the object itself.

Give all your orders for barge movement in 2- or 3-foot increments, wait until all movement has stopped, and then move again. There is no better way to lose the affection of the barge foreman than to make him seesaw his barge back and forth on the anchors. You will have to make the final adjustment in the position of the sled over the pipeline by holding onto the back rollers at the bottom of the claw and straddling the pipe with your legs. When the claw is directly over the pipeline, lower away. You must be sure that your hose is clear and that your hands and feet are not between the claw and the pipeline.

SETTING THE SLED ON A MUD BOTTOM

If the bottom is soft mud, the claw will penetrate it and the sled will settle until the pontoons are resting on the bottom. On some sleds on a mud bottom, you must be careful not to get under the back horizontal roller as the sled is lowered away. When you are sure the claw is inserted over the pipeline, back away a few feet while the sled is lowered the rest of the way. Watch out for the hoses that go to the sled. They generally lead into the back of the sled, and if there is too much slack in them when you are setting the sled, there is a good chance that you will get a rap on the head from them as the sled is lowered.

SETTING THE SLED ON HARD SAND OR CLAY

On a hard sand or clay bottom, you might have to set the sled with the jets running. This is in order to blow a hole for the claw to fit down into; otherwise, the claw would fetch up on the hard bottom and the sled would tip, because the claw hangs 3 to 6 feet below the pontoons and the sled cannot rest level until the pontoons are squarely on the bottom. It might sometimes be necessary to set the claw just over the pipe, and then climb up on the catwalk for 4 or 5 minutes while the pumps are turned on and a large hole is blown. After the pumps are shut off, you can go back under the sled to check it and be sure it is properly on the line.

BLOWING A HOLE AT THE RISER

When starting out at a riser, it is important to get the sled as close to the riser as possible. The material directly under the riser-tube turn will have to be jetted out by hand, and the closer to the riser you can set the sled, the less hand-jetting will be required.

Some jet sleds are equipped with backward-pointing jets as well as forward-aiming ones. These back jets are controlled by a valve on the sled and are used specifically for a situation like this. The sled is set, the valve is turned to activate the back jets, and a hole is blown under the riser-tube turn. This is generally a two-way valve. When the back jets are operating, the forward ones are shut off and vice versa. After the hole under the riser is blown, the valve is set for the forward jets and the barge moves out.

In the absence of back jets on the sled, a large hole will be blown as close to the riser as possible, the sled will be pulled ahead making ditch for about 15 or 20 feet and the pumps will then be shut down while the material under the riser-tube turn is hand-jetted.

CONSTRUCTION AND OPERATION OF A JET NOZZLE

Years ago, jet nozzles for hand-jetting used to be carefully and expensively machined out of bronze with one forward aperture for the working-pressure stream and a series of radial holes on the back end for the counterbalancing stream. No matter what position the nozzle was held in, one of these back jets was invariably scouring the diver.

Once again, some unnamed innovator assembled a bunch of inexpensive pipe fittings into a T, using reducers and short pipe nipples, and created the jet nozzle which is the standard in the industry today. The large volume of water carried by the hose is divided at the T and forced out of the opposing pipe nipples at increased pressure. The nozzle can be swiveled on the threads of the T, and there is generally a stop welded onto it to prevent screwing it all the way off of the hose. No matter how much pressure is forced through the nozzle, it remains perfectly balanced because there are two exactly equal and opposed streams of water coming out of it.

Manipulating a jet nozzle is not exactly a science, but there are a few elementary rules that apply. A hose with 300 to 500 lbs. of water pressure in it cannot be regarded as supple. Take the time to lay your hose out on the bottom in such a way as to make your work easier. Hose with pressure in it naturally wants to straighten out. Lay out your hose with a large bight, so the tendency of the hose to straighten will be forcing the nozzle in the direction you want to travel. When the pressure is turned on, all the air in the hose and the pumping system will be forced down to the nozzle. The hose, which is generally 3 to 4 inches in diameter, will become very buoyant, especially if long lengths are used, and this buoyancy is enough to pluck you off of the bottom. A short piece of line tied between the nozzle and something solid on the bottom will save you from this embarrassing situation when the pressure is first turned on. How you hold or handle the nozzle is a matter of personal preference and comfort, but the jet can blow the patches off of your wet suit or the galoshes off of your feet and the counter-jet can rip the mask off of your face and strip the tape from your hose assembly, so be sure you have the nozzle secure and aimed properly before calling for pressure.

Jetting is always easier if you have a current to carry away the loose material. Jetting sand without a deep hole to drive the material into is almost useless unless you also use a small air lift. When jetting in mud it is always better to plan to get your work done in one pass. That is, if you need a trench 3 feet deep, cut your trench 5 feet deep on the first slow pass, rather than 3 feet deep with the probable necessity of having to pass over it again. Do not let go of the jet nozzle with the pressure running. Not only is there the possibility of accidentally coming in contact with the lost jet stream, but an unmanned jet will have the tendency to keep burrowing deeper and deeper into the bottom, and in a short time it may be impossible to recover it.

PROPER HANDLING OF THE JET HOSE

When your time is up, shut off the pressure and lay the nozzle over the top of the pipe. Do not take a round turn on the pipe with the hose. The foreman would not appreciate having to blow another dive just to retrieve the jet. Jet hoses in deep water are difficult for the topside crew to retrieve. An excellent and simple way to recover the jet hose is to fit the system with an air connection and valve. When the job is completed, the hose can be filled with air; this will float it to the surface and eliminate a lot of bull work.

The long lengths of jet hose required in deep water are very difficult for the diver to handle, especially in a current. It has been suggested that a tap be made into the jet water system right on the jet sled. Fifty or 100 feet of hose could be stored on the sled on a rack, or more ideally on a reel, with a valve at the jet nozzle. This would eliminate current drag and a number of other problems.

INSPECTING THE JOB

When the material under the riser has been jetted, the machinery will be started, and the sled will be pulled along the pipe for about 200 feet. A diver will be sent down to check the ditch, and if it is O.K., the sled will be picked up and the barge will be backed up to the structure again. If the riser did not go down of its own accord, an inspection will be made to be sure there is no material under the riser or pipeline. When making this inspection, the diver must be careful not to get under the riser or pipeline; it could drop suddenly.

USE OF COME-ALONGS

If there is a good ditch, the crew will rig up come-alongs to the top of the riser and jack it down. Sometimes an air tugger is rigged between the barge and the top of the riser with a horizontal pull, and alternately taking a strain and slacking off on the tugger sets up a vibration in the riser. This, together with the come-alongs jacking down should lower the riser to grade. When the riser is down, all the riser-clamp cap bolts must be tightened with an impact wrench. This is the final underwater work done on the riser before inspection and acceptance by the client. Be sure to tighten every bolt, even the ones on the jacket-leg side of the clamps before you pass the job as completed. On making the final inspection of the riser, be sure to report any damage to the weight coating that might have occurred while the riser was being lowered. Any such damage will have to be repaired with epoxy (Chapter 11).

When the riser is completed, the claw will once more be set on the pipeline, and the barge will move out on its jetting operation. The diver's duties will then be periodic checks of the pipeline and the ditch.

CHECKING THE DITCH

The frequency with which the ditch is checked, as the barge moves along on the jetting operation, is determined by the barge superintendent, and depends upon the type of bottom, the type of equipment being used and in some cases, the tides. If the claw has electronically monitored rollers, the superintendent might be satisfied with a check every four hours or so. On hard bottom, using a claw without electronically monitored rollers, a nervous superintendent might insist upon a check every half-hour. In areas of exceptionally strong tides, such as the British North Sea, the sled and ditch can be checked only every six hours, between tides.

On almost every jet barge afloat, the custom is for the diver to make these periodic checks while the jetting and suction pumps are running, and the barge is moving ahead over the pipeline. I do not believe that the potential for accident faced by the diver while he is anywhere near this equipment when it is operating justifies the few minutes of time saved. However, it is an established practice and it is not likely to be altered until the insurance companies correlate their accident reports and insist that the practice be stopped.

Precautions. As stressed previously, always keep to the down-current side of all hoses and rigging, and always pass along the sled on the down-current side. If you ever have to cross from one side to the other in front of the sled, always do so above the bridles. There is a chance that slack in your hose will be caught up under the sled pontoons, or in the suctions if you cross under the bridles or pulling cable. Always dive with a tight hose, with an absolute minimum of slack. The chance of a fouled hose (and a consequent free ascent) is always present when diving on a jet sled. Wear a bailout bottle and an inflatable life vest. Use a helmet when diving on a jet sled to protect your head from swinging cables and hoses. Some insurance companies insist that a helmet be worn for all work from a jet barge, and, hopefully, this will soon be a standard dictum. The use of a helmet in preference to a mask has another advantage. When the pumps are running, the level of noise created by the screaming jets is so high that communication with a mask is impossible. A helmet and the air space within it offers considerable insulation from this noise, and communications, though poor, are possible.

Checking the Sled and Hoses. When making these periodic inspections, there are a few specific items that the foreman will want checked, and in most cases the entire inspection can be done in a 6- to 10-minute dive. Usually, the foreman will want to know the relationship of the apex of the sled-pulling bridle to the pipeline, to be sure the barge is directly over the pipeline and pulling in the right direction. If the descending line is attached to the apex of the bridle, it is helpful to tie a 10-foot piece of ½-inch line so that it is hanging from the same point. Slide down the descending line until you reach the bridle, being sure to stay on the down-current side. Drop off the bridle, holding on to the 10-foot piece of line, and find the pipe. Holding the line taut, determine the position of the bridle over the pipeline; depending on the depth of water, the bridle apex will be found 4 to 15 feet above the bottom. If the sled is pulling normally, the apex will never be more than two or three feet to one side of the pipeline or the other.

If, for some reason, you are unable to find the pipe from the end of this short line, the pumps must be shut down while you locate it. This line is kept to a maximum length of ten feet, or shorter than the bridles, so there will be no danger of wandering into the jets in front of the sled if you become disoriented. After you report the position of the bridle, shinny back up the line to the bridle, and hand-over-hand it down the down-current leg until you reach the sled. Pass to the back end of the sled, either alongside the pontoon, or up on the catwalk, and move to the middle, keeping careful check of your hose slack.

The foreman will want to know how the sled hoses are hanging, and you will be able to either see or feel them from this point. If there is no visibility, and you are feeling around, remember the hose knuckles, and keep your hands away from them. If the hoses are too tight, they will be crimped, shutting off the water flow with the possibility of damage to them. If they are too slack, they will be dragging in the ditch behind the sled with equal potential for damage. Frequently, when setting the sled, the lifting cables will be all slacked off, but the sled will still be hanging erratically or at a bad angle. The most common cause for this is that the foreman forgot to slack off sufficiently on the hoses, and they are holding the sled up.

Checking the Pipe. After checking the hoses, you will have to drop off the back of the sled and check the pipe and the ditch. A 5- or 6-foot piece of line tied off at the center of the back catwalk will make this easier for you. In any event, hanging off the back of the catwalk by your hands should enable you to reach the pipe with your feet. Drop down and straddle the pipe with your legs; never drop below the pipe in this position, because that is where the suctions are. Reach in and check the distance between the back horizontal roller and the top of the pipe. This is as much of the sled as you will be able to check with the machinery running. As you feel the sled pulling ahead, release it and move back down the pipe for at least two joints behind the sled. Most pipe will have sagged sufficiently to be lying in the bottom of the ditch as much as two to three joints (or 80 to 125 feet) behind the sled. As you move back along the pipe, check it carefully for damage to the weight coating. Where the pipe touches the bottom, the foreman will want a pneumo on the top of the pipe, on the bottom of the ditch, and on the natural bottom outside of the ditch. When taking readings at the bottom of the ditch, do not get under the pipe, because it will be settling constantly as the sled moves ahead. For the natural bottom reading, be sure you are at least ten feet to one side of the ditch. When these readings have been taken, you can return to the sled. Some divers follow the pipe back, but this has its disadvantages—as the sled moves forward, the side of the ditch will be constantly caving in, with the possibility of trapping the diver or his hose. It is just as easy to get out of the ditch, on the down-current side, and return to the sled by following the furrow that the sled pontoon made in the bottom. When you reach the descending line, be sure your hose is clear and ascend. Should you become lost, confused or disoriented in any way during your inspection of the ditch, immediately call for the sled to be stopped and the pumps shut down.

On sleds without a catwalk, there will be pipe bracing connecting the two pontoons. Follow the pontoon until you locate the aftermost brace, and hold onto it as you move into the center of the ditch to locate the pipeline.

Encountering Valves for Future Tie-ins. If there are no valves or crossing pipelines, your job on a jet barge will be a fairly simple one, primarily limited to these periodic inspections. However, it would be rare indeed to lay or dredge a pipeline today without encountering a crossing pipeline or cable, or a valve for a future tie-in. A valve will be marked by a buoy and/or a sonic pinger. When the barge approaches a valve, a dive must be made from the bow to locate it. The valve will be covered with some type of guard or shield to protect it from fishing nets, and this guard will have to be removed before the valve is buried. The guards are held in place with bolts, and because of corrosion and possible damage to threads, the quickest way to deal with these bolts is with a cutting torch. After the valve cover has been removed and brought on deck, the diving operation will be shifted back to the stern.

Because the valve will project above the pipeline two or three feet or more, the ditch must be cut that much deeper for 200 or 300 feet on both sides of the valve in order to provide the proper amount of cover over the valve. Three hundred feet or so before the valve is reached, the sled is lifted off the pipeline and brought to the surface. The claw is lowered within the sled to provide the extra depth of ditch, and the sled is reset on the pipeline. It is important to pull the sled up as close as possible to the valve without touching it, to limit the amount of hand-jetting. This is a tricky operation because if the sled is accidentally pulled into the valve, the resulting damage would cost thousands of dollars to repair.

After the sled has been reset on the pipeline, the diver carries the end of a small line that leads to the surface and makes his way forward under the barge until he finds the valve. After tying the small line to the pipeline about ten feet ahead of the valve he returns to the surface. With a man tending this small line and taking slack out of it, the barge will move ahead jetting the pipe until the small line is tending straight up and down at the stern of the barge.

The diver then descends and guides the sled a foot at a time until the ditch has been jetted as close to the valve as is feasible. This can be accomplished in one of two ways. He can tie a small line to the forward end of the valve, with a knot in it two or three feet ahead of the valve. Then, while he sits on the forward catwalk, holding a strain on this line, the pumps are started and the sled is pulled ahead until he reaches the knot. The sled will be left in this position for 5 or 6 minutes, blowing a deep hole, and then the pumps are shut down. Then the sled must be lifted and reset on the line on the other side of the valve. Before the sled is lifted, however, the barge is backed up several feet. The reason for doing this is that the sled, in being towed forward through the resisting mud and sand, will still be in tension on the end of the pulling cable. As it is lifted clear of the bottom, the sled could jump ahead and damage the valve. Be sure that the sled is lifted high enough so that there is no possibility of surface swells causing the bottom of the claw to hit the top of the valve.

When the claw is clear, go around to the back of the sled and walk the barge ahead far enough to clear the valve. Reset the claw on the pipeline as close to the valve as possible. The barge will then pull ahead for about 300 feet, jetting down the pipeline. The sled will once more be lifted to the surface, and the barge will back down over the valve again. The valve is inspected to determine if it is down to grade. If not, the material under it must be hand-jetted. When the valve is down to grade, the claw is reset within the sled, to cut the normal depth of ditch. The sled is reset on the pipeline, and routine jetting of the pipeline continues until another valve or a crossing pipeline is encountered.

Encountering Crossovers. A pipeline crossing provides the most work when working on a jet barge. If the crossing pipeline has been recently laid, it is probably marked by a buoy, and pulling up to it will be the same as pulling up to a valve. If it is an old line and passes under the line that is being jetted, then a survey boat will have placed markers in the approximate location of the crossing. Sometimes crossing lines are scouted out by a survey boat with a Seashell or other type of magnetometer working in advance of the jet barge. In this case the crossing line will be clearly buoyed and the approach to it will be the same as for a valve.

If the exact location of the crossing line is not known, then a diver must be used to scout out ahead of the sled when the barge gets on approximate location. The easiest way to do this is with a 6-foot probe and a small line from the surface. The diver walks ahead (in front of the sled), under the barge, probing with the rod. When he has worked ahead 100 feet or so, carefully probing every inch of the way, he ties the small line off to the pipeline that is being jetted and then returns to the surface. The small line is tended exactly as for working up to a valve and when it is straight up and down at the stern, the pumps are shut down and the diver probes ahead for another 100 feet or so, or until he locates the crossing line. Sometimes, when searching for a buried crossing line, some remnant of the ditch, or at least a furrow, is still detectable, but don't depend on it; use the probe assiduously. If you miss the line and the sled is accidentally pulled into it, thus damaging the line, you can be sure of a seat on the next crew boat. If the lines cross at an angle at or near 90 degrees, the claw can be pulled close enough to blow material out from under the other side of the crossover. If the crossing angle is acute, however, the claw will have to be stopped a long way from the actual crossing, and all the material that cannot be reached with the claw, called the plug, will have to be removed by hand-jetting.

Because a vertical distance of two to three feet must be maintained between crossing pipelines, the claw will have to be set to cut a deeper ditch than normal for about 300 feet on all sides of a crossover. If the bottommost pipe is not buried sufficiently deep to allow for this spacing and for adequate cover over the top pipe, then this bottom pipe will also have to be lowered for 300 feet on each side of the crossing. The procedure is to pull up to within 300 feet of the crossing, pick up the sled and set the claw for the deeper cut, and reset the sled on the pipeline. When the sled has been pulled up to the crossing and a deep hole blown, the sled is raised and the barge turned to line up over the bottom pipeline. The barge must be backed up over the bottom pipeline to a point about 300 feet away from the crossing. After the sled is set and pulled up to the crossing, it is picked up, the pipeline is jumped and the sled reset. The barge then pulls ahead for another 300 feet on the other side of the crossing. When the bottom pipe has been jetted for 300 feet on each side of the crossing, the barge is turned around and lined up over the original pipeline.

Now we have three of the four legs of the intersecting pipelines jetted. The bottom pipe should be in the bottom of the ditch, perhaps six feet under the crossing (top) pipe. The top pipe is still suspended because it has only been jetted on one side of the crossover. If the specified clearance between the two pipelines is 3 feet, sacks full of sand and cement are lowered on top of the bottom pipeline to build up this permanent bridge. These sacks are lowered to the diver in a cargo net. He guides the net into the proper position and slacks the load off sufficiently to unhook the slings from one side of the cargo net.

The net is lifted, dumping the sacks on top of the bottom pipe. The diver neatly stacks the bags, perhaps using three or four cargo nets full of them, until he has constructed a bridge, eight or ten feet long and of sufficient height. When working with sand and cement bags in a cargo net, it is always well to have one side of the cargo-net slings permanently shackled into the lifting cable with the other side moused on the hook. This will prevent the accidental dumping of the load, and all the diver has to do when the cargo net is in position, is to cut the hook mousing and unhook one side of the sling. Before allowing the emptied cargo net to be lifted to the surface, be sure all the bags are out of it. Sometimes a bag becomes entangled in the mesh of the net, or in the slings, only to be dislodged when the net is almost to the surface. Being thumped on the head by a 100-pound sack of sand and cement could result in a broken neck.

When the bridge has been constructed, the sled is set on the last leg of the crossing and the ditch is blown for another 300 feet. The sled is picked up and the barge is backed up to the crossing one more time. Pneumoreadings are taken at the bottom of the ditch, on top of the bottom pipe, on top of the top pipe and on natural bottom. If there is sufficient bridge between the pipelines and sufficient cover over the top of the top pipe, the crossing is complete. Sometimes more bags are dumped on top of the crossing to form a permanently fused saddle for the crossing pipelines.

When all this is completed, the claw is retracted within the sled to cut the normal depth of ditch, the sled is reset on the pipeline and the jetting operation resumes.

Setting the Sled on Buried Pipeline. Setting the sled on a buried pipeline as is done on crossings, can present some problems. The pipeline must be located with the probe, and the claw has to be lowered close enough to the bottom so that it is not touching in the swells. The jet pumps have to be turned on for several minutes to blow a hole and uncover the pipeline. When the pipeline is sufficiently exposed, the claw is set over it.

Setting the Sled in Heavy Seas. Often, when trying to maintain the job-progress schedule, the superintendent will insist that the diver try to set the sled on the pipeline in very rough weather. This is dangerous, both in terms of diver safety and possible damage to the pipeline. In a situation like this, your first responsibility is your personal safety. There are two possible ways to set the sled in heavy seas, neither of them positive insofar as avoiding damage to the pipeline. This, however, is the superintendent's responsibility. One way is to have the sled lowered to within 20 or 30 feet of the bottom. Taking a 50-foot piece of small line with you, descend to the sled. Tie one end of the line to the center of the catwalk at the back of the sled. Drop off the sled with the other end of the line and locate the pipeline. Tie the line around the pipeline and return to the sled. By sitting at the center of the back catwalk and facing aft, you can hold a strain on the small line and maneuver the barge until the line is tending straight up and down between your legs. By timing the oscillations of the sled and using the line as a guide, have the sled lowered away when you think it is over the pipeline.

Another method is to take two 10-foot pieces of small line with you. Tie one line into each lower back corner of the claw and, sitting on the pipeline and using the two lines as a pair of reins, maneuver the sled into position as closely as possible. When you think you have it, lower the sled. Either one of these systems will keep you out from under the sled, but the chances of damaging the line are great with both of them. Damage to the weight coating on pipelines is a common occurrence during jetting operations. Repair of this type of damage is covered in the next chapter.

As we approach the end of the pipeline, if it terminates in a riser at a platform, the procedure with which we began the job must be repeated. Some barges carry a miniature sled on the bow for this purpose. The sled is handled by the bow crane, and eliminates the need for turning the barge around and resetting the anchors. When this riser is lowered to grade at the termination of the pipeline, and the riser clamp caps have been tightened, the job is finished.

SUMMARY OF PRECAUTIONS

In closing this chapter, I would like to reemphasize a few points. Never work around a jet sled with the machinery running if you can possibly avoid it. Always have your hose tended tightly, with an absolute minimum of slack. Always descend and work on the down-current side of the sled. If you ever have to cross over in front of the sled, always do so above and on top of the bridles and pull cable. Beware of the lifting pennants, the lifting block and its cables, the hose knuckles and slack hose behind the sled. Remember to stay clear of the back horizontal roller when setting the sled. Use a helmet in preference to a mask, and use a bailout bottle and inflatable life vest whenever working around a jet sled.

Chapter 11

MISCELLANEOUS DIVING APPLICATIONS IN THE OIL FIELD

Many situations develop in the course of laying or dredging pipelines that require the services of a diver. I have selected a few of the most common ones for discussion, and since we just left a jet barge, let us go over the procedure for repairing a type of damage most often caused during the jetting operation. This is the damage to weight coating and dope coating on pipelines and risers.

The thick concrete shell on large diameter pipelines, called the weight coating, keeps the pipeline anchored to the bottom and simultaneously affords valuable mechanical protection for the dielectric insulation, or dope coating, of the pipeline. When setting risers, especially in rough weather, this concrete frequently gets knocked off in patches due to accidental banging of the riser into the jacket leg or the clamps. When setting the sled in rough weather, sometimes the claw is banged into the pipeline. Anchors also frequently get dropped onto or dragged into pipelines, damaging the weight coating. In any of these situations, it is not the loss of the few pounds of ballast the missing concrete represents that is of concern. What the missing concrete does indicate is that the pipeline has probably sustained impact or a severe bending stress with possible damage to the pipe. If the pipe itself has not been dented, nicked or otherwise damaged, then the concern is for loss of or damage to the corrosion-resisting insulation of the dope coat which is under the concrete and bonded to the metal pipe. If two or three feet of weight coating have been knocked off the pipeline but the pipe itself has sustained no damage, and the dope coating is intact with no bare metal showing, the client's inspector may accept the pipeline with no further repair. If the dope coating has been disturbed and bare metal is showing, the client will insist upon repairs to restore the integrity of the pipeline insulation.

SPLASH ZONE AND OTHER EPOXY COATINGS

Repairs to coating damage are usually effected with Splash Zone or some other type of epoxy resin coating. The various manufacturers of epoxy coatings for underwater use imply that the application of their product is far simpler than actual experiences in the field indicate. Epoxy is an effective and durable material for underwater use, but its proper application is a matter of careful mixing, timing and some skill. Epoxy comes in two and sometimes three containers, consisting of the base, a converter or catalyst and sometimes a hardening or drying agent. Splash Zone, the trade name of the type of epoxy with which I am most familiar, comes in two cans, one containing a black base and the other a yellow converter. The two are mixed in equal parts to form a thick putty-like dark-green paste. They are toxic and should never be mixed barehanded; use rubber gloves. The mixed epoxy can be used barehanded under water. The pot life of the mixed epoxy is about an hour in 70- to 80-degree temperatures, but this time is considerably shortened in cold water. In 50- to 60-degree water the material might remain workable for only 15 or 20 minutes.

For proper bond, the metal surface to which Splash Zone is to be applied must be completely cleaned of rust, scale and marine growth. For small areas of metal which have not been exposed to salt water for a lengthy period, vigorous cleaning with a wire brush will probably be sufficient. For large areas where scale and marine growth are present, the surfaces will be easier cleaned by sandblasting. Oddly enough, the equipment used for sandblasting on the surface is equally effective under water. The only improvisation necessary is the addition of weights five or six feet from the nozzle to keep it submerged so that the diver can hold on to it. Sandblasting hose (generally of large diameter) is very buoyant. The corrosion process is rapid, therefore the epoxy must be applied as soon after sandblasting as possible, and large areas should not be attempted during any one dive. When applied to a vertical surface, epoxy will tend to slump or sag before it has had a chance to set up, and some means of containment is absolutely essential for a good job. In my experience, the best agent for this is Pycoflex or True-X plastic pipe tape which comes in 4-inch rolls. This tape retains some of its adhesive qualities under water, and I strongly feel that any attempt to apply epoxy under water without it or some other suitable binding or reinforcing is a total waste of time.

For coating repairs to a pipeline laying on the bottom, it may be necessary to hand-jet to provide easy access to all parts of the pipe.

Some success has been achieved by spreading a thick coat of epoxy onto one side of a burlap sack. This pad or poultice is then sent down to the diver and he wraps it around the pipe, binding it in place with tape or small line. When using pipe tape, after the affected area has been cleaned, take three or four wraps of tape around the edge of the undamaged section of pipeline, adjacent to the area to be coated. The epoxy will then be sent down readymixed in a bucket, and you can grab handfuls of it and press and spread it onto the pipe, wrapping the tape over the epoxy in a spiral as you go. It is advantageous to secure the beginning and terminating ends of your tape with several round turns of small line, pulled tight and knotted around the pipe. The epoxy is best handled barehanded or with rubber gloves. When you begin to feel the material stiffen, scoop out what remains in the bucket and clean the bucket out as best you can before sending it back to the surface for another batch. Before it sets, the epoxy is soluble in water and is easily removed from your hands or rubber gloves by rubbing them together. After it has set, only a paving breaker could remove it, so avoid getting it on your wet suit.

For vertical applications such as risers or jacket legs, it is helpful to tie a piece of ½- or ¾-inch rope around them to form a base or support on which to build up the epoxy. Most epoxy coatings will be hard in 3 to 4 hours and completely set up in 24 hours.

The use of epoxy for weight-coat repairs to large areas of pipeline is difficult, expensive and of dubious efficacy. I have often wondered why pipeline companies do not use simple prefabricated sheet metal forms and "Subac," or other underwater cement, for these repairs.

WHEEL JOBS

Sometimes the tugboat or crew boat catches a line or a cable in the propeller. If the boat just picked up the end of the line or cable in the wheel, and the other end is still on board, you may not have to dive. Hook the bight or free end of the line to the deck crane, and with the clutch of the affected propeller disengaged, take a strain with the crane. The chances are good that the cable or line will unwind itself and come free. Otherwise, you will have to go to work. If it is a cable, have the cutting equipment hooked up before you go overboard. If you can get the barge electrician to shine a floodlight overboard you will not need a hand light, thus leaving both hands free. Whenever you do have to carry a handlight of the floating "Ikelite" or "Dacor" type, it is generally easier to secure it to your hose with a short lanyard about six feet above you. Thus the light is always available but out of your way.

I strongly recommend the use of flippers for this type of work as they are a big help in getting you under the boat. A full wet suit and no weight belt helps to stabilize your position while working under any vessel because the buoyancy of the wet suit holds you up against the underside of the boat. A crew boat is more dangerous to work on, especially in a running sea, because of its shallow draft. It bounces around much more violently than a tugboat. Then, too, on a crew boat there is no place to wrap your legs or to brace yourself. Your safety demands that you keep one hand above you to push yourself away from the boat as it drops between the waves. On a tugboat, the diver can generally get forward of the propeller and lock his legs around the skeg. A fouled cable will be in great tension, so be careful while burning so that individual strands do not snap out and hit your mask or hands. Fouled propellers are never the same, so study the job carefully before you start burning. Also (and this is extremely important) don't be satisfied that the clutch is disengaged. Insist that the main engines be turned off before you go into the water on a wheel job. Invariably, parts of the cable will be jammed between the propeller hub and the stern bearing, and great care must be exercised to cut this portion free without burning the shaft or the bearing housing. This is difficult to do when working with one hand on a wildly pitching boat.

Several razor-sharp knives are needed for rope (Manila, nylon or poly); common butcher's knives (not stainless) are adequate. While the diver is using one knife, a deck hand can sharpen the other with a file. For heavy rope the most effective tools are a 3- or 4-inch carpenter's wood chisel and a 3- or 4-pound mason's striking hammer, but wood chisels are rarely on offshore barges. Many types of stern bearings are water-lubricated, so after you have finished the job, it is important to inspect carefully to be sure no fine strands or threads are jammed around the shaft and in the bearing. At the same time, it is well to give a report on the condition of the propellers—bent blades, nicked tips, etc. Check the rudder also.

Be sure your tender stands in a position where there is no chance for your hose to get caught between the boat and the side of the barge.

SUCTION STRAINERS

Every barge or large vessel has suction strainers that occasionally need cleaning or perhaps removal and replacement of the screens. For cleaning, bend a 3/16-inch welding rod into a "J" at one end and an "L" on the other. The "J" inserted into one of the screen holes provides a simple and effective way to hold yourself in place. Flippers are a big help on this type of job also. A 4-inch stiff-blade putty knife will remove all the barnacles and heavy growth, and a touch-up with a wire brush should complete the job. If a screen must be removed, a 2-pound coffee can, tied to a 5- or 10-pound magnet attached to the side of the barge or vessel makes an excellent receptacle for tools, nuts, washers.

When working under a vessel, bend a welding rod with a "J" on one end and a closed eye on the other. Tie a piece of 1/4- or 3/8-inch line in a deep loop to the eye in the welding rod. Hook the "J" into the screen and adjust the loop so that you can sit in it and conveniently reach your work. A 25- or 50-pound magnet also works well for this application.

Needless to say, with very powerful pumps the machinery must be shut down before you approach the suction screen.

SEARCHING

Searching for lost objects under water is probably one of the most frustrating and generally unrewarding tasks a diver is called upon to do. Success depends mostly upon Lady Luck. First, determine as nearly as possible at what point the object was lost overboard. Taking the current into consideration, heave a weighted line overboard in the same position and with a 30- to 50-foot search line tied to the weight, go to work. A wrist compass (if you have enough visibility to see it) will help to determine when each circular sweep has been completed. Pay out 4 or 5 feet of your search line, and make successive sweeps in opposite directions. On a muddy bottom use flippers and try to swim all over the area of search without disturbing the bottom. Investigate all holes and craters if the lost object is very heavy. In this situation a probe may be necessary.

INSTALLING ANODES

The inhibition or control of environmental corrosion to offshore structures and pipelines by cathodic protection systems is a highly specialized offshore service which frequently requires the use of divers. Most modern pipelines are laid with zinc bracelets fabricated directly onto the pipe at carefully engineered intervals along the pipeline. These are designed for maintenance-free corrosion protection of the line for periods of up to 40 years.

A number of older pipelines are protected by anode beds laid some distance away from the pipeline, connected to each other and to the pipeline with insulated copper cables. The diver's work on anode beds is the inspection and occasional replacement of anodes. The only tool normally required for this is an adjustable or crescent wrench to remove the nuts from the studs embedded in the anodes. Occasionally, cable ends may have to be trimmed with cable cutters and new end lugs attached with a "Nico-Press" or similar type of crimper.

The addition or replacement of anodes on structures involves considerably more diving work. To begin with, the corrosion engineer uses a diver to help him with numerous tests required for the design of an adequate corrosion-protection system. The engineer may require temperature readings at various depths at the structure. He may also want water samples from various depths to determine salinity, alkalinity and other chemical characteristics. The easiest way to collect these water samples is by using a number of baby food jars with the covers numbered from one to the number of samples required. Distribute the covers by successive numbers in various pockets of your coveralls and carry the jars in a cloth or net sack. Starting at the bottom, hold one of the jars inverted over the exhaust of your mask or helmet or over the end of your pneumo hose. When the jar is full of air, tip it right side up, screw cover No. 1 on tightly and place it in the bag. Repeat this operation at each specified depth, using successively numbered caps.

Sometimes a diver is required to hold or place the sensing end of a variety of wire-connected instruments at various locations on the structure while the engineer takes his readings at the surface. These will range from instruments to measure electrical currents, resistivity of the water, etc., to taking sonic measurements of the thickness of the metal at various locations on the structure. Diver's tools most frequently required for this type of work are a scraper and a wire brush to remove marine growth and scale at the locations of these measurements.

Fig. 45. A worn-out anode, badly in need of replacement. Note the oysters and other shellfish on the structure bracing, a good reason for wearing coveralls. Photo: Aqua-Salvors, Inc.

The bull work begins after the cathodic protection system has been designed. The number of anodes on a structure depends on the structure size, the corrosive and electrical properties of the water in which it stands, and the design life of the protection system; it can range from two or three to fifty or sixty anodes on one structure. The size of the anodes, from a few pounds to several tons, is dependent upon the same considerations, but anodes will generally weigh between 100 and 500 pounds. Anodes are usually made of zinc, magnesium, or aluminum. These anodes, commonly cylindrical or rectangular bars, are attached to structure bracing at specified locations with bolted clamps. It is important that the area of attachment on the brace be carefully cleaned of all growth, scale, and rust. After the clamp bolts are secured with an impact wrench, the electrical connection between the anode and the structure is further insured by fillet welding, between the clamps and the brace. The required length of bead is carefully specified in the design drawings, but it is generally about four inches per clamp.

Because this type of work is usually done from a supply boat or a small barge, the plan of operation should always be based upon establishing air tuggers or other lifting gear to handle the anodes directly on the structure. This eliminates all extraneous motion, and lowering and pulling the anodes into place is a straightforward rigging job, requiring air tuggers and snatch blocks.

SETTING A STRUCTURE

The design and construction materials for the many types of permanent offshore structures have changed through the years as these platforms have moved further offshore into increasingly deeper water. In the early days and close inshore, these platforms were erected on timber pilings. In many parts of the world cylindrical prestressed concrete pilings are still used for platform foundations. For a time, because of the exceptionally corrosive qualities of the waters of Lake Maracaibo in Venezuela, oil well and production platforms were erected on aluminum jackets. The majority of the structures being installed today in the offshore waters of the world are of the tubular-steel jacket-leg template design. These structures are fabricated ashore and then towed to their offshore location aboard huge deck barges.

A jacket template consists of multiple legs—four, eight, twelve or more—depending on the depth of water and the type of service of the platform. The legs are of large-diameter pipe, from 24 to 40 inches or more, increasing in diameter and wall thickness with the depth of water. They are connected and reinforced by a network of horizontal and diagonal braces made up of smaller-diameter pipe. A jacket for service in 300 feet of water can weigh as much as 4000 tons. The legs, except for structures in very shallow water, are battered; that is, angled so that the area of the base is much greater than that of the top. The bottom of each jacket leg is sealed with a heavy rubber diaphragm and the tops are also temporarily sealed. This results in the entire jacket being hollow and also watertight. From 10 to 20 feet up from the bottoms of the jacket legs, at the first row of horizontal bracing, timber planks, called mud boards, are fastened diagonally across each corner of the jacket to form a supporting mat. These mud boards are designed to come in contact with the ocean bottom and support the jacket when it is initially set on the bottom. The legs on one side of the jacket will be fitted with massive timbers, bolted along their entire length to form skids. The skids are liberally greased and when the barge is on location, hydraulic rams are used to jack up one end of the jacket, launching it off of the barge. If the engineers' calculations are correct, the jacket will float. It is then maneuvered under the boom of a mammoth derrick, and the lifting slings are hooked up. As the derrick exerts a lift on the top end of the jacket, the bottom will sink until the jacket is suspended vertically. The diver will have previously made a bottom inspection at the platform location, to be sure there are no pipelines or other obstructions there. The derrick barge will then winch itself into exact position.

Running down the outside of each jacket leg, but on the inside face of the structure, will be a 2- to 4-inch pipe, called the grout line. Each pipe terminates in a valve somewhere in the vicinity of the first or bottommost level of horizontal bracing. These valves are generally connected to the surface with long control rods but, frequently, because of damage or mechanical problems with the rods, the diver is called upon to open and close these valves. Initially, the valves may have to be opened to flood the jacket to give it negative buoyancy so that it can be set firmly on the bottom. Approach all valves under water with extreme caution and whenever possible place yourself behind the valve before attempting to operate it. Needless to say, the greater the depth, the greater the head of pressure, and the consequent greater suction and danger of an underwater valve. A 3-inch valve at 200 feet would instantly suck the flesh from your hand or arm down to the bone.

If the structure is not going to be set over an existing conductor pile or pipeline, and if the valves on the grout pipes operate properly from the surface, then you will not have to make a dive during the actual setting of the jacket. When the jacket is in place with the valves open and the jacket flooded and resting solidly on the bottom, you must go down to make an inspection of the mud boards. The engineers will be concerned with how they are resting on the bottom. Be extremely cautious when approaching each jacket leg in case the jacket is not entirely flooded. Occasionally, because of an uneven bottom, one or more of the mud boards are not resting solidly on the bottom. In this event the project engineer may require that the space between the mud boards and the bottom be filled with sandbags. This is just plain bull work. You can greatly reduce the amount of labor required if you take care to dump each cargo net of bags as close to the jacket leg as possible. Build up a solid wall of bags along at least two sides of the mud boards before venturing in underneath them. Always be sure to pass over the top of the lowest horizontal braces before passing to the inside of the structure, to keep your hose from being trapped if the jacket settles.

If the structure is to be placed on top of an existing pipeline, timber saddles will be built on the underside of the two opposing bottom horizontal braces as a guide. In this case, two divers, one on either side, are required to line up the structure. At the diver's direction, the structure is turned or twisted with air tuggers on the surface. Perfect telephone communications are essential for this operation, and you must be careful to keep yourself and your hose out from under the structure.

If the jacket is to be set over an existing conductor pile or well-head, the pile will probably be marked by a buoy. The buoy is unhooked

Fig. 46. (Top) Massive drilling platform "jacket" being towed to its location in the Gulf of Mexico. (Bottom) Drilling platform "jacket" being set by a huge derrick barge. Photos: J. Ray McDermott & Co., Inc.

and the diver feeds a winch or a tugger cable down through the appropriate bell guides in the jacket. He attaches the cable to the top of the conductor pile, which might be 10, 12, or as much as 50 feet off of the bottom. This cable is set taut and used as a guide cable. The diver must ride the bottommost bell guide and stab the conductor.

Fig. 47. Drilling platform upper deck being set in place on top of "jacket" by two huge derrick barges. Photo: Brown & Root, Inc.

Remember to give your directions for the movement of the structure in precise measurements. Remember also that it takes a few minutes for all motion to stop after heaving ahead or back or from side to side on the anchor winches. Be sure that all motion has stopped and that the conductor is well centered under the bell guide before you give the order to lower away. When the jacket is set firmly on the bottom, tubular steel piles are inserted into each jacket leg and driven firmly into the bottom.

On very large structures in deep water, frequently there are jacket legs that do not come to the surface. In this situation, a diver must stab the piles, called skirt piles, that fit into these jackets. The pile-driving operation can take from days to weeks, and the diver normally has nothing to do during this phase; sometimes he is sent ashore until the grouting operation.

When all the piles have been driven to grade or refusal, they are solidly welded to and around the tops of the jacket legs. Hoses are then hooked up to the top of the grout pipes and cement grout is pumped under high pressure to fill the annulus ring between the pile and the jacket leg. Engineers will have calculated how much grout and how much time will be required to fill the annulus on each pile. The diver may have to watch at each valve at the bottom of the grout pipes to report when grout is emerging and to close the valve. He may also be required to take a sample of grout for surface inspection and testing by the engineer. It is important that this sample be taken with a covered can or jar and that the receptacle be completely filled with grout to eliminate dilution by sea water.

MECHANICAL REPAIR SLEEVES AND CLAMPS

The sea beds of the world are crisscrossed and interlaced with tens of thousands of miles of pipelines, from 2 inches to over 48 inches in diameter. All of these pipelines are subject to damage from a wide variety of causes, as well as pitting and holing from corrosion. When pipeline leaks and failures occur close to major ports or construction centers, repairs are usually accomplished by pipeline or derrick barges. The pipeline is cut; the two ends are brought to the surface and a repair section welded in. Then the large loop of repaired pipe is carefully laid back on the bottom. In deep water, the ends of the line are lifted to the surface, the damaged section cut out and a carefully measured repair section is welded in, with both ends of the pipeline fitted with flanges. The two flanged ends are lowered to the bottom where they are made up by divers. In extremely deep water, the entire repair job may be effected using a dry underwater welding hut (Chapter 14).

In remote areas, many types of pipeline damage can be satisfactorily repaired from a boat or small barge, using one of a number of patented repair sleeves and clamps. Several companies manufacture these sleeves and clamps. For representative illustrations, I have selected a few of the products of the Pipeline Development Company, or Plidco.

LOCATING LEAKS

The most annoying and difficult-to-find type of leak is a pinhole leak caused by internal or external corrosion, especially if the leak occurs on the underside of a pipeline under the mud. The general area of the leak can be determined by gas bubbles or oil droplets on the surface. The exact location may not be determined until after many hours of careful underwater inspection. A further complication to this type of job is when the leak occurs in weight-coated pipe. The gas or oil bubbles may be escaping through a crack in the weight coating many feet removed from the actual leak in the pipeline. On high-pressure lines, your sense of hearing is a great aid in locating leaks. Place your ear directly on the pipeline, shut off your air, hold your breath and listen carefully. The hissing or whistling sound peculiar to this type of leak will increase as you approach the leak. Of course, the weight coating must be removed before a repair can be made; marine growth and scale, if present, are removed, too. On a very old or badly corroded pipeline, be careful not to aggravate the damage by overzealous scaling.

The Plidco Smith + Clamp (Fig. 48) is an excellent device for repairing pinhole leaks in otherwise sound pipe. Spring the clamp apart and push it over the pipe. Pull the ends together and screw up on the draw-bolt nuts until the clamp fits loosely on the pipe. The draw-bolt ends are tapered to fit into the ears on the clamp so that the draw bolt will not fall out after a slight tension is applied. Rotate or slide the clamp on the pipe and center the pilot pin over the pinhole. Push the pilot pin into the hole as far as it will go and screw in the force screw until the packing cone touches the pipe. Securely tighten the draw bolt to hold the clamp firmly in place and remove the pilot pin. Tightening the force screw with a wrench will flatten the packing cone against the pipe, sealing the leak.

SPLIT SLEEVES

For large-surface areas of pitted or corroded pipe, or for long splits, Plidco split sleeves can be fabricated for any diameter pipe or specified length. Figure 49 shows a 54-inch-long split sleeve for 18-inch diameter pipe. It was installed on a split pipeline with a working pressure of 1250 psi in 100 feet of water in the Gulf of Mexico. These sleeves have patented steel girders around all edges to prevent the gaskets from being displaced when the sleeve bolts are tightened, and they are designed for a recommended working pressure of 1000 pounds. It is imperative to thoroughly clean all marine growth, scale and pipe dope from the area to be covered by the split sleeve. In the case of splits or cracks in the pipeline, it is advantageous to drill a small hole at each end of the split or crack to keep it from propagating under the sleeve. Locate each end of the crack and mark it with a heavy hammer and center-punch to start your drill accurately.

These sleeves, depending on the pipe size, can weigh in excess of 1000 pounds. The larger-size sleeves require lifting equipment and careful rigging to install. If working from a small barge or boat in

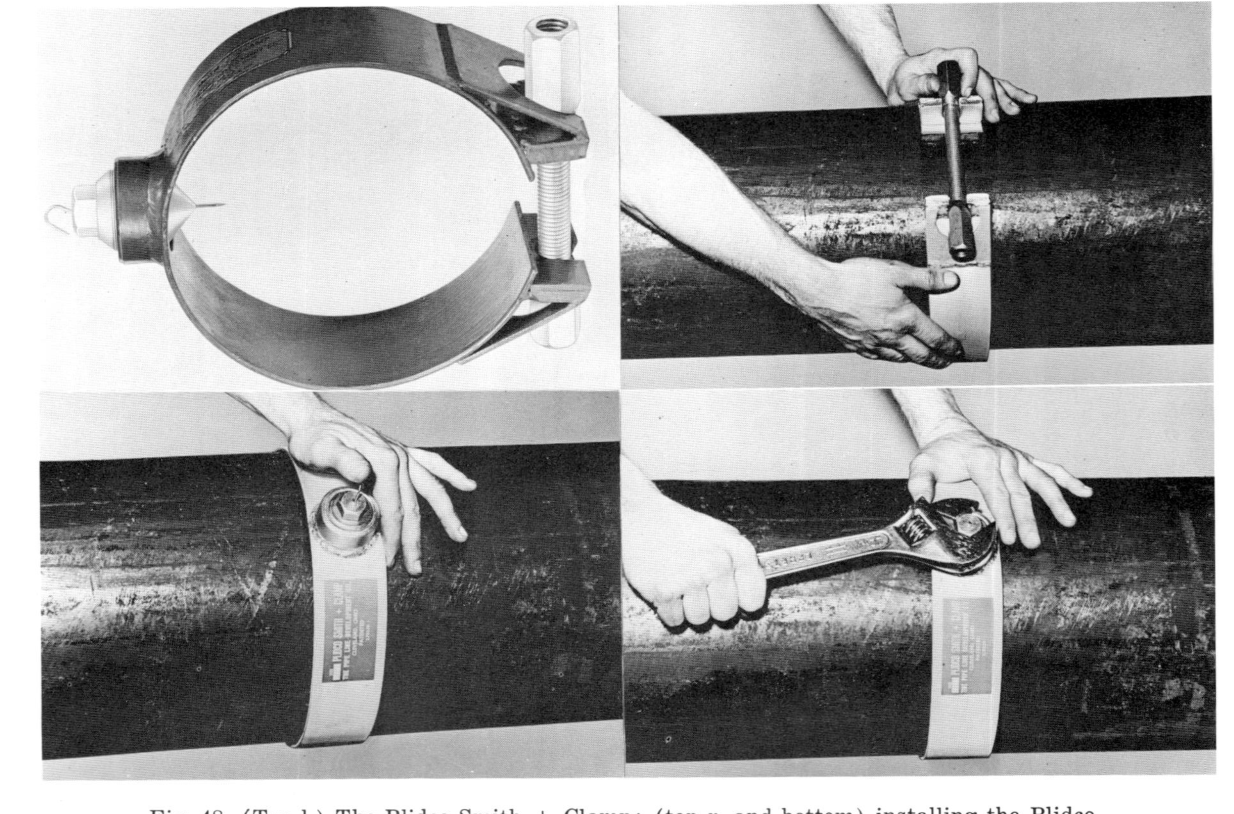

Fig. 48. (Top l.) The Plidco-Smith + Clamp; (top r. and bottom) installing the Plidco-Smith + Clamp. Photos: The Pipe Line Development Co.

unsettled seas, each half of the sleeve can be rigged to a small pontoon or buoyancy bag. Buoyancy should be calculated to leave each piece 20 to 30 pounds negative for easiest handling by the diver. After the pipe has been cleaned, assemble the two halves loosely around the pipe. Adjust the sleeve over the split or leaking area and securely tighten the bolts with an impact wrench. Be sure to tighten the bolts evenly, and, on long sleeves with many bolts, check and recheck to be sure all the bolts are tight.

WELD +® ENDS

Extensive damage to a pipeline often can be repaired without heavy floating equipment by using Plidco (Pipe Line Development Co.) weld +® ends. In the case of a bad hole or buckle, the damaged section must be stripped of concrete and cut out. The cuts must be as square as possible, so a burning guide is mandatory. After the cuts are made and the damaged section removed, the cut ends are dressed with a pneumatic grinding wheel to remove all slag and burrs, both inside and out. The two ends of pipe then must be lined up vertically and horizontally. It is often advisable to build saddles of sand and cement bags to hold the pipe ends in alignment. Make careful measurement between the cut ends. If the replacement section, or pup, is less than ten feet long, this measurement can best be made with a telescoping measuring rod made of ¾- or 1-inch pipe. The rod should be fitted with a locking nut to hold the measurement after it has been taken. It should also be fitted with a flat strip of plate on each end, welded at perfect right angles to the measuring rod. These plates should be six inches longer than the diameter of the pipe. When the pipe ends have been dressed and lined up and the measurement taken, a pup, or replacement section of pipe can be cut about 2 inches shorter than the measurement. The next step is to measure the length of the weld + ends. These are stocked in diameters from 1½ to 36 inches, and they are 16 inches long in the larger sizes. (They can be fabricated for any diameter or length desired.) If the weld + ends you are using are 16 inches long, measure off 8 inches on each cut end of pipe and mark this measurement clearly (Fig. 49). The weld + ends are then slipped over the ends of the pipe and pushed back clear of the cuts. The repair pup is lowered into place and lined up. The weld + ends are pushed back over the ends of the repair pup, and stopped at the 8-inch marks on the pipeline ends. This divides the coupling equally between the two sides of the ends. After the weld + ends are positioned, pull up the clamping and thrust screws tightly with a wrench, exerting about 100 foot-pounds of torque. These screws adjust and equalize the space between the pipe and coupling and resist end-pull of the pipeline. The thrust screws on the ends are then tightened and pushed against a forged steel thrust ring which compresses the pack-

Fig. 49. (Top, l. to r.) A Plidco split sleeve for repairing 18″ diameter pipe; measuring off the pipe for installation of a Plidco Weld + End. (Bottom, l. to r.) Tightening up the clamping screws on a Plidco Weld + End; tightening up the thrust screws on a Plidco Weld + End. Photos: The Pipe Line Development Co.

ing ring with tremendous pressure, causing packing to flow out between the pipe and the coupling. When the thrust screws are pulled up tight the joint is made and a pressure-tight seal is completed. The installation is now complete unless welding is desired. If welding is specified, the clamping and the thrust screws are cut or burned off. The coupling ends are fillet-welded to the pipe; all screws are then seal-welded. The fillet weld should be about 1¼ times the pipe-weld thickness (Fig. 50).

Plidco also manufactures flanges that fit and seal over pipeline ends in the same way as the weld + ends. These flanges are frequently used to repair tanker loading-terminal hoses where they connect to submarine delivery pipelines. They are effective only on straight pipe, and will not work well on oval or out-of-round sections.

INSPECTION DIVING

From time to time, every diver is called upon to inspect and report on the condition of some kind of underwater installation. The most important consideration during an inspection dive is to take your time, make accurate observations, and if possible, recheck them. Memory plays strange tricks, especially in the zone of nitrogen narcosis. Explain your findings carefully and clearly to topside while you are actually making your inspection, so they can be recorded. An invaluable device for inspection diving is the tape recorder, hooked into the diver's communications system. Orienting directions and dimensions are always of extreme importance, and a wrist compass and some type of measuring device are indispensable. Measurements can be taken with a small line knotted every 6 inches, or with a stick notched every inch. Where there is visibility, a steel tape will do. As soon after the inspection as possible, while memory is still fresh, translate your findings into a comprehensive written report, complete with sketches and drawings.

LIVE-BOATING

A number of diving companies specialize in inspection work. Their principal clients are the oil companies and gas transmission companies actually buying the pipeline or installation.

Inspection divers are hired to follow a pipe-lay barge or a jet barge and check every foot of laid pipe to be sure it meets specifications. Pipeline inspection divers generally work from their own boat, either a crew boat or a small supply boat, and the majority of their work is done "live-boating."

During this operation, the boat is not moored, but underway, following closely behind the diver, keeping station on the diver's surfacing air bubbles as he walks along the bottom and checks the pipeline. Live-boating can be extremely dangerous and there are a few cardinal rules governing the practice that must never be broken for any reason, with the single exception of lifesaving. The greatest potential hazard is the fouling of the diver's hose in the boat's turning propellers. Therefore, live-boating should never be done without a thoroughly experienced boat skipper and a seasoned tender. Because the skipper relies solely on the diver's surfacing bubbles to maneuver his boat and keep safely on station, live-boating must never be done at night, or in conditions of poor visibility, or when surface conditions are such that the diver's bubbles are obscured by white caps or sea-foam. I know of no boats used for live-boating equipped with baskets or guards around the propellers, and this is unfortunate, because these guards would greatly reduce the danger of a fouled hose. Because of this ever-present danger, I firmly believe that live-boating should never be done unless the diver is wearing a bailout bottle and an emergency flotation device, regardless of the depth. A boat equipped with a flying bridge would also increase the efficiency of the operation by giving the boat operator a much clearer view of the diver's bubbles.

Whenever possible, dives made from a live-boat should be planned so that no decompression stops in the water will be required. This is because maintaining the diver at accurate depths for proper decompression is extremely difficult from a drifting or unmoored vessel. If water stops are anticipated, a heavy hang-off line (at least 100 pounds) should be rigged prior to the dive. A suitable diving ladder is imperative and because a small boat is likely to roll considerably, the ladder should be long enough so that the diver can board it with ease 5 or 6 feet under water. All live-boat line tending must be done from the bow, with the hose coiled there and the telephone and diving manifold located there. The compressor is likely to be set on the stern, out of the tender's sight and hearing. Since the small internal combustion engines used on compressors are more susceptible to malfunction on a small rolling boat because of the constant stirring of sediment in the fuel, a gauge indicating the air-supply pressure should be mounted on the bow within easy visibility of the tender. Both the diver and tender must become accustomed to working with an absolute minimum of hose slack. In a current or in deep water, the closer over the diver that the boat can operate, the less drag there will be on the diving hose. The boat operator must be constantly alert to avoid positioning his boat directly downcurrent from the diver making it possible for a bight in the diver's hose to be carried to the screws.

When the diver is ready to come aboard, the engines should be shut off.

A free descent is very difficult when live-boating, because of the problem of orientation when the diver reaches the bottom. For thi reason, the diver usually makes his descent down a riser, a jacket leg or a buoy line. The first move is for the captain to position his boat as accurately as possible with regard to the wind, the current and the expected direction of the diver's travel. Then he maneuvers his boat as close as possible to the diver's means of descent. Perhaps the most common mistake made by divers working from a live-boat is jumping overboard before the boat is close enough to the descending line buoy. Have the boat skipper put the boat as close as possible before jumping overboard, and save yourself a long, exhausting swim, or the possibility of being swept away by the current, only to have to climb back aboard and start all over again. Before the diver jumps, he must also be sure that the tender has fed enough slack overboard so that he can reach the water safely. The tender must be careful not to have a foot or leg in a bight of hose when the diver jumps. It can lead to serious injury.

The foredeck of a small boat is quite unstable and precarious even in light seas, and both the diver and tender must guard against being accidentally pitched overboard. A live-boat must have some kind of handrail at the bow for safe operation.

As soon as the diver has a hold on the buoy line, the captain must back the boat away to a point where he has a clear view of the diver's bubbles. As the diver descends, he must take careful note of the set of the current and lead of his hose so that he can move out in the proper direction when he reaches the bottom. As noted in Chapter 9, observing the position of the flaps on the field-joint tins is an excellent way to determine direction; they will invariably point the same way along the entire length of the pipeline. The field-joint tins are spaced approximately every 40 feet along the pipeline, and also provide a means of fairly accurate measurement of distance traveled. Count them as you move along the pipeline and if you should encounter damage, you can establish the location with a measurement from a fixed point in addition to marking it with a buoy; for example: "The pipeline has extensive damage to the weight coating, 63 joints from the riser."

WALKING ALONG THE PIPELINE

The method of locomotion along the pipeline is a matter of diver preference. Some divers, though strong swimmers, prefer to use flippers and divide their energy output between steady stroking with the flippers and pulling themselves forward along the pipeline with their hands. Where there is no current and scant hose drag, this is an effective method for covering long distances. Of course, the diver must know how to pace himself, expending energy in a smooth, steady flow. The flippers are also an advantage in gaining the buoy line when the diver first enters the water.

Fig. 50. (Top) An inspection diver jumping from the bow of a crew boat to make his descent down a jacket leg. Note the handrails on the bow and the recompression chamber on the stern. Photo: Author. (Bottom) A diver welding up a Plidco Weld + End. Photo: The Pipe Line Development Co.

In deep water and with strong currents, the most feasible method is for the diver to be heavily weighted and planted firmly on the bottom. With the hose leading over one shoulder, the diver leans forward against the drag, straddles the pipe and plods along, perhaps pulling himself along the pipeline with his hands, as well. Rubber boots and work gloves are musts for this type of work, because the rough surface of the weight coat will quickly shred hands or wet-suit booties. The field-joint tins and their bands present a potential for serious cuts, because they rust quickly and their edges are razor-sharp. In poor visibility the rate of progress over the pipeline will be very slow as the diver must actually feel along every foot of pipeline in his search for damage. Some minor damage to the weight coating is tolerated by virtually every pipeline owner. However, the determination of acceptable limits of weight-coat damage must be made by the chief inspector. Your job is to report any and all damage you observe in the weight coating with the exception of minor cracks and spalls. Concrete broken away from the pipeline in chunks is an indication of possible serious damage; inspect such areas very carefully. You must determine if the dope coating is damaged, exposing bare metal, or if the pipe itself is nicked, creased, dented, or out of round. If the dope coating is damaged, any exposed metal will have to be coated with epoxy. If the pipe itself is damaged, the weight coating and dope coating surrounding the damaged section must be removed to allow for a thorough inspection. Molded impressions of the pipeline damage may be required. The easiest way to get an accurate impression of narrow gouges and shallow nicks is to lay a sheet of ¼-inch-thick soft lead over the damage and beat it in with a ball-peen hammer. For larger indentations, a mold can be made with stiff modeling clay, although this mold will be very delicate and lose its shape if not handled with extra care. Plaster of Paris can be used also. It is mixed with a minimum of water and sent down to the diver in a plastic bag. A useful material for making mold impressions is Splash Zone, or other epoxy. It is well to smear the damaged area lightly with Vaseline or grease before applying the Splash Zone to prevent it from adhering to the pipe. After this has set for about two hours it can be removed, giving a perfect impression of the damage.

Photography is an excellent tool of the inspection diver, but on pipelines on or under the sea bottom, conditions of light and visibility generally preclude its extensive use. Underwater photography is a speciality in itself and will not be treated in this book.

Underwater television and video tape, where visibility permits, are excellent tools for the inspection diver. Little expertise is required by the diver to operate an underwater television camera. Matters of focusing and angle are controlled by directions from topside. Practically all that the diver must remember is to position himself comfortably so that the camera can be held steady. All movements of the camera must be made as slowly and steadily as possible.

PLACING BUOYS

Any damage or questionable areas on the pipeline must be buoyed. When your time is up after walking along a pipeline, place a buoy where you stopped so that the next diver can begin his walk at that point. Placing buoys is a frequent chore of inspection divers and not as simple as it may at first appear. In water 100 feet or so deep, the easiest method of getting the buoy line to the diver is to put the bow of the boat directly over the diver, take a heavy strain on the hose and slide the buoy line down the diver's hose on a shackle. The diver must hold firmly to the pipeline as the slack is taken out of his hose or he will be plucked off the bottom. A simple way for him to brace himself against the vertical pull on the hose during this operation is to pass a bight of the hose under the pipeline and hold onto it on the opposite side. Thus a substantial strain can be taken on the hose without pulling on the diver. When the shackle reaches the diver, the crew topside secures the buoy to the upper end of the line and throws it overboard. The diver unties the buoy line from the shackle and passes it under the pipe, pulling out all of the slack before tying it off. A round turn should be taken around the pipeline before tying off the buoy line. The diver must be careful not to pass a turn of buoy line over his hose, thereby securing himself to the pipeline. If the shackle should hang up while it is being slid down the hose, the tender must slack off the hose to allow the diver to pull in the slack until the shackle is reached. When using the hose for a messenger line like this, it is important to unhook the shackle as swiftly as possible, so that the hose slack can be recovered immediately.

When the buoy line is being sent down the hose, the line tender must stand as far away from the hose tender as possible to prevent the buoy line from taking turns around the hose. After the buoy line is secured, the diver must be sure that he is not fouled on it before he moves out. One disadvantage to this method of sending down a buoy line is that if more than one buoy is to be set, and the walk along the pipeline is to continue, the fairly expensive shackle will have to be discarded. In order to slide easily over the hose, the shackle should be large and consequently, too heavy for the diver to carry along with him on his walk. Another way of sending buoy lines to the diver in deep water, or if more than one buoy must be set, is to have the diver tie the end of a small line to his belt or 5 or 6 feet above him on his hose. This line must be tended the same as the hose; slacked off and taken up as the diver walks along. The diver must be careful not to make any complete turns, thus wrapping this line around his hose. When he is ready for it, the buoy line is tied into this messenger line and the diver pulls it down until he reaches the end of the buoy line. As soon as the buoy line is unfastened, the slack in the messenger line is pulled aboard.

One consistent problem encountered in walking pipelines, especially in deep water, is the amount of time required for the diver to dig under the pipeline in order to secure the buoy line. In sand or clay and with large-diameter pipelines, this could require 5 minutes or more. To reduce the amount of time required for this operation I use a very simple device. I make up a number of rods from 4 to 8 feet long, depending on the diameter of the pipe. These rods are made from either ½-inch black iron pipe or ⅜-inch malleable rod, with a small closed eye on each end. While walking the pipe, I use the rod as a combination probe and cane. When my time is almost up, I bend the rod into a "U," then push this "U" down, over the top of the pipe and rotate it until both eyes are looking upward. The buoy line can then be tied into one of the eyes, or if more security is desired, between both eyes across the top of the pipe. The next diver unties the line, and rotates the "U" rod until he can lift it clear of the pipe. Then he straightens out the rod and continues on his walk until it is time for him to set his buoy, whereupon he repeats the operation. Using this method, a buoy line can be passed around a 24-inch-diameter pipeline in a minute or two, where digging a hole might require 10 minutes or more.

INSPECTING CROSSOVERS

Pipeline inspection divers are frequently called upon to check pipeline crossovers. Where pipelines are laid over each other, the contract specifications will clearly indicate how much spacing or vertical distance there must be between the crossing pipelines. On large lines this spacing may be as much as two or three feet, and maintained by sacks of sand and cement stacked between the pipelines. On smaller lines the distance may be only a few inches, with spacing and insulation provided by rubber sleeves banded to the pipelines at the point of crossing. Often, old truck tires are used for this purpose. It is the inspector diver's job to check that the crossings are installed according to specifications. The ultimate responsibility for accepting or rejecting a job not in strict accordance with the specifications belongs to the chief inspector. Your job is to report conditions as they exist, and not to make judgments of acceptability. *Report conditions as you find them.*

INSPECTING BEHIND THE JETTING OPERATION

When inspecting behind a jet barge, the diver is primarily checking the amount of "cover" over the pipeline or the depth of the ditch. The required cover over the pipeline can be from 3 to 10 feet, depending upon its location. Soon after a jet barge has made a pass over a pipeline, the sides of the ditch begin to cave in, covering the pipeline. As a result, detecting damage that may have occurred during the jetting operation is frequently impossible. The diver must walk along the ditch, keeping track of the pipe and measuring the depth of cover with a probe. In deep ditches he must be constantly on the alert against being caught in a cave-in.

INSPECTING RISERS

Inspecting risers is a much more straightforward operation. The diver will be checking to see that there is no damage to the riser or to the weight coating, that the clamps are installed at the proper elevations and that all nuts and bolts are tight. Starting at the bottom, the diver ascends from clamp to clamp, inspecting the riser either visually or by feel, depending upon the conditions of visibility. At each clamp, pneumo readings are taken to check the elevation. Riser clamps that are secured to structure braces must be carefully checked to be sure that the clamp is perpendicular to the riser and that the riser is bearing evenly upon the entire inner surface of the clamp. Almost all clamps are provided with rubber insulation where they bear on the riser and on the structure members. You must check to be sure that this rubber has not been displaced. The ultimate objective of all inspection work is to be sure that the client is getting what he has paid for, a workmanlike job installed to specifications. There are times when the inspection diver can correct minor discrepancies, thereby hastening the completion of the job and ultimately saving money for the client. This opportunity occurs most frequently while you are inspecting riser clamps. As you can readily admit, generally it is more difficult to do a job than it is to check the job to see if it has been done properly. When inspecting riser clamps, carry a wrench to see that all the nuts and bolts are properly tightened. If you occasionally encounter a loose nut, don't set up a hue and cry about sloppy work; take the minute or two required to tighten the nut yourself. This type of simple magnanimity will redound to your credit in speedy acceptance of your client's installation. This is not to suggest that it is the inspector diver's job to finish uncompleted work. If the character of the job is sloppy and unworkmanlike, it should be rejected. But if the difference between accepting and rejecting a job lies in the simple act of tightening a nut or two, or rearranging a couple of sand bags, then by all means take the trouble to do it yourself.

Remember, take your time; be certain of your observations. Report conditions as you find them. Report your findings over the telephone. Draft a written report of your findings, together with sketches and drawings. The three essential characteristics of a good inspection diver are experience, judgment and integrity.

Chapter 12

DIVING FROM A DRILLING RIG

The enormous financial and technological complex known to us today as the Offshore Oil Industry began in the early 1930's, when wildcatters and exploration crews began drilling from barges in the shallow marshes of southern Louisiana, adjacent to dry land producing acreage. A drilling mast or derrick was erected over the stern or side of a conventional deck barge, and the barge was pumped full of water until it rested firmly on the mud of the marsh bottom in water depths of six to eight feet. Drilling was then conducted with typical dry-land techniques. These rusty makeshift barges proved to be the genesis of our modern, huge submersible drilling rigs that sit on the ocean bottom in water as deep as 100 feet.

When the existence of oil reserves under the waters of the marshes and bayous was proven, stationary platforms were erected on timber pilings, and these rickety structures slowly leapfrogged out of the marshes into the ever deepening offshore waters of the Gulf of Mexico in search of more oil. Since that time, exploration drilling and production of offshore oil and gas has become a worldwide multi-billion-dollar industry.

The problems of open-sea deep-water drilling seemed insolvable for many years until, in 1948, four oil companies (Continental, Union, Shell and Superior) pooled sufficient finances to form the Cuss Group with the express purpose of solving these problems. In 1953, this forward-looking, financially courageous group, operating from a small converted Navy patrol ship, the *Submarex,* began experimental drilling operations in deep water off the coast of California. The results of these operations were far from conclusive, but the wall of technical and economic possibility had been breached, and all subsequent advances in the art-science of deep-water drilling date from this pioneering attempt.

Today, offshore drilling is conducted from stationary platforms or from mobile rigs of four basic configurations. The fixed stationary platform, erected on the spot on pilings or a jacket template, is perhaps the most efficient structure for offshore drilling operations, but a fixed platform is very expensive, especially in deep water, and the cost is usually not justified, except in areas of proven reserves. As of this writing, the deepest water in which a permanent platform has been erected is 340 feet.

Two of the four types of mobile or movable drilling rigs are termed bottom-supported, and the other two are known as floaters. These terms apply only during actual drilling operations, they all float when moving from one location to another.

Fig. 51. A service boat stands ready near the three-legged jack-up rig *Intrepid.* Photo by Jim Thomas; Zapata Norness, Inc.

BOTTOM-SUPPORTED RIGS

Bottom-supported rigs, depending upon their design, have a predetermined limitation as to the depth of water in which they can drill, whereas "floaters" can theoretically drill in the Mindanao Deep.

A jack-up rig is a barge with three or more vertically adjustable legs protruding through it. On location, the legs are lowered to the bottom and the barge is then jacked up above the surface of the water to a height of thirty feet or more. This results in a stable platform, unaffected by surface waves or currents. Because of the massive weight of the legs, necessary for strength, the stability of the barge is affected when the legs are in the retracted or raised position for towing.

This factor precisely limits the length of legs possible for a given barge size, and consequently, the depth of water in which a given barge can drill. Three hundred feet appears to be the maximum, practical depth of water for a jack-up rig. On some jack-up rigs, the legs are spudded in individually, and the barge or platform is then leveled or adjusted to compensate for an uneven bottom. On other jack-ups, the legs are connected to a mat under the barge, and this entire mat is lowered to the bottom, resulting in a larger area of contact with the bottom and greater stability for the drilling rig.

Large submersible drilling rigs are built on a series of large-diameter caissons, connecting the platform deck to a mat or to huge horizontal cylinders which form the underwater hull of the vessel. The length of the caissons determines the depth of water in which the rig can drill, and the maximum depth is about 100 feet. When this type of rig is ballasted and setting on the ocean floor, it provides a very stable platform, with the cylindrical caissons offering minimum resistance to surface waves and currents.

A diver has little to do with the actual drilling operation aboard stationary platforms or bottom-supported rigs. When these rigs are operating, they are firmly fixed in place, with no relative movement between the rig and the well hole, and the drilling procedure is similar to that when drilling on dry land, with the wellhead and the blowout preventers at the surface.

Sometimes divers are called to inspect the bottom for junk or obstructions prior to ballasting a submersible rig or lowering the legs or mat on a jack-up rig. On an uneven bottom, they might be used to sandbag voids under a jack-up or submersible mat. There are always inspections to be made, zincs to be attached, and lost tools and other objects to be recovered for these types of rigs.

Fig. 52. (Left) A fixed, stationary oil well platform drilling in the Gulf of Mexico; (right) the *Louisiana*, a semi-submersible rig, drills offshore West Africa.
Photos: Zapata Norness, Inc.

FLOATERS

Floaters, drill ships and semisubmersible rigs provide a great deal of diving work. A drill ship is exactly that, a ship of conventional or catamaran hull, usually selfpropelled, with a derrick and drilling equipment. On the *Submarex* the mast or derrick was positioned over the side, but modern drill ships have the derrick amidships where it

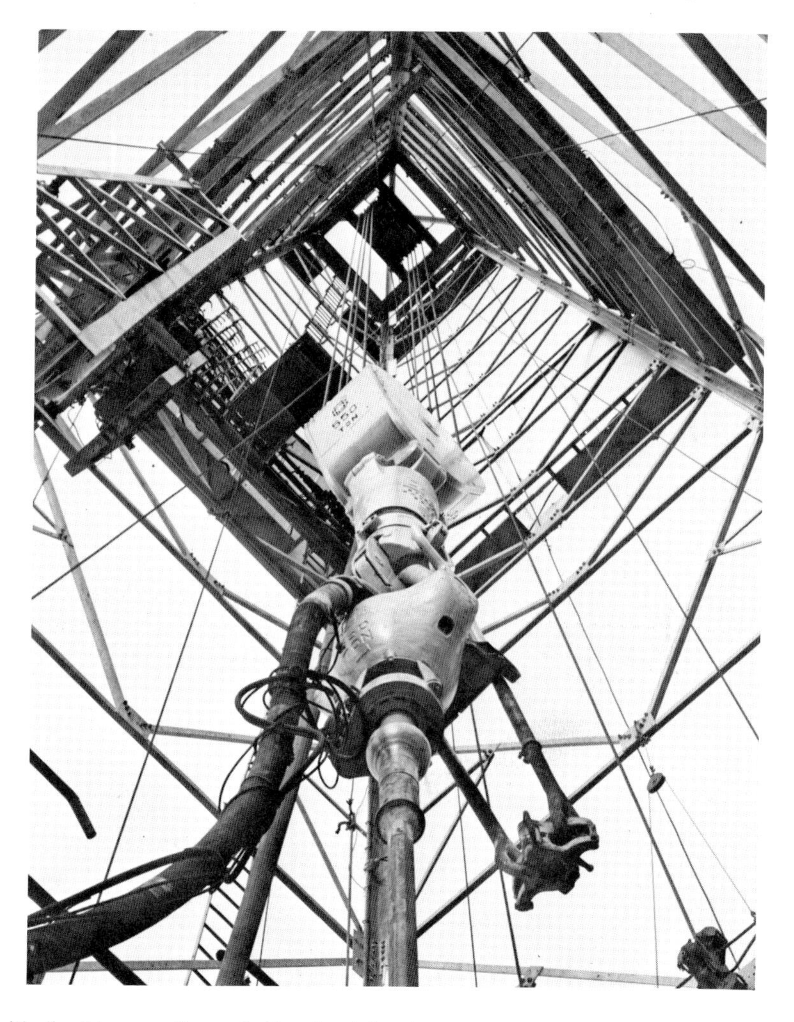

Fig. 53. Looking up through the derrick of a drilling rig. Going from bottom to top, we first see the flat-sided drill pipe, or "Kelly." Above it is a 550-ton Gardner-Denver swivel. Connected to the swivel is the hose through which the drilling "mud" is pumped. Above the swivel is the traveling block. Atop the derrick is the crown block. Photo: The Gardner-Denver Company.

affords greater stability, and all operations are conducted through a large hole in the center of the ship, called, for some unknown reason, the moon pool.

A semisubmersible rig, like a submersible rig, is also built on huge caissons. On location, it is ballasted to its drilling draft, sometimes as much as 90 feet, and, like an iceberg, it gains its stability from the huge mass of it under water.

Drill ships and semisubmersible rigs rise and fall with the waves during the drilling operation, and they rely upon efficient anchoring systems to maintain their lateral position over the well.

SUBSEA DRILLING SYSTEMS

Floaters always use what is called the subsea system when drilling; that is, the conductor, casing strings, wellhead, and blowout preventers are firmly positioned on the ocean floor.

Subsea drilling systems were designed so that the entire operation, from spudding in to producing oil, could be accomplished without the use of divers. It works fine on paper, but with drilling rig costs of $25,000 a day and more, most drilling rig companies carry divers and, sometimes, deep-diving systems aboard, especially when working in remote areas. This is a form of low-cost insurance for the times when the blueprints fail to work out.

Drilling is a precise, highly technical trade with hundreds, perhaps thousands, of tools and hardware components and employing dozens of varying techniques. Drilling is not within the intended scope of this book, but an elementary discussion of the process will help you to better understand the possible duties of a diver when a well is drilled by the subsea technique.

The one universally recognized piece of oil-drilling apparatus is the mast, or latticework derrick, and it is as essential offshore as it is in Oklahoma. A derrick is usually tall enough to accommodate three joints of made-up drill pipe above the drilling floor. Since a standard length of drill pipe is 30 feet, most derricks are 90+ feet tall. At the top of the derrick is the crown block and through its sheaves are rove the many parts of cable that connect the draw works, or high-speed winch, to the traveling or lifting block. This traveling block is used to raise and lower the drill pipe in the well hole during the drilling operation, and it is also the lifting medium for handling the heavy wellhead and blowout preventer and all other hardware required in a subsea system.

During drilling operations, a swivel connector is attached directly to the traveling block. Depending from this swivel is a length of flat-sided drill pipe, called a "Kelly," and below this is attached the remainder of the drill pipe, or the drill "string." At the bottom of the drill string is the drill bit, the tool that does the actual cutting. Placed in the exact center at the base of the derrick, on the drill floor, is the rotary "table." This table, powered by the draw works, has a flat-sided chuck to fit the Kelly and thereby transmit a rotary motion to the drill string, while allowing the drill string to move freely up and down through the table in a vertical plane. The drill string is suspended by the traveling block and it is carefully lowered as the drill penetrates the earth's crust, making a hole. As each successive 30 feet of hole is drilled, the string is broken below the Kelly, and a new length of drill pipe is added. When the drill bit becomes dull, the entire string must be withdrawn from the hole. This is usually done three joints at a time, with 90-foot lengths of drill pipe standing upright inside the derrick. When the bit has been replaced, these 90-foot lengths of drill pipe are screwed together and lowered back into the hole, and drilling is resumed.

The simplest topside drilling operation requires lubrication and the frequent removal of chips or the material being drilled; both of these requirements are critical when drilling through the earth's crust. These functions are accomplished by constantly pumping and circulating drilling "mud" through the swivel, down through the hollow drill pipe and out of holes at the drill bit. This mud, a carefully controlled blend of special clays, chemicals and water, lubricates the bit and washes the cuttings back to the surface on the outside of the drill pipe. It also cements and stabilizes the inside surfaces of the hole. At the surface, the cuttings are screened out and the mud is recirculated. By controlling the density of this mud, the drilling engineer is able to hold back internal pressures in the well if they develop.

We will continue our oversimplified description of drilling an oil well, and move into the ocean environment.

DRILLING AN OCEAN-BOTTOM WELL

After the drilling rig has been anchored, the first step in drilling an ocean-bottom well, using the subsea system, is to set the conductor. The conductor is heavy wall pipe, usually 30 inches in diameter and 100–250 feet long. Setting the conductor is perhaps one of the most critically important preliminary steps. The conductor pipe provides the first large-diameter opening or hole in the sea floor, and one of its purposes is to hold, or maintain that hole in the usually soft-surface mud or sand. The length or depth of the conductor depends entirely upon the surface geology at the well location. The conductor can be drilled, driven or jetted into place. It is usually further stabilized by cementing it in place. The cement is pumped down the drill pipe and prevented from returning up the conductor by a "pack-off shoe" at the bottom of the conductor, and it is thereby forced up around the outside of the conductor, bonding it firmly to the soil through which it passes. This is important, because the conductor has to withstand

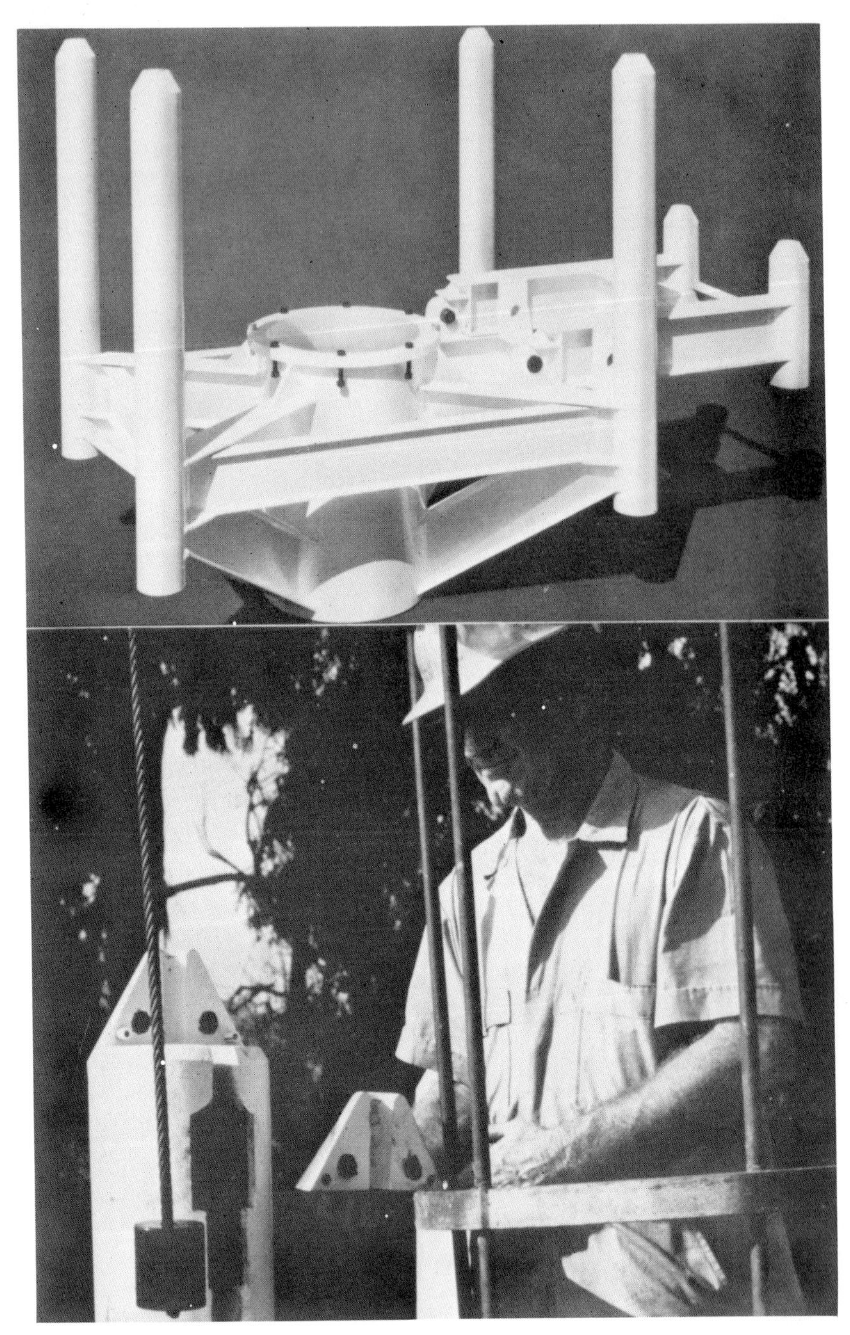

Fig. 54. (Top) A permanent guide base. Note the guideposts and the two auxiliary guideposts for the TV system. Photo: Cameron Iron Works, Inc. (Bottom) Replacing broken guidelines. The diver manually removes the top half of the post assembly as shown, thereby allowing full penetration to the post for insertion of a replacement guideline complete with spelter socket. Photo: VETCO Offshore Industries, Inc.

the potentially great lifting forces of the natural gas pockets that are apt to be encountered during the drilling operation. A stable conductor is additionally important, because it must support the great weight of the well casings to be subsequently run, the weight of the wellhead and the blowout preventer stack. It must be able to resist the vertical pull of the guide cables and the lateral forces of the drill string and marine riser, as the floating drilling rig oscillates over the well hole in the surface waves.

At the top of the conductor pipe is a conductor housing, a machined fitting designed with internal threads to accept and support the next string of well casing to be run, which is usually 20 inches in diameter.

At the top of the conductor housing is a hub. Either a hydraulic connector or a mechanical clamp attaches the blowout preventer stack to the hub. Mounted just below the hub of the conductor housing is the permanent guide base, Fig. 54. When the conductor is set in place and cemented, the conductor housing and guide base sit at or near the ocean floor and together represent the wellhead, or ocean-bottom foundation of the well. The guide base usually has four vertical guideposts extending above it and frequently two additional small posts for a TV camera guide system. If you are ever assigned to a drilling rig as a diver, ferret out the blueprints and drawings of the guide base, because you will be doing most of your work on and around it.

There is a ⅝- or ¾-inch cable attached to the top of each guidepost, and these cables run up to the drilling rig on the surface where they are spooled onto air tuggers, set to maintain the cables in constant tension. These cables are used to guide down to the wellhead the blowout preventer stack, the TV camera and various other tools and components, including the marine riser.

These cables frequently break, and replacing them is a common chore for a diver working from a drilling rig. The method of attachment of the guide cables to the guide base guideposts varies with the manufacturer, so study the blueprints. However, the usual method utilizes a spelter or zinc socket on the end of the cable. The socket is slid into a slot in the side of the guidepost, and is further secured with a split cap at the top of the post. The split cap is anchored in place with large countersunk allen screws.

After the guide cables are rigged, the blowout preventer stack is lowered through the drill floor on a string of drill pipe.

THE BLOWOUT PREVENTER STACK

The blowout preventer stack is a huge piece of machinery consisting of a number of valves and rams, Fig. 55. Its purpose is to prevent the unexpected pockets of gas encountered during the drilling operation from escaping out of the well, with the consequent danger of explosion and fire. The cemented conductor casing seals the initial well hole to the surrounding earth, and the blowout preventer stack attaches to the top of the conductor housing with a metal gasket between them, thereby producing a pressure-proof seal. After the conductor is in place, all of the drilling operations are done through the blowout

Fig. 55. (Left) A 16¾″ blowout preventer stack on top of a permanent guide base. Note how the four guideposts slide into the funnels of the B.O.P. guide structure. The hydraulic control pod is at the upper right-hand corner of the B.O.P. (Right) How the B.O.P. would look to a descending diver. The large diameter white pipe in the center is the marine riser. The two black pipes going into the funnels are the choke and kill lines. Hydraulic control pods, upper left and lower right. Photos: VETCO Offshore Industries, Inc.

preventer stack. If a gas pocket is encountered while drilling, the gas escaping up the drill pipe is controlled by increasing the density of the drilling mud being circulated in the well. The gas is prevented from escaping up around the conductor pipe because it is firmly cemented to the surface formations surrounding it. The blowout preventer rams prevent it from escaping up around the outside of the drill pipe. A blowout preventer usually has three to five hydraulically-actuated rams, and at the first indication of escaping gas, these rams are closed around the outside of the drill pipe. There is generally at least one blind ram in a preventer stack, so that the hole can be sealed in the event of a gas eructation when there is no drill pipe in the hole. A stack will also have at least one bag type preventer. This is a doughnut-shaped packing unit designed to close around irregular parts of the drill string, such as the mating joints. A number of these preventers of various shapes are stacked vertically, atop each other, giving the whole assembly the name of blowout preventer stack.

The various hydraulic controls for the rams and connectors of the blowout preventer (B.O.P.) are manifolded at a point at the top of the stack, and a detachable "pod" connects the blowout preventer stack and the hydraulic control hoses running in a bundle from the surface. The connections from the choke and kill lines are at the top of the preventer. Mud or cement can be pumped into the well below the preventer stack through these lines.

Blowout preventer stacks vary considerably in design and the diver should carefully study the prints and drawings of the particular stack being used on his job. There are several jobs that a diver will be called upon to perform on a blowout preventer stack, and further discussion of them follows.

There is a guide structure around the B.O.P., usually consisting of four posts and their connecting framework. The posts are slotted to receive the guide cables, and the cables are retained inside of the posts by hinged gates. The lower ends of the posts are funnel-shaped, and as the B.O.P. is lowered to the bottom, sliding down the guide cables, these funnels slip over the guideposts of the guide base at the wellhead on the ocean bottom. When the B.O.P. is landed on top of the conductor housing, the hydraulic connector is actuated, locking the B.O.P. firmly to the wellhead. This operation is usually monitored by television, with the camera mounted on a frame that slides up and down the TV guide cables, controlled by an air tugger. On occasions when the TV is inoperative or when visibility is too poor to rely on it, a diver is used to guide the B.O.P. into place.

After the B.O.P. is landed and locked into place, the rams are closed and the connection between the B.O.P. and the conductor housing is pressure-tested by pumping fluid through the drill pipe. If this connection leaks, the B.O.P. must be raised a few feet and the steel gasket between the B.O.P. and the conductor housing may have to be replaced by a diver. On some types of connector, this steel gasket is held in place by four setscrews.

Frequently, when the B.O.P. has to be lifted off of the wellhead (either to replace a gasket or at the termination of the job), the hydraulic connector will not release. Various types of connectors are provided with manual overrides for such a contingency, and this is usually a job for the diver.

The manual override can be a pair of cams, or a series of pad eyes to which cables are attached to provide an upward lifting force, thereby releasing the connector. The override might be a series of countersunk screws that will have to be backed off. I was on one job where even the mechanical overrides failed to operate and the connector had to be burned off with a torch under water. It was an extremely delicate and critical job, for if the hub on the conductor

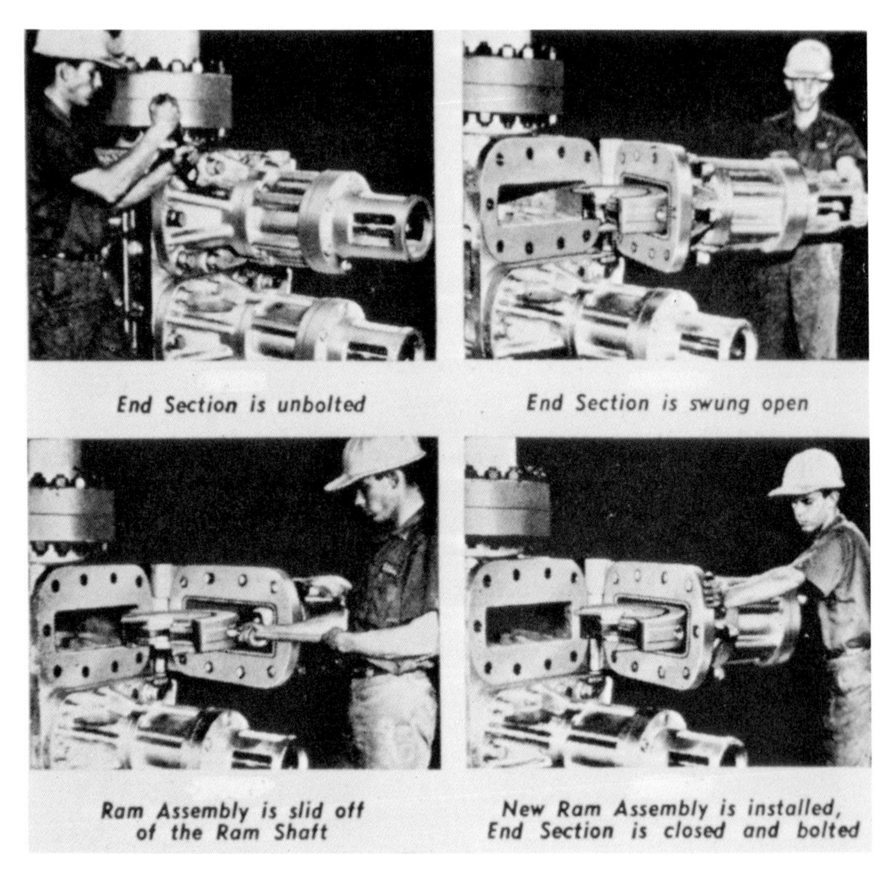

Fig. 56. Changing rams in "Type LWS" preventers. Photo: Shaffer Tool Works.

housing had been damaged by the torch, the whole well would have been ruined. Again, it is imperative for the diver to consult the manufacturer's drawings for proper tools and procedures before any work on a B.O.P. or wellhead is attempted.

Quite often the rams of a B.O.P. will have to be changed under water by a diver, either because they leak, or because of a change in drill pipe sizes. Changing rams is generally a simple nuts-and-bolts job, but the manufacturer's drawings are essential. Figure 56 is a group of photographs illustrating the method involved in changing rams in a Shaffer type "LWS" preventer. Figure 57 shows Shaffer blind rams, and rams for closing around pipe.

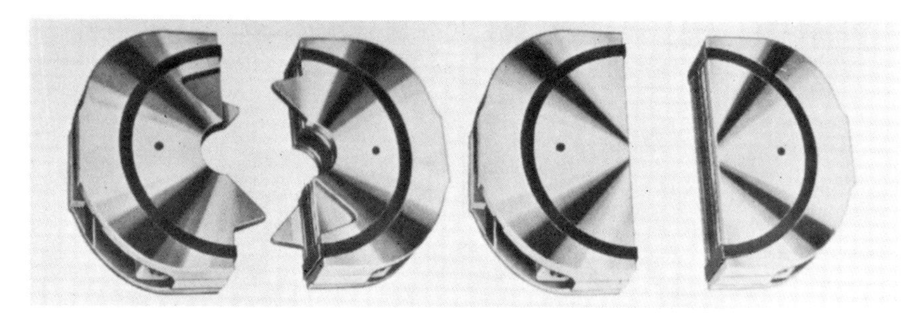

Fig. 57. (Left) Type 60 self-centering ram assembly complete with rams for closing around pipe. (Right) Type 60 ram assembly complete with rams for closing off open hole. Photo: Shaffer Tool Works.

THE MARINE RISER

After the B.O.P. is in place and tested, the marine riser is lowered and locked in position on top of it.

The marine riser is large-diameter, thin-wall pipe, and it extends from the top of the B.O.P. to the surface, just under the drilling floor. The purpose of the marine riser is to guide the casing strings and the drill string into the well, and to provide a conduit for the return of the drilling mud to the surface. The marine riser can be 18–24 inches in diameter, and modern ones have built-in buoyancy so that they are neutral or only slightly negatively buoyant.

Compensation for the rise and fall of the drilling vessel in tides and ocean waves is afforded to the rigid drill string by a telescoping splined joint of pipe, called a bumper sub., and sometimes also by a hydraulic constant-weight compensating mechanism hanging from the traveling block. This vertical motion is compensated for in the marine riser by a telescoping slip joint. The riser is hydraulically locked to the top of the B.O.P. with a connector and a flexible joint, or a ball and swivel socket. This swivel joint allows for horizontal movement of the rig on its anchors, due to wave action. The marine riser is lowered into place on the end of a string of drill pipe, and it is guided into position by a series of arms with funnel-like receptacles on their ends that lock around two opposing guide cables with pin-locked gates. If the TV is not working, or if the drilling rig is slightly off-center over the well, a diver is used to stab the marine riser onto the B.O.P. The marine riser is usually suspended from four cables attached to the underside of the drill floor.

The choke and kill lines are steel pipes which are run down—one on each side of the marine riser—through guide arms and funnels attached to the riser. These lines screw into the top of the B.O.P., and often a diver will be used to guide them through the funnels and stab them into the B.O.P. The choke and kill lines are hung off on the marine riser, just below the slip joint, and from that point they are connected to the drilling rig by hoses.

The hydraulic control hoses are stored on reels on the deck of the drilling rig. They connect to one or two control pods which are lowered to the manifold on top of the B.O.P. with an air tugger. As the pod is lowered away, the hoses are seized to the tugger cable, forming a bundle. The pod is guided into place by arms connected to the guide cables, and although they are theoretically capable of being stabbed remotely, ordinarily a diver is used to do this. There are a series of O-ring sealed projections on the underside of the pod which fit into corresponding holes on the manifold. As the pod is held in correct alignment by the guide frame, the diver has only to push it slightly this way or that, to be sure it is stabbed, and perhaps jump on it once or twice to be sure it is properly bottomed out.

When all of the foregoing has been accomplished, the well is drilled. Generally, a series of casings of diminishing diameters is run; for instance, a surface casing of 20-inch diameter pipe, an intermediate casing of 13⅜-inch pipe, another intermediate casing string of 9⅝-inch pipe and a long string of seven-inch casing pipe. The lengths of each size casing depend entirely upon the type of geological formations being drilled through and the ultimate depth of the well. The purpose of the casing strings is to keep the well hole open through the soft or unstable stratifications of soil being drilled through.

During the actual drilling operation there is nothing for a diver to do unless trouble develops. If you ever pull a job as a diver aboard a drill ship or a semisubmersible rig, there are a number of things to check out and prepare for and also to beware of.

PRECAUTIONS DURING THE DRILLING OPERATION

When rigs do not carry permanent divers aboard, a number of diving companies contract to leave a compressor, chamber and other diving equipment aboard the rig in order to insure getting the job if need for a diver develops. This saves the time and difficulty of loading and unloading this heavy equipment for each job but it also presents a few problems. Two to three months might pass between diving jobs and the equipment will have been untended and uncared for during this time. If you are sent on a job in this situation, hook up and test out all the equipment as soon as you board the rig. The compressor should be lubricated, fueled and run, and the chamber should be opened up, aired out, hatch seals checked, etc. Idle equipment deteriorates rapidly, especially at sea, and it is your responsibility to assure yourself that it is safe and workable.

On most drill ships and all semisubmersible rigs, the deck from which the diving operations are conducted will be too high above the water for a ladder and the diver will have to use a stage, probably powered by an air tugger. There should always be two stages to enable a standby diver to get into the water in case of emergency, and the tugger cables should be long enough to send the stage to the bottom. The tugger cables should be carefully inspected before use and new ones rove, if they are worn or rusted. The stage is an ideal carrier for all of the diver's tools and materials, and you should spend a little time with the rig welder and fabricate convenient lockers and racks on it. The stage is generally lowered with a messenger line shackeled to one of the guide cables.

The diving compressor will usually be remote from the diving station, especially on semisubmersible rigs. Therefore, it is imperative that a large-volume tank and pressure gauge be located at the diving station.

Before you make a dive, study the plans and prints of the system carefully, and be sure you understand perfectly what you are expected to do.

After you have dressed into your gear, be very careful when moving around the edge of the moon pool or the opening in the cellar deck. A fall from that height while in your gear would probably be fatal.

The stage should have a seat, and if the weather is rough, some type of safety belt or line tying you into the stage is recommended. When looking down from a height of 30 or more feet to the surface of the water, it is difficult to judge the height of the waves. Passing through the air-water interface, when surface wave conditions are turbulent, can be dangerous. If you are using a deep-sea rig, keep your suit inflated to the absolute minimum.

Regardless of the type of gear you are using, it is always wise to wear an extra belt that can be removed and hung on the stage after you are under water. When the stage approaches the top of the waves, hold on tight and have the stage dropped quickly 20 feet or so. If you are using a diving mask, be sure that it is firmly strapped on beforehand.

If the stage cable is not long enough to take you all the way to the bottom, be sure that you are lowered 20 or 30 feet below the surface turbulence before you leave the stage. Remember how you exited the stage so that you can return the same way and avoid a round turn of your hose on any of the guide cables. As you descend on one of the guide cables, be careful of your hands on frayed and prickly wire. It is extremely important not to spiral around a guide cable during your descent. The guide cables are apt to be heavily greased, so try to keep your diving suit away from them.

When diving from a semisubmersible rig, your tender is standing so high above the surface of the water that he can do very little to help you if you are swept off the descending cable in a strong current.

It is important to wear a very stout harness; if you are swept away, your tender will be able to haul you bodily to the surface with help from the drilling crew.

If your stage will reach the bottom and the current is quite swift there, exit the stage under one of the horizontal stage bars, so that if you are swept away, your tender can hold your slack tight and you will still be able to pull yourself back to the stage. Sometimes in deep water and a strong current, the drag of your hose will be so great that you will have to tie your hose off to the stage. If this happens, allow yourself 20 feet or so of slack—at least enough to reach the wellhead—before you tie off your hose. Whenever you have to tie off your hose, do it with a few rope yarns or light line, or tie the securing line through one of the seizing tapes of your hose bundle, so that a heavy pull from the surface will part the securing line or rip through the tape in case of an emergency.

On semisubmersible rigs, the recompression chamber will be located at least one and probably two decks above the diving station. In most cases, the time required to pull the stage up to the surface from the last water stop, undress the diver, and then rush him up several long flights of stairs to the chamber will be more than the recommended interval of five minutes between the diver's leaving the last stop in the water and reaching 40 feet in the chamber. Because of this, dives from this type of rig should be kept as short as possible, and never over the optimum schedule. If optimum or longer dives become necessary, the decompression schedule should be routinely increased to the next longer or deeper table.

You must be just as careful when leaving the water as when entering if the surface water is rough. Hold on tightly to the stage and have it lifted more rapidly as you approach the surface.

After you complete a short diving job on a drilling rig, be sure that any equipment to be left behind is properly stored and covered.

Many offshore drilling rigs carry bells and deep-diving systems such as the Divecon "Seachore," or the Reading and Bates Seashell. These will be described in Chapter 14.

When a well is completed it is either abandoned, temporarily abandoned or completed and put on stream. In the case of a dry hole, the well is plugged with cement and the wellhead removed by cutting the casings and conductor 10–20 feet below the bottom surface. A well might be temporarily abandoned awaiting the construction of a platform, at which time the casing will be extended to the surface and the well will be completed there aboard the permanent structure. When a well is temporarily abandoned, it is plugged with cement and after the B.O.P. is removed, a cap is installed on top of the wellhead. This cap can be run on the end of a string of drill pipe and locked in place hydraulically, or it can be diver-installed. A diver-installed cap is held in place by a Cameron or other type split clamp with two or four securing bolts. After the cap is installed, the guide cables are unhooked and the well is marked for future completion with a pinger, or the guide cables are gathered together in a bundle and secured to a buoy.

If the well is to be completed using a subsea completion system, the production tubing strings are run into the well and secured in a special hanger inside the seven-inch casing. The B.O.P. is removed, and a special subsea Christmas tree is lowered to the wellhead. A Christmas tree, Fig. 58, is the assembly of well operating valves.

Fig. 58. A "Christmas Tree" or submarine wellhead being lowered to the bottom. This is the valving and control unit for the oil well. Note the guide funnels at the bottom of the "Christmas Tree," which will fit over guideposts on the ocean bottom. The production flow lines, through which the oil will flow to shore, will be connected to the flanges at the lower right-hand side of the picture. Photo: Cameron Iron Works, Inc.

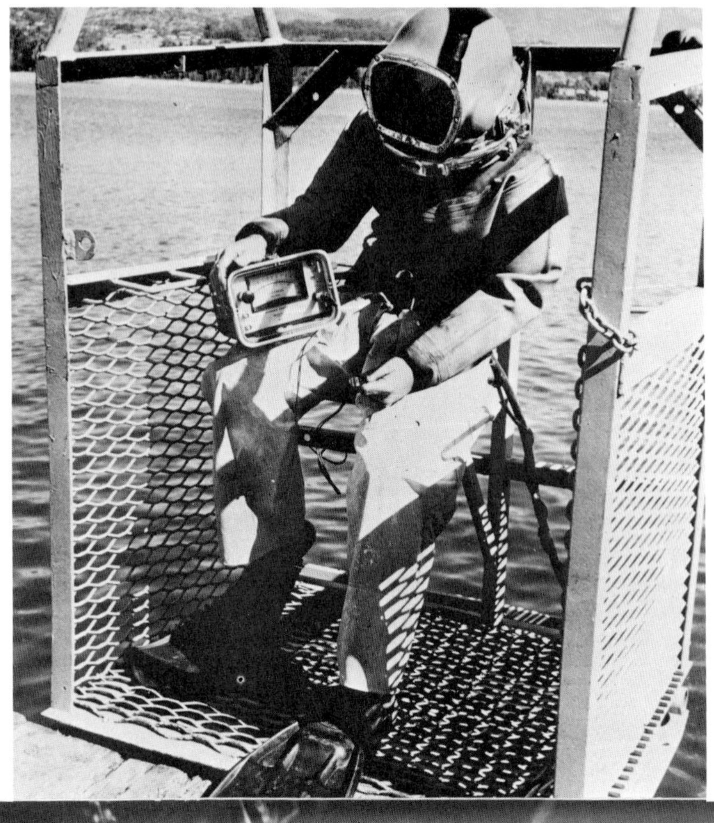

Fig. 59. (Top) A rugged diver's stage. The diver is wearing a Ratcliffe helmet. Photo: California Divers, Inc. (Bottom) The author working on a damaged "Christmas Tree," wearing hose-supplied Mark VI diving gear. Photo by Ron Church; courtesy of Westinghouse, Inc.

The tree attaches to the wellhead with the same type of hydraulic connector as used for the B.O.P., or with a diver-installed mechanical clamp. After the tree is installed, the production flow line pipes are attached to the flow line loops of the tree. The valves on the tree are usually of the fail safe type. Hydraulic pressure is remotely maintained on the valves to keep them open. If the hydraulic pressure is removed intentionally or not, springs automatically close the valves.

The hydraulic control lines are run together with the production flow lines to an operating station either ashore or on a remote platform. Some new and experimental subsea wellheads have the valves controlled by acoustic signals and also electrically with nuclear power sources.

Christmas trees in diver depths usually have manually-operated valves also. The hydraulic control lines and the production flow lines are generally of small diameter, from one to four inches in diameter, and they can be made up to the Christmas trees either remotely from the drilling rig or by divers.

I recommend that you write to the various manufacturers of subsea drilling and completion equipment for literature and brochures of their systems if you anticipate diving from a drilling rig. (See the Appendix for addresses.)

Chapter 13

MIXED-GAS DIVING

DEPTH LIMITS FOR COMPRESSED-AIR DIVING

The maximum depth for compressed-air diving was reached by the U.S. Navy in 1915, during the salvage of the submarine *F4,* which was sunk off of Honolulu.

When we consider that ten short years earlier the Navy had no formal decompression tables, the successful salvage of this submarine, requiring dives to 304 feet with standard compressed-air-supplied deep-sea diving gear, represents an extraordinary achievement and a milestone in the history of diving. For many years this depth was considered to be the absolute limit for man, dressed in a flexible rubber suit, to perform useful work.

There are three or four interlocking physiological conditions that establish 300 feet as the approximate limit for compressed-air diving. Almost every diver is noticeably affected by nitrogen narcosis at depths beyond 200 feet and the man who can perform constructively and intelligently at 300 feet is rare indeed.

The density of compressed air at 300 feet is so great that it imposes severe resistance to breathing and reduces the mechanical efficiency of the lungs. Because of this, the diver's potential for performing work at this depth is considerably less than half of what it is on the surface. Because the lungs are unable to ventilate properly, the amount of CO_2 in the diver's blood remains dangerously close to the toxic limits, even with the most efficient diving equipment.

The one single factor which positively prohibits compressed-air diving beyond 300 feet is the high partial pressure of oxygen in air at that depth. Susceptibility to oxygen poisoning or toxicity is highly variable between individuals, but at depths beyond 300 feet a dangerous reaction is assured, even from people with a proven tolerance to oxygen.

EXPERIMENTS WITH OTHER GASES

To extend the practical limits of deep diving, it was necessary to construct synthetic mixtures of breathing gas in which the partial pressure of oxygen could be carefully controlled. There are only three known inert gases that have a possible use as the diluent or carrier gas for oxygen in a diver's breathing mixture. They are nitrogen, hydrogen and helium. Nitrogen, used with lower percentages of oxygen, will eliminate the threat of oxygen toxicity; but for deep diving, the problems of nitrogen narcosis and unacceptably high breathing resistance still remain.

Hydrogen, the lightest gas known, is an attractive possibility, but its use as a breathing gas poses many problems because it is explosive when mixed with oxygen.

Arne Zetterstrom, a Swede, conducted extensive experiments using hydrogen-oxygen mixtures for deep diving, but unfortunately he was killed in an accident which was unrelated to the use of hydrogen during a dive to 520 feet in 1944. Zetterstrom's diving stage was accidentally hauled directly to the surface after this dive, and he died, probably from the effects of explosive decompression.

Experiments with hydrogen-oxygen mixtures are currently being conducted by several diving and hyperbaric investigators, and because hydrogen is only half as dense as helium, it holds definite promise for extending diving depths beyond the expected limits for helium.

USE OF HELIUM IN DEEP DIVING

In 1919 Dr. Elihu Thompson, a world-renowned physiologist, suggested in a letter to the Bureau of Mines that the use of helium as a replacement for nitrogen in divers' breathing-gas mixtures would increase the depth of diving 50 percent or more over that with compressed air.

The discovery of the existence of helium forms a curious and interesting footnote.

During a total eclipse of the sun in 1868, P.J.C. Janssen, from a vantage point in India discovered a theretofore unobserved spectrum while studying the chromosphere of the sun. During subsequent observations with a spectroscope, Sir Norman Lockyer became convinced that this new spectrum was produced by an element as yet unknown on the earth. He named this new element helium, from the Greek word for sun, *helios*. The existence of helium on earth was not proven until 1895, 27 years later.

Helium is extracted from natural-gas wells in Kansas and Texas and it appears that the only commercial concentrations of this gas in the world are in the United States.

Helium is characterized by many properties which make it ideal for mixed-gas diving. It is chemically and physiologically inert, nontoxic and nonflammable. It is second only to hydrogen in lightness and, because of its low density, it effects a considerable reduction in breathing resistance over air or nitrogen-oxygen mixtures. It is colorless, odorless and tasteless. Helium diffuses rapidly and after neon it is the least soluble of all known gases.

Helium has some disadvantages. It is a relatively rare gas and therefore expensive. Because of its diffusibility and low density, it manages to find escape routes from circuits and systems that are leak-proof with other gases. Helium distorts the human voice to near unintelligibility for reasons that have yet to be adequately explained, and it is also a marvelous conductor of heat. This latter fact results in the requirement of significantly more thermal protection for the diver than would be necessary for an equal exposure to cold water when compressed air is the breathing medium.

The disadvantages of helium are minor, and they can be coped with technologically within an acceptable financial and practical boundary when we consider the many advantages—scientific, military and commercial—of being able to dive to great depths.

Shortly after Elihu Thompson's suggestion to the Bureau of Mines regarding the suitability of helium for deep diving, that agency began extensive experiments in collaboration with the U.S. Navy culminating in today's routine use of helium-oxygen mixtures.

My intention is not to present a chronologically precise history of the development of mixed-gas diving, but rather to point out a few of the highlights in order that you may be better able to appreciate the contributions of your predecessors.

DEVELOPMENT OF RECIRCULATING SYSTEMS

In 1927, Elihu Thompson made a second contribution to deep diving by suggesting that the very expensive mixtures of helium and oxygen used by divers could be conserved by recirculating the gas mixture through a filter to remove "impurities and effete gases," and by renewing the oxygen content.

Around the same time, an Englishman, Sir Robert H. Davis, motivated by an understanding of the insidious effects of carbon dioxide upon divers in deep water as well as by an appreciation of the low volumetric capacities of the air pumps then in use, designed a helmet and back pack assembly to be used for deep diving with compressed air. This unit recirculated a portion of the diver's air through a canister containing a chemical carbon-dioxide absorbent, using the venturi eductor principle. Divers using this unit breathed purer air, free of carbon dioxide, and they also profited from another, unexpected benefit. The incidence as well as the severity of symptoms of nitrogen narcosis were markedly reduced among divers using this rig.

English divers have long considered a depth of 240 feet to be within the limits of routine compressed-air diving. From extensive personal experience I know that there are few American divers capable of working efficiently at this depth. The interrelationship between high partial pressures of carbon dioxide and the effects of nitrogen narcosis has been pondered without resolution for many years. I am forced to inquire whether this seemingly greater proficiency of English divers in depths over 200 feet is not directly attributable to the efficient carbon-dioxide scrubbing systems used on their equipment during deep air dives.

In any event, the carbon-dioxide scrubbing system designed by Davis in 1928 for deep compressed-air diving required only a change in the size of the injector-nozzle orifice to make it suitable for helium-oxygen diving; this because of the different densities of air and helium. The "New," "Total Concept," "Revolutionary" semiclosed-circuit recirculating rigs for helium-oxygen diving, marketed today by Clark, Advanced Diving, Aquadyne and others, all owe a considerable debt to Mr. Davis's first model.

The U.S. Navy built its recirculating system onto a modified standard deep-sea helmet, and the CO_2 absorbent canister was attached directly to the back of the helmet. The breathing gas, supplied through a hose from the surface at a minimum pressure of 100 psi over bottom pressure, is forced through the venturi nozzle, located at the top of one side of the canister. The high-pressure stream of gas flowing past an opening in the helmet and then entering the canister creates a slight pressure differential. This sucks the used air from the helmet and forces it through the CO_2 absorbent in the canister and then discharges it back into the helmet, free of carbon dioxide. Oxygen is constantly replenished by the stream of incoming gas which operates the venturi. Excess gas escapes through the exhaust valve. The volume of gas consumed during a dive with semi-closed recirculating diving equipment is generally about one-fifth of that required for an equal dive on open circuit. If the recirculating system should malfunction, the diver can go on open circuit simply by closing the recirculating valve and opening the control valve.

HELIUM-OXYGEN MIXTURES

The Navy helium-oxygen diving helmet weighs about 103 pounds and it is an extremely awkward piece of equipment to use. Nevertheless, over the years an extraordinary amount of good work has been done with it.

The first large-scale operational use of helium-oxygen mixtures for diving by the Navy was during the salvage of the submarine *Squalus,* sunk in 1939 in 240 feet of water.

The first civilian use of helium-oxygen for diving was by Max Gene Nohl, one of the founders of Desco. On December 1, 1937, Nohl descended to 420 feet in Lake Michigan, wearing on his back high-pressure cylinders containing a helium-oxygen mixture. This dive established a new deep-diving record.

The U.S. Navy reached 440 feet in 1938, and repeated the accomplishment in 1941 during open sea dives to the sunken submarine *Conger.*

The advancement of helium-oxygen diving technology was delayed by World War II, and the next step into the depths was made by the

British Navy in 1948, when a diver was lowered to the stunning depth of 540 feet in the open sea.

Fig. 60. Max "Gene" Nohl dressing in for the first record-setting civilian mixed-gas dive. Photo: Dr. Edgar End.

In almost all of the foregoing deep dives, the depths had been earlier reached or exceeded in various hyperbaric test facilities ashore.

AUTHOR'S RECORD-BREAKING DIVE

The British record of 540 feet in the open sea was not broken until 1967, when the author, working for Westinghouse, Inc., made the first open-sea working dive in 600 feet of water in the Gulf of Mexico. This dive was possible only because of saturation diving techniques, which will be covered in Chapter 14.

CIVILIAN ENTRY INTO MIXED-GAS DIVING

The U.S. Navy was the world-recognized authority on helium-oxygen diving until the early 1960's. At that time, civilian diving companies, spurred on by the requirements for deeper diving by the offshore oil operators, began experimenting with diving gas mixtures and developing new equipment and techniques. By 1970, the amount of helium-oxygen diving experience and expertise gained by the civilian diving community during the previous decade far outweighed what the Navy had accumulated in the preceding 35 years. The Navy now looks to the industry leaders of the commercial diving field for deep-diving equipment and techniques.

This is not meant to belittle the Navy's outstanding contributions to deep diving. Until about 1960, the Navy did more deep diving and deep-diving research than any other agency or group in the world. Since that time, however, the requirements of the offshore oil industry for deep-diving capabilities and their willingness to pay for them have resulted in a level of civilian diving activity undreamed of ten years earlier.

The civilian entry into the field of mixed-gas diving effectively began with the drilling of oil wells off the coast of California in the late 1950's. Because of the characteristics of the California coast, the first wells were in an average depth of 250 feet of water, although they were only a stone's throw from shore. They were in a depth range accessible to compressed-air divers, but the work performance of the average diver was generally disappointing mainly due to the effects of nitrogen narcosis. Eight or ten divers were required over a period of days to complete work that one man could easily have done in an hour or two in shallower water.

Most of these early California offshore divers were recruited from the local abalone business, and because their trade required them to spend long hours under water, day after day, they formed a group of divers acknowledged to be among the best in the world. However, abalone diving is seldom practiced in water deeper than 100 feet, and these men were unaccustomed to the effects of nitrogen narcosis.

In 1962, Dan Wilson, at that time an abalone diver and now president of the diving company, Sub-Sea International, began experimenting with a mixed-gas diving system. In September of that year, using a standard Japanese abalone helmet modified by the addition of a demand-breathing system, Wilson made a record-breaking civilian mixed-gas dive to 420 feet in the open sea.

As a result of the unqualified success of this dive, Phillips Petroleum Company contracted with Mr. Wilson to supply mixed-gas diving services aboard the drilling barge *CUSS I*, operating in 250 feet of water.

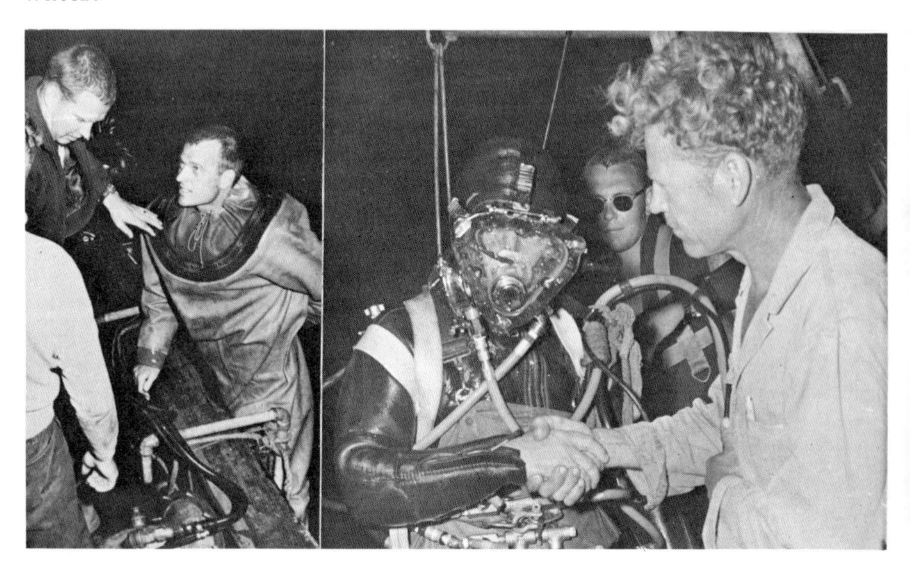

Fig. 61. (Left) Dan Wilson climbing aboard the support vessel after his record-breaking mixed-gas dive to 420 feet. Photo: Dan Wilson. (Right) Norman Ketchman, after his record-setting mixed-gas dive in the Gulf of Mexico, shaking hands with Norman Knudsen, who duplicated the dive the following day. Mike Hughes, current president of World Wide Divers, is in the background. Photo: Norman Ketchman.

The other west coast diving companies, in order to maintain competitive parity with Mr. Wilson, were forced into mixed-gas diving. The age of civilian mixed-gas diving was born.

Shortly after this dive, Gerry Todd, a partner of Associated Divers, Santa Barbara, California, in collaboration with Richard Kirby and other members and employees of the firm, designed a semi-closed-circuit recirculating system and mounted it on a standard light-weight Japanese diving helmet. This basic unit, which today bears the name of Kirby-Morgan Gas Hat, provided the major tool for all early west coast mixed-gas diving, and it remains a standard in the industry.

At this same point in history, commercial offshore oil-field diving had been an established industry in the Gulf of Mexico for more than a decade. All of the existing wells at that time were easily within the efficient range of compressed-air diving, but floating rigs were already probing the Gulf bottom in water depths of 300 feet and more.

With an eye to the future, Norman Knudsen and Norman Ketchman, partners in Gulf Coast Diving Service, planned a record-setting demonstration dive. In October of 1963, in a dive designed by mixed-gas specialist Peter Edel, Norman Ketchman descended to a depth of 335 feet. This dive established a depth record for the Gulf of Mexico and also marked the first application of mixed-gas diving in that area. Ketchman used a modified Desco mask and this fact has a significant bearing on the subsequent development of mixed-gas diving gear in the Gulf of Mexico.

DEVELOPMENT OF MIXED-GAS DIVING GEAR

Better than 90 percent of all diving in the Gulf was done with either a Scott or a Desco mask, with the Desco in more widespread use.

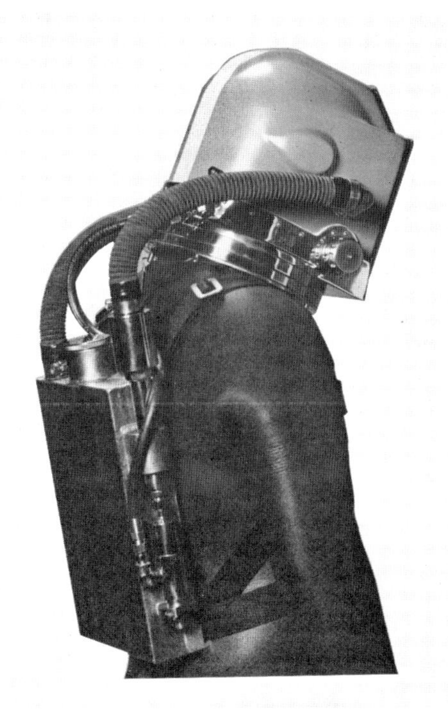

Fig. 62. Swindell, or Advanced mixed-gas diving equipment. Photo: Commercial Diving Division, U.S. Divers.

Few Gulf Coast divers were experienced in the use of hard-hat gear, and because of this, the Kirby-Morgan Gas Hat received only limited use in the Gulf.

Taylor Diving and Salvage Company, perhaps the largest commercial diving company in the area, began its mixed-gas diving opera-

Fig. 63. Back view of Aquadyne mixed-gas diving equipment, showing the CO_2 scrubber canisters and cylinders of emergency reserve gas. Photo by Brooks Institute; courtesy of General Aquadyne, Inc.

tions using a modified Navy Mark VI semiclosed-circuit recirculating diving rig. Instead of a mouthpiece, the unit was equipped with a Desco mask fitted with an oral-nasal mask, and rather than high-pressure scuba cylinders, the gas was supplied through an umbilical hose.

The Mark VI is a very efficient breathing rig, but the breathing bags worn on the chest interfere with the diver's ability to work. The unit also requires rather delicate adjustment of the breathing bag inflation and of the exhaust valve, and it is therefore far from idiot-proof. Flooded CO_2 absorbent canisters are a common occurrence when this rig is used by unpracticed hands.

In 1966, the David Clark Company, a manufacturer of space helmets, assessed the expanding potential of the inner-space industry and developed, with technical collaboration from Taylor Diving, the first of the line of semiclosed-circuit recirculating mixed-gas diving systems to use a lightweight plastic helmet and a back-mounted CO_2 scrubber canister.

The Clark system was soon followed by one from the Advanced Diving Equipment Manufacturing Company and another from Aquadyne. These three similar semiclosed-circuit recirculating systems are probably used in more than 60 percent of all mixed-gas diving in the world's oil fields today.

A type of equipment rapidly gaining in popularity for mixed-gas diving is the open-circuit demand mask. It was long thought that a demand system would be unsuitable for deep diving because of increased breathing resistance, but the new masks use an oral-nasal to cut down on circuit dead space and they are also equipped with a manually-loaded diaphragm so that inhalation resistance is virtually eliminated.

The Kirby-Morgan Band mask is perhaps the most popular of this type, followed by the Aquadyne and Advanced Diving Equipment Company's new mask. Phil Widolf is also currently marketing a new diving mask for mixed-gas diving.

Gas consumption is undeniably higher when using a demand mask compared to a semiclosed-circuit recirculating system, but all of the semiclosed-circuit rigs are inherently awkward to work in. A diver wearing a mask is unquestionably more versatile and capable of performing more work faster than his brother in a semiclosed-circuit recirculating diving rig. In my opinion, the increased work potential of a diver wearing a mask will justify the cost of the extra gas involved, in most circumstances.

DECOMPRESSION TIMES REQUIRED

When helium was first proposed for use in deep diving, it was thought that greatly reduced decompression times would ensue, but this has not been the case. Decompression times with mixed gas are as long or longer than they are with air. The amount of time required for safe decompression is directly related to the partial pressure of the inert gas or gases at the depth of the dive, and the length of time of exposure.

The higher the percentage of oxygen in the breathing-gas mix, the lower the partial pressure of the inert gas and the shorter the decompression. To illustrate: A 30-minute dive to 190 feet on compressed air would require a total decompression time of about 64 minutes, according to the U.S. Navy standard air decompression table. A dive to the same depth for the same time using a helium-oxygen mixture of 79 percent helium and 21 percent oxygen (this being the same ratio of oxygen to inert gas that is present in compressed air) would require a total decompression time of 92 minutes, according to the U.S. Navy helium-oxygen decompression tables.

This is an increase in decompression time of 28 minutes. The same dive with less oxygen would increase the decompression time further; to wit, the same dive with a mixture of 85 percent helium and 15 percent oxygen would require a total decompression time of 97 minutes.

In order to shorten decompression times, some diving contractors have, in the past, been led to conduct dives with a partial pressure of oxygen dangerously close to the toxic limits. The Navy had at first calculated a maximum safe oxygen partial pressure of 2½ atmospheres absolute, for a maximum exposure time of 30 minutes, but, because of subsequent additional experience, they reduced this to 2 atmospheres and then to 1.6 atmospheres. The percentage of oxygen in a diver's breathing mixture must be reduced with increasing depth and increased times of exposure. It is very well to plan a dive for a 30-minute period using the maximum safe allowable oxygen partial pressure in order to shorten decompression, but it might be something else again to get a diver unfouled and on his way up in that period of time.

The work rate or level of exertion is another definite causative factor in oxygen toxicity and because this is so highly variable between individuals, it virtually defies tabulation or the drawing of curves. Because of these unpredictable elements, I personally feel more secure diving with lower oxygen partial pressures even with the resultant longer decompressions.

Another important consideration concerning the percentage of oxygen in a diving-gas mixture comes to bear during the ascent from deep dives over 300 feet. In this case a mixture of 90 percent helium and 10 percent oxygen might be used. This low percentage of oxygen would be fine at depth, but it would not result in a sufficiently high partial pressure of oxygen for proper decompression during the shallower stops. A mixture containing 10 percent oxygen is theoretically acceptable for breathing in depths beyond 36 feet. However, the level of physical exertion is unpredictable and lean mixtures at shallow depths could result in anoxia and unconsciousness, especially if the diver is exerting himself. Some companies eliminate this risk by shifting the diver to a richer mix, normally about 20 percent O_2, during the ascent and the water-decompression stops.

USE OF COMPRESSED AIR

A number of companies shift the diver to compressed air when he reaches 100 feet during the ascent, and they also carry out all of his water-decompression stops on compressed air. After a short period at 40 feet, the diver is surfaced and he completes his decompression schedule breathing oxygen at 40 feet in the chamber.

I prefer this technique to any other for several reasons. To begin with, there is a tremendous psychological lift when you get back to breathing air. You feel as though you have returned to the world and this feeling is reinforced by the return of your normal speaking voice. You also begin to feel warmer, and although the change in your body heat is probably not very substantial, you at least feel that you will be able to endure the remainder of your water stops. I also feel that the elimination of helium from your system is much more rapid when the breathing gas is shifted to air at the 100-foot level. Granted, while your body is eliminating helium, it is also taking up nitrogen, but the decompression times at the shallower depths are more than long enough for the proper decompression of this nitrogen. My views on this matter are not the result of an educated understanding of gas physiology, but are purely empirical. I have never suffered from the bends or decompression sickness after a mixed-gas dive when I have been decompressed on compressed air from the 100-foot level. But I have frequently suffered from "hits" after similar dives when I was decompressed on mixed gas up to the 50-foot level. It has not been found practical to shift to compressed air at depths greater than 100 feet because of the rapid onset of nitrogen narcosis.

My preference for this system of decompression is further reinforced by the disastrous consequences of an attack of oxygen poisoning to a diver in the water. I don't believe that the resultant shorter decompression times in the water justify the risk involved in administering oxygen to a diver in the water. The Navy warns that the diver must be at complete rest while breathing oxygen in the water, and this is quite often an impossible requirement to comply with operationally, especially in rough water or strong currents.

When oxygen is used for decompression in the water, the diver normally begins breathing it at his 50-foot stop. During rough weather, because of waves and the rolling of the diving vessel, the diver's actual decompression stop depth could easily fluctuate up to ten feet or more, possibly putting him dangerously close to the oxygen toxic limits.

Finally, if it is available at the control panel, there is always the possibility of the diver being accidentally shifted to oxygen while he is at depth. The chance of pure oxygen leaking through valves and entering the breathing-gas circuit, although remote, is still a possibility with some control panels or "racks" currently in use.

HELIUM-OXYGEN DECOMPRESSION TABLES

The U.S. Navy was the first agency in the world to compute a set of helium-oxygen decompression tables and, to my knowledge, these tables are still the only ones published publicly. However, the extensive use of these tables by civilian divers, especially in the depth ranges of over 250 feet and exposure times of longer than 45 minutes, has led to an unacceptably high incidence of the bends, or decompression sickness. Because of this, most major diving companies have developed their own helium-oxygen decompression tables. Unfortunately, because of the great expense involved, most of these companies consider their tables proprietary and hold them secret for exclusive in-house use. As a result, you as a diver will be using the decompression tables and procedures of the particular company employing you.

The computation of helium-oxygen decompression tables is very complex, as is the design and selection of different gas mixtures for specific dives and objectives. Both of these functions are the rightful responsibility of a trained physiologist or mixed-gas diving technician, and they will not be treated in this book. For futher insight into the technical aspects of helium-oxygen diving, I refer you to the U.S. Navy Diving Manual, 1970.

The use of nitrogen with helium-oxygen mixtures, called a trimix, has definite value although it further complicates the decompression calculations. Nitrogen is far less expensive than helium and when it is used in a percentage that results in a partial pressure of less than three atmospheres (the concentration at which the first symptoms of nitrogen narcosis usually appear), it will reduce the cost of the dive and increase the diver's speech intelligibility.

Helium-oxygen diving procedures differ somewhat from those used for compressed-air diving, and they also differ slightly with the varying types of diving gear used.

HELIUM-OXYGEN DIVING GEAR

Helium-oxygen diving is synonymous with deep diving and this means long exposures and long water-decompression stops. Because of this, and due to the increased body-heat loss experienced when helium-oxygen mixtures are used, the diver must be dressed very warmly. With deep-sea gear, all the woolens that you can wear and still be able to stuff yourself into the diving dress are recommended.

With wet-suit gear, if a hot-water-heated suit is not used, a 1/8-suit should be worn under your standard 3/8- or even 1/2-inch suit. If a source of hot water is available, this can be run through ordinary tool hose secured to the diving stage or hang-off bar. When the diver reaches the stage at the beginning of his first decompression stop, he can insert the end of the hose under his wet-suit jacket and the warm water will provide instant relief from the numbing cold.

It is important for the diver to carefully check over all of his diving gear before any dive, and it is imperative with helium-oxygen gear.

The hoses and their connections between the helmet or mask and the CO_2 absorbent canister, as well as all plugs, caps, and fittings, must be carefully checked for leaks. Although Baralyme® (a registered trademark of National Cylinder Gas Co.) and Soda-Sorb® (a registered trademark of Ohio Chemical Co.) are not as dangerously caustic as earlier CO_2 absorbent chemicals, they must be kept dry to scrub CO_2 out of the breathing mixture properly. While diving, the presence of milky-white spray or liquid in your helmet or mask is an indication of a leaking gas circuit and wet CO_2 absorbent. If this occurs, you must immediately go on open circuit and abort the dive. The decreased efficiency of wet absorbent chemical can quickly lead to dangerous levels of CO_2. Be sure the CO_2 absorbent chemical in the canister of your diving rig has been changed before you begin your dive.

Although the amount of chemical absorbent contained in the canisters of most diving rigs is sufficient for dives of from two to ten hours, it is wise to change the chemical before each dive. The efficiency of chemical CO_2 absorbents decreases rapidly with lower temperatures, and this is another good reason for fresh absorbent chemical at the start of every dive.

The speed of descent when diving with helium-oxygen mixtures must be moderate (the Navy prescribes 65 feet per minute) because the rack or diving-panel control operator increases gas pressure during the descent through a regulator, maintaining a pressure of 100 pounds over the diver's ambient depth pressure. In a rapid descent, the diver could outstrip his gas supply, leading to a possible squeeze. In some recirculating rigs, especially during deep dives, a gas pressure of 100 pounds over bottom pressure might not be enough for adequate ventilation. If you feel the least bit unusual during your descent or dive, don't hesitate to call for more gas pressure or to go on open circuit.

During any dive, with any type of gear, it is always wise to stop work occasionally and go on open circuit briefly to flush out your helmet or mask. Don't worry about gas consumption. A completely phony mark of pride with some divers is the small amount of gas they use during a dive. To produce work, your body needs oxygen and the efficient elimination of CO_2. Both of these requirements can only be met by sufficient gas flow. If you feel you are not getting enough to breathe or you are uncomfortable in any way, stop work and go on open circuit until you feel normal. With the lightweight helmets that employ a neck dam or seal, if the helmet raises and lowers like a piston with each breath you take, check your exhaust adjustment and then call for more gas pressure. If necessary, crack your free-flow control valve slightly.

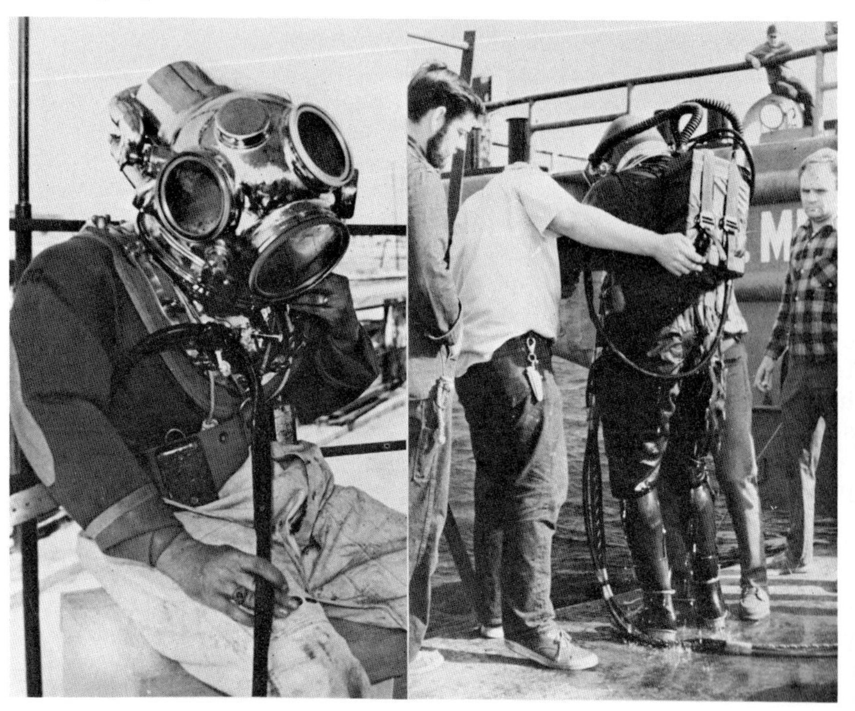

Fig. 64. (Left) Diver dressed out in deep-sea gear and Kirby-Morgan mixed-gas helmet. Photo: California Divers, Inc. (Right) A diver wearing the Swindell, or Advanced mixed-gas equipment. Note the back-mounted CO_2 absorbent canister. Photo: Author.

Carbon-dioxide poisoning hangs as a dangerous and ever-present specter over all mixed-gas and deep-diving operations. Know its signs and symptoms. I repeat, if you feel the least bit unusual at any time, go on open circuit and ventilate freely until you feel normal. If ventilating does not improve your condition, abort the dive.

Frequently, the effects of CO_2 will not be felt until the ascent. For this reason, it is a wise routine to ventilate freely for about a minute before you begin your ascent.

When you are diving with helium-oxygen mixtures, you must speak slowly and articulate each word carefully because of the "Donald Duck" effect, or voice distortion.

An involuntary blowup from the great depths of helium-oxygen diving would almost certainly be fatal, so if you are wearing deep-sea gear, pay close attention to your suit inflation and your exhaust valve. Wear plenty of weight on your belt and heavy shoes or ankle weights.

If you are wearing the type of equipment with a back-pack-mounted canister, be sure the canister is snugly secured to you because a loose canister not only is annoying but also can throw you off balance.

Because of the inherent awkwardness of most helium-oxygen diving equipment, a stage should be used whenever possible for lifting the diver into and, more importantly, out of the water. A rugged harness or safety belt is just as important with helium-oxygen diving gear as it is with other types.

Because of its lighter density, helium will come out of solution and form bubbles much more rapidly than nitrogen. This requires a much deeper first stop when diving with helium than would be the case with compressed air. In long, deep helium-oxygen dives, the first stop could be well over 200 feet deep.

This tendency towards rapid bubble formation completely rules out the possibility of emergency bailouts or hauling the diver rapidly to the surface in case of equipment malfunction. The diver must make his water stops if he is to survive and some emergency auxiliary gas supply must be provided during every dive.

To my knowledge, Cal-Divers was the first company to hook up the pneumofathometer hose to a mixed gas supply so that the end could be inserted into the diving rig in the event of a primary gas supply failure. This practice has already resulted in the saving of a diver's life. Additionally, Cal-Divers rigs the descending line and the diver's stage with a hose and a valve so that on the bottom, or during decompression, the diver is always close to an auxiliary gas supply. These are excellent practices and I would like to see them adopted by all diving companies. The standby diver, operating from a completely independent gas source, should be equipped with an emergency gas supply which he could bring to a stricken diver. This could be as simple as a "T" and a valve and a short whip leading off of his own gas supply hose.

RATE OF ASCENT

The Navy prescribes differing rates of ascent from the bottom to the first stop, depending upon the oxygen partial pressure at the depth of the dive and the exposure time. The Navy rate of ascent between stops is 60 feet a minute. It is almost impossible to comply with ascent rates during actual diving operations to the second, and if the ascent is not faster than 60 feet a minute. I believe these rates are not critical.

As already stated, during very deep dives with lean oxygen mixtures the gas supply will probably be shifted to a richer mix for the ascent and the water decompression stops. If your company uses compressed air for all or part of the water stops, you will probably be shifted to air at 100 feet or at the first water stop if it is shallower than 100 feet. As explained earlier, this shift is not usually made any deeper than 100 feet because of the possible rapid onset of nitrogen narcosis. When you reach the proper depth, you will be told to ventilate or go on open circuit. With deep-sea gear, close your recirculating valve and open your free-flow control valve. It is important to keep your exhaust valve open by pressure on the chin button or dump valve to keep ahead of the increased gas flow and prevent an involuntary blowup. Because of the difference in densities between compressed air and the helium-oxygen mix, the sound of the gas flowing into the helmet will undergo a marked and readily identifiable change when the air reaches you. A positive test can be made by speaking or counting off. The return of your normal speaking voice will be instantaneous as soon as you are breathing compressed air.

The Navy and a number of commercial companies carry out the water decompression stops on a helium-oxygen mixture up to the 50-foot stop, and then shift to pure oxygen. In this case the procedure is practically the same as above. When you reach 50 feet, you will be told to ventilate or to go on open circuit. Open your control valve wide, paying close attention to your suit inflation, until the return of your normal speaking voice signifies that you are breathing oxygen.

Now you are told to recirculate. To conserve oxygen, close the control valve and open the recirculating valve. On some recirculating rigs, such as the Clark or Swindell, the recirculating system does not have a separate diver-controlled valve, but is in constant operation. With these rigs, ventilating and recirculating is simply a matter of opening and closing the main control valve.

Oxygen decompression in the water can be very hazardous unless the diver is at complete rest. For this reason, before oxygen decompression in the water is attempted the diver must be provided with a comfortable, secure stage so that he is not required to exert himself in any way.

When diving out of a bell, oxygen decompression can be safely started at the 60-foot stop, resulting in a shorter decompression period. This is because the diver is accompanied by an attendent in a bell, and also, at the first signs or symptoms of oxygen poisoning, the diver can remove his oxygen mask and breathe the chamber atmosphere.

RECOMPRESSION

The onset of decompression sickness following a helium-oxygen dive is usually very rapid, often occurring just after the diver has left the 40-foot water stop and before he reaches the surface. In this situation, the diver must be rushed into the recompression chamber as quickly as possible. It is my opinion that even if complete relief is achieved before the diver gets to 40 feet, he should be taken down to 60 feet and given at least a 20-minute period of oxygen breathing before the normal decompression is resumed. Often, after leaving the 40-foot water stop the diver will experience only mild symptoms (such as slight pains in the arms or legs) and neglect to report them. When he reaches 40 feet in the chamber, if he experiences complete relief he will probably still not report his symptoms and so receive only the normal decompression. Usually this situation results in a recurrence of symptoms, either during the final ascent from 40 feet or shortly after the diver leaves the chamber, thus requiring a longer treatment schedule than would have been necessary if the diver had reported his symptoms when they first occurred. *Report all symptoms immediately, no matter how slight.* After a helium-oxygen dive, stay close to the chamber for at least three hours.

Unfortunately, the use of proper recompression practices and routines has been slow in catching up with clearly known physiological facts. For instance, many dives are conducted from the surface to depths of 400 feet or even more, with on-the-job recompression facilities limited to compressed air and treatment tables for 165 feet. The U.S. Navy treatment Table IV might be suitable for cases of decompression sickness contracted during compressed-air dives to about 200 feet or so, but this table is of dubious value in treating cases contracted during deeper helium-oxygen dives. On all helium-oxygen diving operations, the recompression facilities must be at least equal to the depth of the dive, and the chamber must be supplied with helium-oxygen breathing mixtures. Because of the great expense of these mixtures and because of the large volumes of gas that would be required to properly ventilate a chamber during a protracted treatment, chambers used on helium-oxygen diving projects should be fitted with carbon-dioxide scrubbing systems.

If a diver is stricken with the "bends," or decompression sickness, after a deep helium-oxygen dive (say, to 400 feet), he might conceivably have to be recompressed to the depth of his dive or even deeper to achieve relief. Therapeutic decompression following such a recompression might well have to be according to a series of slow linear bleed rates similar to those employed for decompression from a state of total saturation. These rates could be as slow as four feet an hour, and involve days. For this as well as other reasons of obvious safety, I strongly recommend that helium-oxygen diving from the surface be limited to a depth of 250 feet. Beyond that depth, diving operations should be conducted with a diving bell and a mating deck decompression chamber. The deck chamber should be equipped with CO_2 scrubbers and all other fittings and accouterments for the safety and comfort of the diver. Bell diving and saturation diving are discussed in the following chapter.

HELIUM-OXYGEN DIVING EQUIPMENT

Helium-oxygen breathing mixtures are usually supplied in premixed banks of cylinders, rated according to the percentage of oxygen in the mixture. These Heliox banks are hooked into a diving control

Fig. 65. AIRCO diving gas mixing and control panel, built for International Underwater Contractors. Photo: AIRCO Central Research Labs.

panel or console, sometimes called the "rack." The rack is also hooked up to supplies of compressed air and oxygen, and through a series of valves, gauges and pressure regulators, the rack operator controls the pressure, volume and the type of gas flowing to the diver. The rack operator is a trained technician with the grave responsibility of the diver's life at his fingertips.

Dives of differing exposure times at different depths require a fluctuating selection of oxygen partial pressures for diver safety and shortest decompressions. Premixed cylinders of gas with a fixed percentage of oxygen afford a rigid depth-exposure limitation.

AIRCO, one of the world leaders in commercial gas products, has designed an automatic gas-mixing console for diving with which the optimum partial pressure of oxygen for a given dive depth and time can be automatically selected and mixed from banks of pure gases. This device promises numerous advantages for mixed-gas diving and I expect it to be an industry standard in the near future.

The recirculating rigs used in the commercial diving industry today are called semiclosed circuit rigs. That is, unlike open-circuit rigs (i.e., standard diving helmets and masks, where ventilation is provided by a constant flow of gas that is exhaled into the surrounding water after breathing), these rigs recycle a portion of the exhaled breathing gas through a chemical carbon-dioxide scrubber and return it to the diver. In this way, a substantial amount of the expensive gas mix is reused, and the amount expended is only about ⅕ of that used with an open-circuit rig.

The ideal rig in terms of gas conservation is the completely closed-circuit rig. With this unit the gas is recirculated, carbon dioxide is scrubbed out and the exact volume of metabolized oxygen is automatically metered into the breathing circuit with theorectically no gas loss. A safe partial pressure of oxygen is automatically controlled and adjusted for the exact depth. Closed-circuit rigs offer many potential advantages, and although efficient units are currently manufactured by Beckman, Bio-Marine and General Electric, these rigs are not yet in use in the oil fields and so we will not discuss them here. Proper operation of these rigs requires special instruction and training from the manufacturers; they should never be used without this training.

Chapter 14

DIVING FROM A BELL; SATURATION DIVING; DRY-ATMOSPHERE WELDING HUTS

DIVING BELLS

The use of diving bells is recorded in ancient history books. Aristotle reports that Alexander the Great descended in one in the year 332 B.C.

In 1687, Sir William Phips, later to become the first governor of Massachusetts, employed an open-bottom diving bell to complete what is still one of the most lucrative salvage jobs on record, the recovery of 1½ million dollars' worth of Spanish gold from a sunken wreck in the Bahamas.

In more recent times, one-atmosphere observation chambers were used in two other remarkable gold recovery operations—the removal in 1922 of 5 million dollars in gold from the liner *Egypt,* sunk in 400 feet of water, and the salvage of 10 million dollars in gold in 1941 from the cargo ship *Niagara,* sunk in 438 feet of water.

Throughout the history of diving European diving companies have developed and used diving bells in all their variations, including open-bottom work chambers for harbor works and submarine foundations, one-atmosphere observation bells for deep exploration, and one-atmosphere articulated armor shells for deep-water salvage.

With two notable exceptions, American divers remained strangely indifferent to the possibilities of the diving bell.

In 1934, William Beebe and Otis Barton descended to a depth of 3000 feet in a cable-suspended sphere, and in 1939, the U.S. Navy, using the McCann submarine rescue bell, saved the lives of 33 men trapped 240 feet below the surface in the sunken submarine *Squalus.*

These were both magnificent achievements, to be sure, but they represent the only significant use of diving bells by Americans from the time of Sir William Phips until the beginning of the 1960's.

In 1928, Sir Robert H. Davis, one of the true fathers of modern diving, designed the Davis Submersible Decompression Chamber. He was motivated by the long, punishing decompression that divers were forced to endure in the water if they spent any appreciable time working in deep water. The original Davis S.D.C. was a steel cylinder, large enough to hold two men, with two inward opening hatches, one on top and one on the bottom. The S.D.C. was put into the water suspended from a cable from a boom and winch, and with a tender inside, it was lowered to a depth of 60 feet with the lower hatch open. Air pressure was supplied through a hose from the surface, to keep the water out and for ventilation. The dive was then conducted from the deck of the surface vessel in the usual way. When the diver's bottom time was completed, he ascended on his own down line and made his deep-water decompression stops hanging from it. When he reached 60 feet, he climbed a short ladder into the S.D.C. and his tender then removed his helmet. The hose and communications cable was unhooked from the helmet and the ends dropped out of the hatch, to be hauled in by the surface crew. The diver commenced breathing oxygen from a mask supplied from a cylinder on the S.D.C., and the tender closed and dogged the bottom hatch. The S.D.C., with an internal pressure corresponding to a sea-water depth of 60 feet, then was lifted aboard the diving vessel and the diver completed his decompression on deck in the relative safety and comfort of the S.D.C.

While this was a considerable improvement over the old method of decompression in the water. Davis realized that drawbacks remained, especially in view of the increasingly longer decompressions required for deeper mixed-gas diving.

In 1931 he designed the three-compartment deck-decompression chamber with mating attachments to receive the S.D.C. under pressure. The diver now was able to complete his lengthy decompression in a comfortable bunk with ready access to food, dry clothes and medical help. The S.D.C. was also liberated for use by another team of divers so that diving operations could be conducted continuously if necessary.

Except for modern instrumentation and differences in hatch design, location and mating methods, the original Davis concept of a submersible decompression chamber and mating deck-decompression chamber continues in use today. Variations of this basic unit represent the principal hardware for all commercial deep-diving and saturation-diving systems.

Incredibly, American divers did not rediscover the diving bell or submersible chamber as a useful commercial tool until 1963.

In that year we again observe Dan Wilson working to advance the state of the art of deep-sea diving. Mr. Wilson and Jon Lindbergh, under the corporate heading of Deep Submergence Systems, Inc., designed and supervised the building of a twin-sphere diving bell, called the "Purisima."

Another submersible chamber predates the Purisima in the American diving bell renaissance, and this was the aluminum cylinder designed and built by Mr. Ed Link in 1962. Mr. Link's diving chamber was a research tool, however. With it he conducted the first successful open-sea testing of the recently evolved theory of saturation diving.

Mr. Wilson's Purisima was a commercial tool, designed and built to support divers working on the deep-water wellheads offshore in California. It contained several innovative features still in use on some present-day diving systems.

The Purisima consisted of two interconnected spheres, one on top of the other, capable of independent pressurization. In practice, a diver occupied the lower sphere and a standby diver occupied the upper one. With both spheres pressurized to one atmosphere, the Purisima was lowered to the bottom as close as practicable to the work area.

After surveying the situation, the diver donned his gear, pressurized the lower sphere and opened the bottom hatches. The standby diver, still under only one atmosphere of pressure in the upper sphere, then observed the diver working in the water only a short distance

Fig. 66. The diving bell "Purisima." Photo: Dan Wilson.

away. If necessary, to supply a little extra muscle, or in case of an emergency, the standby diver could "blow down" his upper sphere, open the connecting hatch and drop down through the lower sphere into the water.

A novel feature of the Purisima was that an engineer could ride the upper sphere in a shirtsleeve environment and observe and direct a working diver over telephones.

After a dive, the diver decompressed within the Purisima in a procedure similar to that first used with the Davis S.D.C.

It was not until 1965, 34 years after Mr. Davis built his first complete unit, that American commercial divers were provided with the proven advantage of a submersible decompression chamber and a mating-under-pressure deck-decompression chamber.

Taylor Diving and Salvage Company, of New Orleans, La., anticipated the application of saturation diving techniques to offshore oil work and built the Mark DCL deep diving system.

The Mark DCL is a sophisticated and, relatively speaking, a commodious and comfortable diving system. The D.D.C., or main deck chamber, is seven feet in diameter and 26 feet long. It is divided into two main living compartments and one small entrance lock. It can house six men in comfort and eight men (not so comfortably) if necessity demands. It has a large service lock for passing in food and supplies and is fitted with toilets, washbasins, bunks, piped-in music, temperature and humidity controls and CO_2 scrubbing and other gas monitoring and metering systems.

The S.D.C., nowadays more appropriately referred to as the P.T.C. (for personnel transfer chamber, or capsule), mates onto one end of the deck chamber with a bolted flange. Access between the two chambers is through a narrow trunk.

The P.T.C. is a cylinder eight feet high and five feet in diameter. It is large enough for two divers and their diving rigs, with two coiled 100-foot-long umbilical hoses inside.

The P.T.C. is positively buoyant and it relies on a heavy detachable anchor connected to it with a long cable bridle to hold it on the bottom. With the P.T.C. floating buoyantly above a restraining anchor, the unit can be set on the bottom and the lifting cable slacked off so that no surface wave motion is transmitted to the P.T.C.

The advantages of diving out of a bell, in terms of diver safety and comfort, are numerous and undeniable. Personal cognizance of these advantages leads me to reiterate my firm belief that diving operations from the surface should be limited to a maximum depth of 250 feet. Beyond that depth, all diving operations should be conducted out of a bell.

The term "diving bell" was first coined to identify the open-bottom work chamber which looked like a bell; it was later applied to bathyspheres and observation chambers. Today, the word "bell" is used for any submersible chamber except submarines, including S.D.C.'s and P.T.C.'s, and my use of the word generally refers to these.

BASIC TYPES OF BELLS

Today's diving bells are of two basic types: those capable of withstanding only internal pressure, and those capable of withstanding both internal and external pressures. There are differences in operation between these two types and in some applications the bell capable of withstanding pressure in both directions has advantages over the other type.

Fig. 67. (Top) The Taylor Diving & Salvage Co., Inc., Saturation Diving Complex. Photo: Taylor Diving & Salvage Co., Inc. (Bottom) The "Dick" Evans, Inc., Saturation Diving Complex. Photo: J. Ray McDermott & Co., Inc.

The two-way bells are most often used from drilling rigs where dives are apt to be deep and of short duration. In operation, this type of bell is similar to the Purisima.

Fig. 68. (Top) Divcon, Inc., diving bell being lowered from a drilling rig. Photo: Divcon, Inc. (Bottom) The "Cachalot" diving bell being lowered to the bottom by a crane. Photo: Westinghouse, Inc.

The diver enters the bell and is then lowered to the bottom, quite often 400 or 500 feet deep, with the bell pressurized to only one atmos-

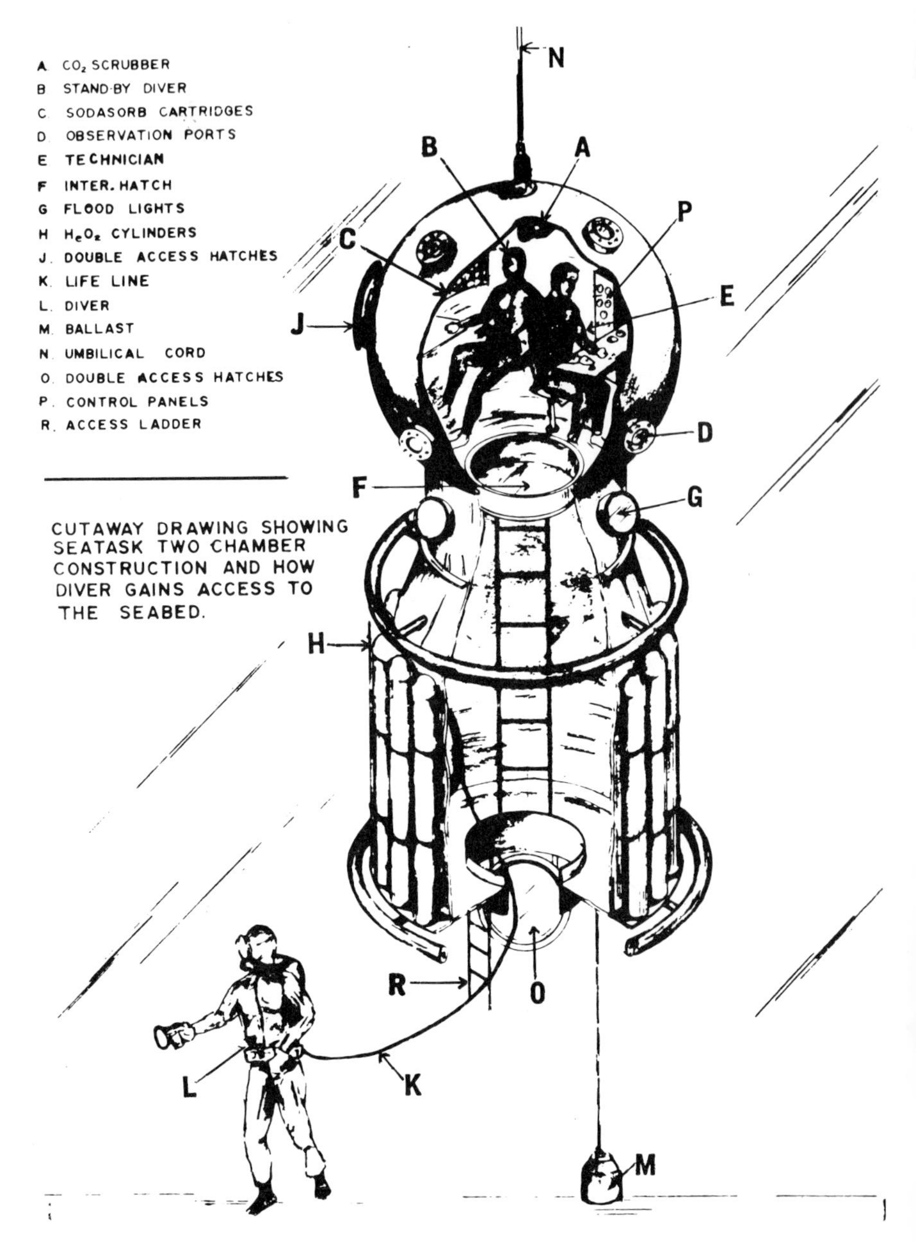

Figure 69. Courtesy of Divcon, Inc.

phere. On the bottom, the diver surveys his task (possibly the replacement of a guide cable), gets into his gear and quickly "blows down" the bell. He then opens both lower hatches (one opening inward and the other outward), and with his hose tended by a standby diver who remains in the bell, he drops into the water. When he has completed his job, he reenters the bell and it is lifted up to the level of his first stop with the hatches open. The inside hatch is then closed and dogged and the bell is lifted to the surface where decompression is completed either in the bell or in a lock on D.D.C.

The principal advantages of this system are that the diver is not under pressure during the descent or while positioning the bell, and his dive will consequently be much shorter. This type of bell can also be used as a one-atmosphere observation bell.

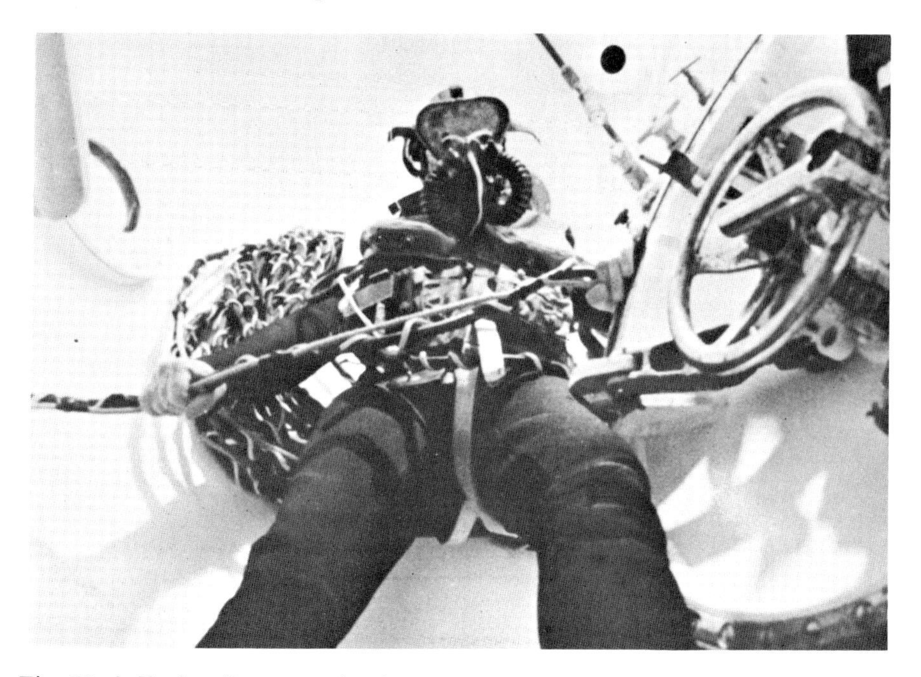

Fig. 70. A Taylor diver, wearing hose-supplied Mark VI gear, ready to exit the bell in 300 feet of water. Photo: Taylor Diving and Salvage Co., Inc.

The Mark DCL bell and others of its type can be put in the water only when they are internally pressurized. The hatches open inward, and if the bell is not pressurized, water pressure would force the hatches off their seats and flood the bell. This type of bell cannot be used as a one-atmosphere observation chamber.

The Mark DCL was the first American commercial diving system comprising an S.D.C. and a mating D.D.C. It was used for the first time in 1965 to put three teams of five divers down to 300 feet in the Gulf of Mexico. The diving equipment first used with this system was hose-fed Navy Mark VI semiclosed-circuit recirculating gear,

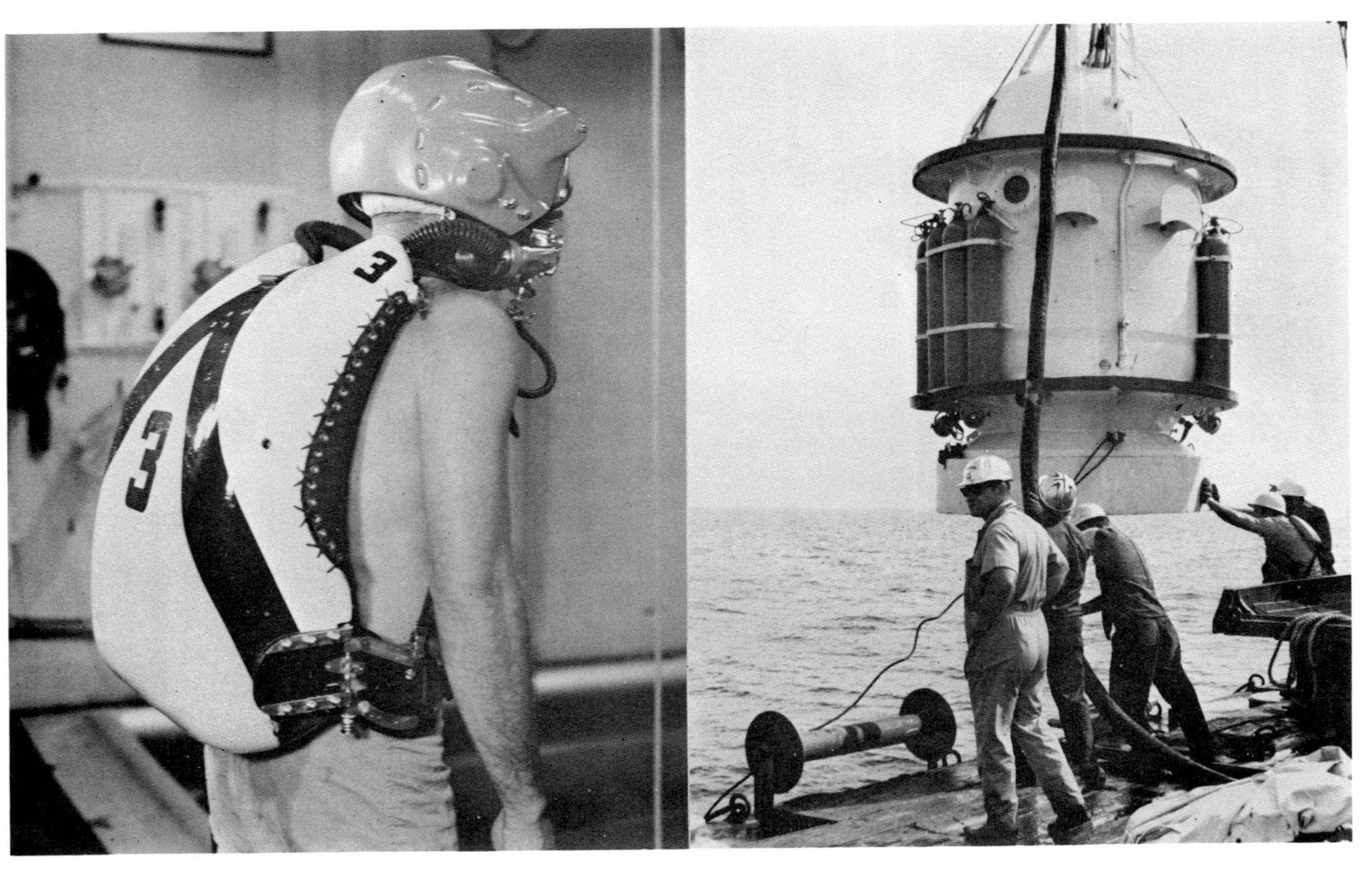

Fig. 71. (Left) The "Abalone" diving rig. Photo: Westinghouse, Inc. (Right) The

but this was soon replaced by the David Clark semiclosed recirculating rig. The gear most frequently used out of diving bells today is a Clark, Swindell or Aquadyne semiclosed circuit rig. For wellhead work or short-duration dives where gas consumption is not a critical factor, the Kirby-Morgan Band mask is the preferred equipment.

A new concept in hose-supplied semiclosed-circuit recirculating diving equipment is Westinghouse's "Abalone," designed and developed by Jerry O'Neill. This rig has been used by the Navy to depths of 850 feet.

The routines and procedures for diving out of a bell are virtually the same whether the dive is a saturation dive or a nonsaturation dive. I will cover these procedures in greater detail in the following discussion on saturation diving.

SATURATION DIVING

Saturation diving was born in 1957 of theories and speculations proposed by U.S. Navy Captain George Bond, at that time director of the Submarine Medical Center, New London, Connecticut.

Captain Bond's initial hypothesis was that after a given time of exposure to a given partial pressure of inert gas, the tissues of the body would become completely saturated with that inert gas. This simple but revolutionary theory meant that after a certain but, as yet, unspecified period of time at depth, a diver's body could not absorb any more inert gas, no matter how long he stayed at that depth. Consequently, after the saturation point was reached, decompression would require the same amount of time whether the diver stayed at depth days, weeks, or even months.

The theory was proven to be valid and the time for saturation of all the body's tissues was established as 24 hours, during a series of animal tests conducted at the Submarine Medical Center under the project name of "Genesis." The results of these tests were further confirmed by a series of human-subject dives conducted at the U.S. Navy Experimental Diving Unit, Washington, D.C., under the direction of Captain Bond, Captain Walter Mazzone and Captain Robert Workman.

Diving had at last entered the space age. From a terminal practical depth of 300 feet for a few short minutes, the door was suddenly opened on the possibility of working for days and weeks at a time at depths of 1000 feet or more.

The tremendous potential of saturation diving for commercial work, scientific investigation of the seas and military applications, led many agencies, private and governmental throughout the world, into a crash program of design and development of deep-diving equipment, hardware and gas mixtures, for use with the saturation technique.

To identify the individual contributions to the present state of the art of saturation diving would require a separate book.

OPEN-SEA TESTS AND COMMERCIAL APPLICATIONS

The first open-sea test of saturation diving was conducted by Mr. Ed Link in September of 1962. Using the previously mentioned aluminum cylinder, Mr. Link put a diver, Robert Stenuit, down 200 feet in the Mediterranean and kept him there for 24 hours.

Following this historic dive there came a long list of deeper and longer experimental saturation dives in the open sea, including M. Cousteau's Con Shelf series and the U.S. Navy's much publicized Sea Lab tests.

The world's first commercial application of saturation-diving techniques was accomplished by Marine Contractors, Inc., of Southport, Connecticut, with hardware, systems and technical supervision supplied by Westinghouse, Inc.

Fig. 72. The author returning from a surface dive. The "Cachalot" diving bell is in the background. Photo: Paul J. Tzimoulis, *Skin Diver Magazine.*

During this first commercial saturation-diving job, conducted in the late summer of 1965, divers were kept at a depth of 200 feet for periods of up to five days while working to replace faulty trash racks at the Smith Mountain dam in Virginia.

The system used on this job, Westinghouse's "Cachalot," was designed and operated under the supervision of Alan Krasberg and Jerry O'Neill. It was similar in physical dimensions and configuration to the Taylor Mark DCL, but there were a number of operating and structural differences.

Perhaps the heart or brain of the Westinghouse system was an oxygen partial-pressure sensor designed by Mr. Krasberg which precisely monitored the PO_2 in the deck chamber, the submersible chamber and in the diver's individual breathing circuits.

Marine Contractors, Inc. used the Cachalot system again the following year, in the first saturation-diving operation ever to be carried out in the Gulf of Mexico.

Two eight-leg well structures, owned by Gulf Oil Company, which were demolished by hurricane "Betsy," were completely removed to a depth of 240 feet by saturated diving crews. The author worked as a saturation diver during this project.

The following year, in June of 1967, the author, again diving out of the Cachalot system, was the first of a team of divers out of the bell at a depth of 600 feet, thus establishing a new world's record for a commercial saturated dive in the open sea. (Hannes Keller, the noted Swiss mathematician and theorist had exceeded this depth on two occasions prior to this Westinghouse "Project 600" dive. In June of 1961 he astonished the world by taking *Life Magazine* reporter Kenneth MacLeish with him on a dive to 728 feet in Italy's Lake Maggiore. In 1962, Keller survived a dive to 1000 feet for a few minutes off of Santa Catalina Island, California. Unfortunately his diving companion died as a result of the effects of this dive. Neither of these remarkable dives was made using saturation principles, nor were they commercial working dives.)

The author's record (600 feet) was short-lived, and in 1970 the open-sea saturation-diving record was extended to 840 feet by a three-man team of Comex divers.

There have been many simulated chamber dives to depths in excess of 1000 feet, and in 1970 the French concern, Comex, also established a world's record in this department of diving endeavor with a simulated dive to the incredible depth of 1700 feet.

Saturation diving today is an established practice throughout the world, and although dives to 600 or 1000 feet are not yet routine, they could be if sufficient work existed at these depths.

PHYSICAL PLANT REQUIREMENTS

The physical plant required for a saturation dive is complex and expensive. The deck chamber has to be large enough to house the team of divers in relative comfort for extended periods, but not so large that it requires excessive amounts of expensive gas to pressurize it.

It must have toilet facilities and a shower. The service lock must be large enough to pass in suits, tools, diving gear, CO_2 absorbent canisters, food, and all the other items needed for support of the divers. Efficient temperature and humidity controls are absolutely essential because high humidity results in skin, ear and respiratory infections, and the helium atmosphere causes a rapid loss of body heat. Beyond

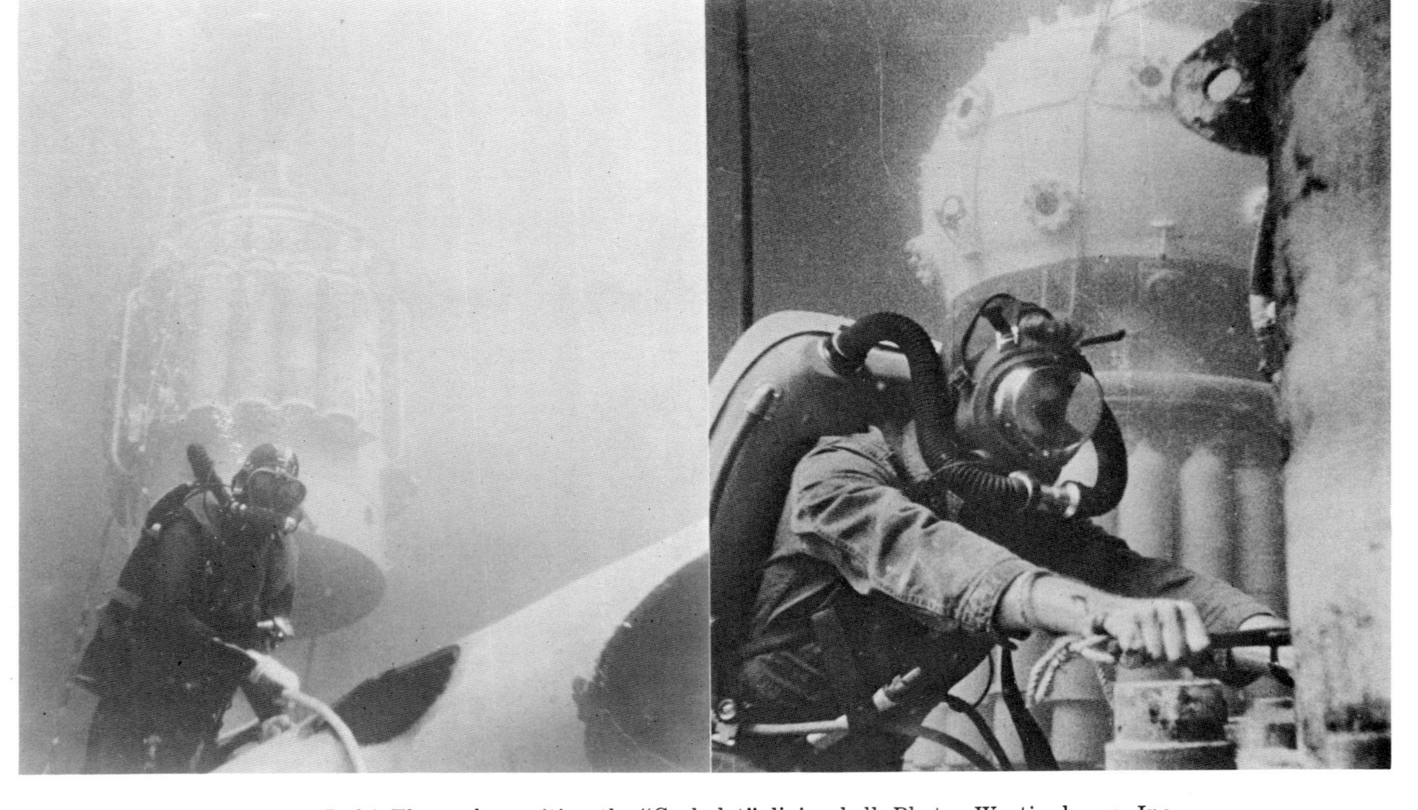

Fig. 73. (Left) The author exiting the "Cachalot" diving bell. Photo: Westinghouse, Inc. (Right) Diver wearing Drager mixed-gas diving outfit. Note diving bell in background. Photo: Divcon, Inc.

the humanitarian considerations involved in providing a warm and comfortable dwelling for the saturated divers, no diver is going to be enthusiastic about putting on a cold, clammy diving suit and going out into the water if his teeth are already chattering.

Atmosphere control within the deck chamber, the submersible chamber and the diver's breathing circuits when they are in the water, is critical, and must be monitored and precisely maintained at all times. This calls for complex equipment and trained technicians.

A partial pressure of 1.2 atmospheres of oxygen might be acceptable for short-duration deep dives, but for saturation diving, a high PO_2 would quickly result in irreversible lung damage. Standard practice has established an oxygen partial pressure of .30 atmospheres as a safe concentration for saturation diving. This is the equivalent of 30 percent oxygen at sea level or the partial pressure of oxygen in compressed air at a depth of 14 feet.

The partial pressure of oxygen within the saturation complex must be constantly monitored and the amount metabolized by the divers must be precisely and automatically replaced.

Carbon dioxide liberated by the divers must be immediately removed from the atmosphere. This is done by a system of blowers which circulate the chamber atmosphere through "scrubbers," canisters filled with carbon-dioxide absorbent chemicals. The blowers also mix the incoming oxygen with the helium atmosphere in the chamber and thereby prevent dangerous pockets of pure helium from forming within the saturation complex.

Fire is a constant threat, in terms of actual combustion damage and also of atmosphere contamination.

Lighting and communications, both audio and televised, are additional requirements of the saturation complex.

The S.D.C. or P.T.C. requires the same life-support systems as the main deck chamber, i.e., gas supply, monitors, sensors, scrubbers, lights and communications, and these systems must be resistant to the damp and cold of the underwater environment and to the rough usage of the P.T.C. during launching and recovery. The long hoses and cables of the umbilical are a constant source of potential system failure.

The P.T.C. must be capable of launching, or at least recovery and mating, during rough surface weather.

Some P.T.C.'s carry the gas mix for the diver's rigs in cylinders mounted externally around the P.T.C., while others depend upon a supply hose from the surface. All P.T.C.'s carry one or more cylinders of gas for an emergency supply.

In addition to the deck chamber, the submersible chamber, the technician's control house with all its complex equipment, and the launching and recovery machinery for the submersible chamber, a saturation diving project must have on hand enormous quantities of gas, enough for projected use, inadvertent leakage and emergency contingencies.

UNDERWATER WELDING IN A DRY ATMOSPHERE

The advent of saturation diving made possible the development of another much-needed technology—underwater welding in a dry atmosphere.

As was discussed in Chapter 7, welding in the wet is unsuitable for the rigid code requirements for high-pressure pipelines, due to brittleness, porosity and other undesirable weld characteristics.

Underwater welding in a dry atmosphere had to wait for the development of saturation diving because of the long periods of time required to complete an acceptable pipeline weld.

Successful underwater welding is now accomplished by enclosing the section of pipeline to be welded (either for pipeline repair, or to weld on a deep-water riser—a technique which Taylor Diving and Salvage pioneered) in a suitable structure, and then evacuating the water.

Because of the danger of fire, the oxygen concentration in the welding hut must be very low, and because nitrogen causes an unfavorable reaction in the molten weld puddle, helium is ordinarily used for the inert gas in weld hut atmospheres.

During the preliminary work, preparatory to welding, the hut is charged with a breathable atmosphere so that the welder and his helper can work without wearing special breathing equipment. For the actual welding, the hut is charged with pure helium and the workers must then wear face masks that are hooked up to a breathable gas supply and discharge the exhaust outside the welding hut. These masks are similar to those designed by Scott Aviation for dumping oxygen outside the chamber during O_2 decompression.

The pioneers of underwater dry-atmosphere welding were Ocean Systems, Reading and Bates, and Taylor Diving and Salvage Co., Inc.

REPAIRING A BUCKLED PIPELINE

Let us now make a saturation dive to 400 feet to repair a buckled 24-inch pipeline.

For our dive and welding repair work, we will use a composite of the diving and welding systems developed by Taylor, Westinghouse, Ocean Systems, and Reading and Bates.

A large number of working man-hours are required for a repair job of this type, so we will saturate six divers to be divided into three two-man teams. Few welder/divers are qualified and certified for this type of work, so a professional welder is usually given enough rudimentary diving training to enable him to get from the P.T.C. to the weld hut, and he is then paired up with a professional diver. Usually, the welders are not saturated until all of the preliminary work is done, such as cutting the pipe, setting the weld hut, and so on.

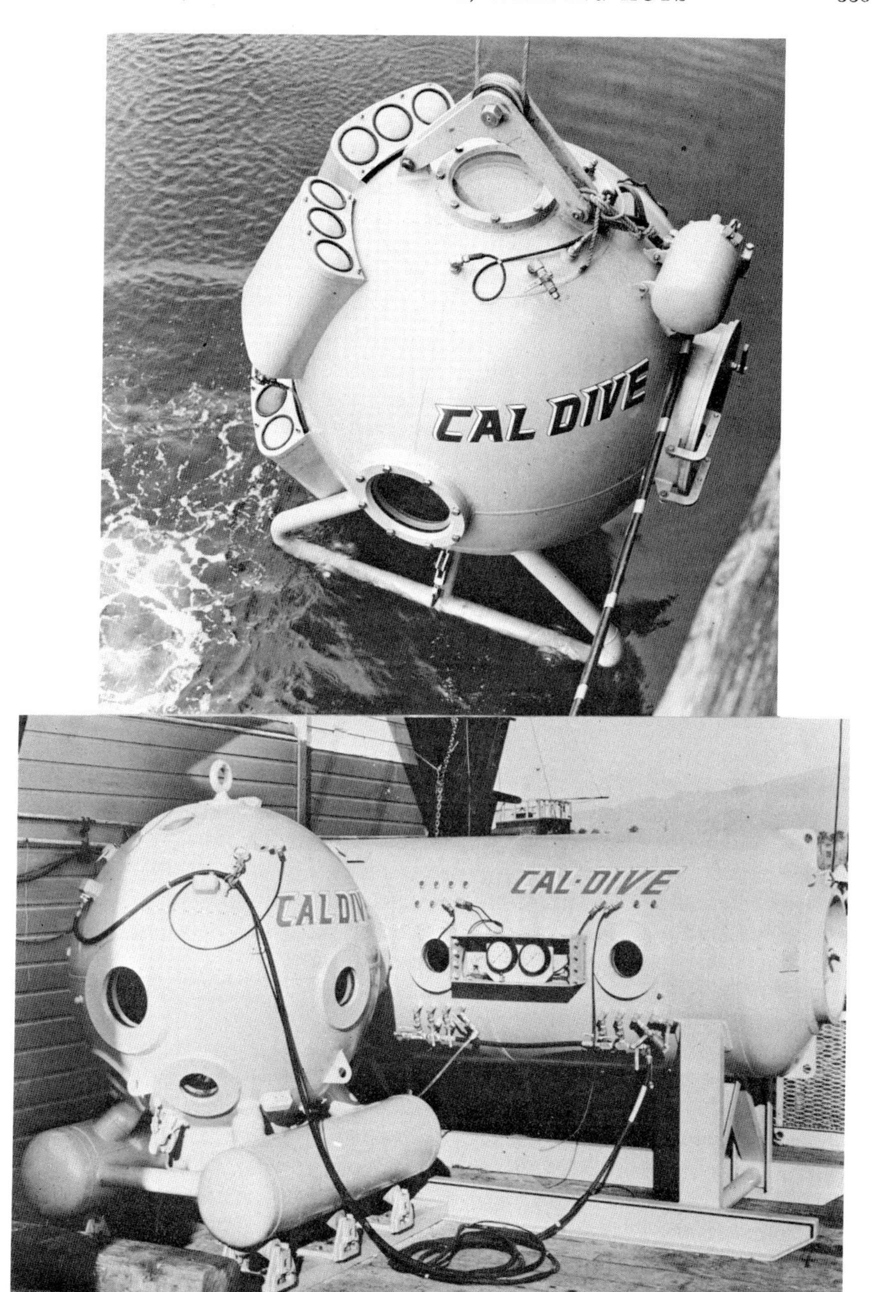

Fig. 74. (Top) A Cal-Dive "bell" going into the water. Note the tanks of breathing gas in groups of three on the left side, and the umbilical hose in the lower right-hand corner. (Bottom) Cal-Dive deep-diving system. Photos: George R. Young and California Divers, Inc.

Before the hatches are closed on the main deck chamber at the commencement of the dive, it is vital to check out all the equipment and tools that will be used: inspect the weld hut, the method of closing the lower gates around the pipeline and making the seals, the location of all the tools within the hut.

EQUIPPING THE CHAMBER

Be sure all the necessary personal gear is in the chamber with you. A saturation dive of this type could last ten days or more, so you will need sufficient clothes, shaving gear, books, etc., to last this long. Items that you need can be locked in to you after the dive has started, but each time the service lock is used, a quantity of expensive gas is lost and the operation of the lock causes a clangor that will disturb your sleeping chamber mates. Properly anticipate your needs and keep the use of the service lock to a minimum.

The long dives made possible by the saturation diving method make efficient thermal protection for the diver essential. The best system so far designed for this purpose is the hot-water-heated diving suit; without this suit, the full potential of saturation diving would be impossible to realize. Select a suit in good repair and mark it with your name to ensure obtaining the same suit for each dive.

Be sure you have your diving knife, harness and weight belt in the deck chamber with you. The most suitable clothing for the chamber is cotton shorts and "T" shirts. Sometimes after a long dive, you might be chilly in the deck chamber, so have at least one set of cotton sweat shirt and pants with you. Comfortable slip-on footgear is also desirable; however, there is a type of inexpensive shower shoe or sandal made of synthetic rubber that must be avoided. The material is of closed-cell construction, and when the chamber is pressurized, the sandals will be compressed to a size much too small to wear. The berths in the deck chamber are generally standard Navy-type pipe berths, and it is difficult to prop up comfortably in one for reading, so try to scrounge an extra pillow for yourself.

If you plan to read a lot, carry your own Birns and Sawyer or Ikelight handlight into the chamber with you. This will enable you to read at night without disturbing your chamber mates. Be sure the light is pressure-rated for the depth of your dive.

During the boredom of the long decompressions following a saturation dive, many divers develop a severe sweet tooth, so provide yourself with an assortment of candies. If you are addicted to tobacco, you will be out of luck as far as smoking is concerned but the yearning for nicotine can be partially assuaged by chewing tobacco.

A supply of basic condiments and snacks should be put into the deck chamber before the dive begins—catsup, mustard, salt and pepper, sugar, bread, peanut butter, jelly, canned Vienna sausage, and so on.

All glass containers must have their caps removed before pressurization or they will shatter and cause a mess. Also, all of the glass bottles in your shaving kit must be opened before pressurizing, and the contents of your shaving kit must be checked over by the dive supervisor. Some types of after-shave lotion or chemical deodorant could dangerously contaminate the closed-chamber atmosphere. Aerosol spray cans are strictly prohibited. The effects under pressure of various medications, such as sleeping pills or diet pills, are not yet known, so it would be wise to forego their use while saturation diving.

Most saturation complexes will have a predive check-off list itemizing such things as fresh CO_2 absorbent canisters in the chamber scrubbers. This list must be carefully checked off, item by item, before the dive.

DESCENT BEGINS

Now we are ready to enter the deck chamber. The hatches are closed and dogged and the long slow pressurization to 400 feet begins.

As previously mentioned, to avoid the danger of oxygen toxicity and lung damage, the oxygen concentration within a saturation-diving complex must be closely maintained at a partial pressure of three-tenths of one atmosphere. This level is reached at the beginning of a dive by pressurizing the chamber to a depth of three feet with pure oxygen, or to a depth of 14 feet with compressed air.

From that point on, the chamber is pressurized to the depth of the dive with pure helium, and the PO_2 is constantly monitored and the metabolized oxygen is automatically replaced.

As soon as helium is introduced into the chamber atmosphere, the divers' voices undergo what is at first a hilarious and then an annoying change. Because of the lighter density and other acoustic qualities of helium, the human voice raises in pitch to a sound not unlike a 45 r.p.m. record played at 78 r.p.m. This so-called "Donald Duck" effect increases with depth, and the tonal quality of some divers' voices is so badly affected that their speech is completely unintelligible. Sound also loses its directional qualities within the chamber and it is difficult to locate the source of a voice when your eyes are closed.

The descent to 400 feet will probably take two hours or more. Experience has shown that when divers are compressed to great depths rapidly, they often suffer from severe pains in the joints, especially those of the ankles, knees, wrists and elbows. A slow and gradual pressurization has been found to lessen and eliminate these joint pains.

Some chambers are fitted with internal depth gauges so that the divers can monitor their descent. After all your gear is stowed, you will have nothing to do but watch the gauge, read, or nap for the few hours it takes to reach dive depth. (Stowage space in a saturation deck chamber is necessarily limited. The best containers for personal gear are small canvas haversacks or knapsacks that can be hung from your bunk.)

Because excursion dives of some duration are possible to depths in excess of the saturated depth, with either no decompression or short decompression periods within the P.T.C., some companies saturate the diving crews 50 or 60 feet shallower than the intended dive depth. This effects some savings in gas and in eventual decompression time.

DIVERS "SADDLE UP"

When the deck chamber reaches the dive depth, the first team of divers is told to get ready, or "saddle up." Most divers strip completely nude before entering the P.T.C., while others, compelled by modesty, wear swim trunks. Motivated not by modesty, but by several practical considerations, I wear a one-piece "shorty" ⅛-inch wet suit. It keeps me warm in the frequently chilly P.T.C. during the descent, and when worn under the hot-water suit it affords protection of delicate skin areas from too-hot water. It also offers some thermal protection in the event of a malfunction in the hot-water system. There is another excellent reason for wearing your own personal suit next to your skin. Quite often on a job, there will not be enough hot-water suits of a particular size to allow each diver exclusive use of one suit. An affliction endemic to saturation divers, and caused by the constant wetness and high humidity, is a fungus jock itch, which can be severe enough to incapacitate a diver or, indeed, a whole crew of divers. After every dive, the hot-water suits are supposed to be scrubbed with soap and water and hung out to dry to prevent the spread of jock itch, but often the exigencies of the job will preclude this being done. Wearing your own personal wet suit under the hot-water suit offers a probable escape from the transmission of this infection. Dry yourself thoroughly after showering at the conclusion of each dive, change your underwear daily and use a medicated talcum powder in your crotch.

After you and your diving partner wriggle through the narrow entrance trunk into the P.T.C., you must carefully check over all of your diving equipment. On the first dive, the diving equipment will have been rigged with fresh CO_2 absorbent canisters by the topside crew before the complex was pressurized. Check it out, anyway. On subsequent dives, the used canisters will be locked out through the service lock and fresh ones locked in. Then you will have to change your own canisters and rig your own diving gear.

Check out the communications between the P.T.C. and topside, and also between each diving rig and topside. Unmating the P.T.C. and lowering it to the bottom requires a lot of work and takes time. When the bell is on the bottom, if the diver cannot exit the bell and get on with his work because of a minor communications problem with his diving rig that could easily have been rectified on the surface, or because he forgot a special tool, the client will be understandably perturbed. Check your gear carefully before the bell is unmated.

Another source of client consternation is the occasional unreasonable length of time it takes for a changeover in diving crews. When it is your turn to dive, be ready. As soon as the bell is mated to the deck chamber and the used-up team of divers has left it, get yourself and your gear into the bell, change canisters and rig your diving equipment. A crew changeover can normally be made in 30 minutes or less, if you are prepared for it. It should never take longer than an hour.

When you enter the P.T.C. for a saturation dive shift, you may be in the bell or in the water for as long as six to ten hours. It is wise to anticipate a few bodily needs for this long stay away from the comforts and conveniences of the main deck complex. To begin with, you are unquestionably going to get hungry, so take some sandwiches, fruit or other snacks into the P.T.C. with you, and most important, a jug of fresh water. A cup of hot coffee is especially refreshing in the middle of your shift, and the best container for this is a Stanley stainless steel thermos bottle. If the P.T.C. is going to be pressurized to a depth greater than the saturation depth, be sure the cork is not in the thermos, or it will be driven in so firmly by the excess pressure that you will be unable to remove it. Paper towels and bath-type towels are handy for wiping condensation off of the chamber viewports and for drying off the divers between dives.

Because of the potentially long stay within the bell, the possibility of a bowel movement must be considered. This situation can be as ludicrous as that of the cat on a hot tin roof unless you provide yourself with a couple of plastic trash bags and a roll of toilet paper.

A small basic tool kit consisting of adjustable wrenches, pipe wrenches, pliers and screwdrivers should be mandatory equipment in every P.T.C. A small first aid kit in a waterproof container and including a plastic airway device for mouth-to-mouth resuscitation should also be mandatory equipment.

There is an additional item which I consider to be absolutely essential for diving out of a bell. The bottom hatch in most P.T.C.'s is fairly small, measuring from 24 to 30 inches in diameter. In the event of an accident, a diver would find it very difficult to pull his injured or unconscious mate into the P.T.C. For this reason, there should be a pad eye welded to the overhead, inside the P.T.C., directly over the center of the hatch. Hanging from this pad eye should be a Handy Billy, or block and tackle. Each diver should wear a stout harness of the type described in Chapter 2, with a ring mounted on the back of it, between the shoulder blades. In the event of an accident, the block and tackle could be hooked into this ring, and it would be fairly easy for a diver to haul his incapacitated mate back into the bell.

Everything within the P.T.C. has been checked out and we are ready to go. The entrance trunk hatches at the P.T.C. and the D.D.C. are closed, and the pressure in the trunk is bled off by the topside crew, sealing the hatches. While the P.T.C. is being unbolted and unmated, the divers get into their hot-water suits.

Some diving systems use a crane or derrick to put the P.T.C. into the water and others use a permanent gallows frame. The descent of the P.T.C. is apt to be very rough and jerky, especially if it is being lowered by a crane—so, hold on tight; some P.T.C.'s have seats and safety belts for the divers just for this purpose. The roughest part of the trip will be the first 20 feet or so, going through the surface waves.

Some P.T.C.'s have internal and external depth gauges so that the divers can monitor their descent, while others use only a pneumogauge reading out on the surface. Sometimes internal controls are provided so that the divers can add or vent gas into or from the bell, while others are controlled completely from the surface.

When the bell reaches the depth of water equal to its internal pressurization, the bottom hatch will pop open. It must be lifted up on its hinges and securely fastened in the open position. Serious accidents have occurred from this hatch slamming shut on a diver.

With the P.T.C. anchor sitting firmly on the bottom, the lifting cable and umbilical are slacked off so that no surface wave motion is transmitted to the P.T.C. If there is water in the bottom of the bell, the internal pressure must be increased slightly to evacuate it. Often, when the anchor is on the bottom, the hatch will not pop open until a little internal pressure is bled off.

Earlier systems had divers' hoses coiled up on the outside of the bell, with the ends fastened to the underside of the bottom hatch. When the hatch was opened, the divers would pull in the umbilical ends, hook up their gear, exit the bell, and uncoil the hoses from the racks as needed.

It has since been found that with the hoses coiled up inside of the bell, much less time is required to get the divers rigged up and into the water. Although the hoses take up a lot of the limited space within the bell, they are much better protected against snagging and accidental damage. Usually, only one diver at a time exits the bell, and his mate stays inside and tends his hose. With the hoses coiled up inside, this tending is much easier. Most bells are provided with an emergency standby mask, so that the tender/diver can slip it on in seconds and get into the water to help his mate in an emergency. You may need more weight to hold you down when wearing only a mask, so be sure that there is a weight belt in the P.T.C. to use with this emergency mask.

When the bottom hatch is open, there is just a narrow rim around it for footing, and you must be careful not to slip and fall while moving around inside the bell. Because of the limited space and the yawning hatchway, dressing into any of the semiclosed-circuit recirculating rigs can be somewhat difficult. The canisters for these rigs are sometimes mounted on a bracket, just above a folding seat. The diver can sit down and adjust the canister harness while its weight is still supported by the bracket.

EXITING THE BELL

When the diver is in all of his equipment and it has been checked out, he can drop on down through the hatch. With heavy or awkward equipment, exiting the bell is facilitated by bleeding off a little internal pressure and allowing two feet or so of water to flood into the bottom of the bell. After the diver has left, the bell can be blown dry.

There are a few surprising deficiencies in some of the otherwise sophisticated diving systems in use today. One is in the communications circuitry. The diver generally cannot talk directly to the tender/diver in the bell just a few feet away from him, but must relay messages through the topside telephone talker.

Another deficiency is in the external lighting on the diving bell. Most diving contractors minimize or completely ignore the emotional and practical advantages to the diver of proper exterior lighting. When the diver drops through the hatch into chilly, pitch-black nothingness, he is a little reluctant to release the anchor and get on with his work. On the other hand, if the diving bell is ringed with floodlights, the diver is immediately assured that no monsters are lurking around the bell, waiting to swallow him; and furthermore, the lights are apt to illuminate his work area, leading to immediate orientation and the other advantages inherent in being able to see what you are doing.

Usually, a guide line or down line is led from the surface, through a shackle on the side of the diving bell and thence to the work area. Even with this guide line, relative orientation between the barge on the surface and the bell, 400 feet below it, is often difficult or impossible. This orientation is very important if the diver is required to work with a deck crane, or if he has to move the bell to a more advantageous position. A small compass mounted inside the bell would be invaluable for estimating this orientation.

WORK BEGINS

The first task on the present job is to remove the concrete weight coating from the damaged section of the pipeline, and then to burn out this damaged piece. All the tools necessary to do this work are secured to the P.T.C. anchor.

As you drop out of the hatch, cold water will quickly seep into your suit through the numerous zippers. You simply close the hot-water bypass valve, usually mounted on the right hip of your suit, to immediately flood the suit with delightfully warm water. Keep the location of this valve in mind, because occasionally scalding hot water can be inadvertently pumped down to you. This bypass valve is your only control of the hot water system. In one position, the water flowing to this valve discharges outside of your suit; in the other position, it discharges inside your suit. If the water is too hot or too cold, adjustments must be made to the heating systems on the surface. The water flows into your suit and is directed to various parts of your body by rubber tubing. The water escapes from the suit at the neck and wrists. Suits generally come from the manufacturer with booties fabricated onto them. When you are out in the water, warmer water within the suit will rise and escape around the neck, but in the diving bell, or when you are trying to climb up into it, the water will fill the legs of your suit making it difficult to move. Many divers cut off these factory booties and use their own footgear. If this is not done, holes should be punched in the booties of new suits to allow the water to drain from them.

Drop on down to the anchor and check out your diving gear. Be sure your gas flow is adequate and that your breathing circuit is not leaking water. When you are assured that everything is working properly, follow the guide line to the pipeline and make an inspection of your project. If the bell is too far away from your work, have it picked up and moved closer.

When you are in the water during a saturation dive, never ascend higher than your saturated depth, because of the dangerous possibility of bubble formation. As an easy reference, never ascend higher than the top of the diving bell, unless under specific instructions from the dive supervisor.

Because the concrete weight-coating surface is so rough and irregular, it is difficult to seal the welding hut properly where the pipeline passes through it. For this reason it is often desirable to remove all of the weight coating from a length of pipe slightly longer than the length of the weld hut. When this is done, the weld hut seals can be made around the smooth bare surface of the pipe. Huts can vary from eight to 15 feet long, and this is a lot of weight coat to remove. It is very strenuous work.

Mark the center of the damage, then measure off slightly more than half the length of the weld hut in each direction and mark these two points.

A device that has recently been used with some success to remove weight coating is called a water blaster. This is a short length of heavy-duty steel pipe with handles and a trigger and a tiny jet orifice the size of a pinhole. The pipe is connected with a hose to the surface, and water at a pressure in excess of 10,000 psi is forced through the tiny orifice, creating a tiny but very powerful stream of water.

This stream erodes the concrete along a narrow kerf, making it possible to cut the weight coating into sectional pieces. In addition to the water blaster, you will need a pair of side-cutting pliers to cut through the chicken wire reinforcing, and a crowbar (or wedges and a heavy sledgehammer) to loosen the pieces of concrete.

Needless to say, the water blaster is a dangerous tool and if the stream is directed at any part of your body or your equipment, serious injury or damage could result. Hold the blaster firmly and aim it carefully. Fortunately, the trigger is spring-loaded and the jet will shut off when it is released. A high-pitched piercing scream attends the water blaster when it is operating. If you lack visibility, avoid any impulse to put your finger over the jet to determine if it is working. The result would be the same as putting your finger over a shotgun muzzle and pulling the trigger. There is a backward-directed compensating jet on the water blaster and you must be careful of this too. The handles are so arranged that the blaster can be held conveniently with this back jet shooting out over your right shoulder.

When all the concrete has been removed, the damaged section of pipe must be burned out. Let us assume that the damaged section is four feet long in this case.

The mastic, or dope coating, must be removed from the lines of cut. To assure square cuts, use some sort of burning guide as described in Chapter 7.

Though the product, oil or gas, has been removed from the pipeline with pigs and the line flooded with sea water, cautiously approach this first penetration of the pipeline with your burning torch unless there is an actual hole in the pipe directly where you are working.

After you have poked this first hole into the pipe, be sure no water flows into it before proceeding with your cut. One of the best ways to determine this is by listening. Place your head against the pipeline a few feet away from the hole, hold your breath and listen. If water is flowing into the pipe with any marked velocity, you will be able to hear it.

Be sure the diving bell is not directly above you when you are burning, or that a current is not carrying the bubbles produced by your burning upward into the bell's open hatch. The oxygen and other gases liberated by the burning process could dangerously contaminate the diving bell atmosphere.

When the cuts are completed, the severed piece of pipe is lifted out of the way by the deck crane and we proceed to the next phase.

LINING UP SEVERED PIPELINE TO EFFECT REPAIR

We now have two open ends of pipeline, approximately four feet apart. They must be perfectly lined up with each other so that the weld hut can be put in place and the repair piece fitted in and welded.

The line-up is accomplished by means of a massive steel girder, equipped with hydraulic actuated clamps and rams. This line-up frame can be 60 to 80 feet long and weigh 50 to 100 tons.

Two points, equidistant from the center of the cut and corresponding to the length of the line-up frame, are measured off on the pipeline and guide lines or cables are shackled on there. These cables are led up through pad eyes on each end of the line-up frame, and then set taut. The line-up frame is then lowered slowly to the bottom, sliding down these guide cables. The frame is maneuvered into final position by the divers, and then set on top of the pipeline.

The line-up frame will have at least four sets of hydraulic clamps and rams. The diver, operating hydraulic controls mounted on the line-up frame, closes the clamps over the pipeline, and then by careful manipulation of the hydraulic rams, he forces the two pipeline ends into alignment. When this has been done, the weld hut is lowered, also on guide cables, into a fabricated position in the center of the line-up frame. The lower edges of each end of the weld hut are fitted with shoes corresponding to the exact shape and size of the top half of the pipeline. When the weld hut is in position, these shoes will be resting firmly on top of the pipe. The sides of the weld hut will extend as much as three or four feet below the bottom of the pipeline, and the slotted ends of the hut must also be closed off and sealed to this depth so that the hut can be blown down, leaving the pipeline in the dry. The end closures are made by hinged gates (also fitted with shoes) which are swung under the pipeline and then jacked up into place and locked. As the weld hut is lowered, a breathable gas mixture is forced into it so that it is kept dry and free of water. When the end gates of the hut are locked in place, the divers enter the hut and do the rest of their work on the inside.

WORKING INSIDE THE WELD HUT

From this point on, the divers, when exiting the P.T.C., wear face masks (either demand or open-circuit) for the short trip over to the weld hut. Entrance into the hut is through an enclosed U-shaped trunk, so that the divers can come and go even if the hut sinks down into the mud. Inside the weld hut they remove and hang up their masks and do their work breathing the weld hut atmosphere.

After entering the hut, the divers must install a final seal around the pipeline ends, where they pass into the hut. This final seal is often accomplished with thin sheets of rubber, with holes cut in them slightly smaller than the pipeline diameter. A sheet of rubber is slid over each pipeline end and slid up to the end bulkheads. These rubbers are clamped around the pipe with steel bands and then fastened to the end bulkheads by means of studs and stiffening strips of metal.

After the end seals are in place, hydraulic "pigs" or balls are inserted several feet into each pipeline and expanded by pumping them up with a hand pump. These balls seal the pipe ends to prevent water from running out of the pipeline when the water level in the hut is lowered. After the end seals and pipeline pigs are in place, the weld hut is blown down, leaving the pipeline exposed and dry. The bottom of the hut is fitted with folding gratings so that the workers have solid footing.

The weld hut is a completely equipped, open-bottom work chamber. All of the tools, equipment and supplies necessary to complete the welding repair work are located in lockers and compartments within the weld hut. The only item not on hand is the actual "pup," or repair section of the pipe, because this must be carefully measured and prepared on the surface.

The weld hut is also provided with lights, communications, a TV camera and a hydraulic panel for running tools such as grinders and wire brushes.

When the hut is blown dry and all the leaks are sealed, the next chore is to trim the pipe ends carefully. Because of the extreme danger of high-pressure oxygen within the weld hut, this job cannot be done with a burning torch. It is usually accomplished with a hydraulic saw which is fastened to and which travels around the pipe on two or more flat chains. The finished cut, which can take up to an hour to complete, leaves the end of the pipe perfectly square and properly beveled for welding.

The next and extremely important step is the exact measurement between the pipe ends. It is best to do this with a telescoping jig with squared end plates. When this jig is in place, the pipeline ends can be moved into final and perfect alignment by means of the hydraulic rams on the line-up frame.

The pup joint is cut according to this final measurement and prepared with bevels on the surface. It is then lowered by crane to a position at the base of the weld hut and unhooked. A diver must go into the water to do this. If the skirt of the weld hut is embedded in the mud, access for the pup must be hand-jetted.

Slings leading from two chain falls hung from the inside overhead of the hut are led out and hooked onto the pup. The floor gratings at the bottom of the hut are raised and the pup is lifted up inside. Then the pup is washed off to remove any mud and lifted into its final position between the pipeline ends. If the pup is too long, within very narrow limits, it can be ground down within the hut to make it fit. If it is impossibly long, or too short, it must be remeasured and a new one prepared. When the proper size pup is in place, it is locked into position with standard pipe line-up clamps. It is now ready to weld.

Each weld could require as many as five passes, and on a 24-inch pipe, this could take two welders as long as ten hours. Before the welding begins, the hut is charged with pure helium and the welders must wear special breathing masks as described earlier. When the welds are completed, they must be wire-brushed and x-rayed. If flaws are detected, they must be ground out and rewelded.

The standard arc-welding procedure with flux-coated rods is not used for dry-atmosphere underwater welding; instead, uncoated weld-

Fig. 75. (Top) Underwater dry welding habitat (center), and massive hydraulic pipe line-up frame. (Bottom) Underwater dry welding habitat being lowered to the bottom. Photos: J. Ray McDermott & Co., Inc.

ing wire is used. In some processes, this wire is fed automatically from a spool, and in other processes it is hand-held and fed into an arc created by a tungsten electrode in a technique similar to gas-torch welding. Instead of flux, the arc is shielded by a carefully controlled flow of argon gas which comes out of the tip of the welding stinger and engulfs the arc and the weld puddle. The various processes used for this type of welding are TIG welding (for tungsten-inert gas) and MIG welding (for metallic-inert gas). This method of welding is sometimes called short-arc welding, or dip-transfer welding.

Proficiency in this type of welding requires training and practice; it is a complex trade specialty. Certification will certainly lead to extra work for the diver who possesses these special skills.

INSPECTION OF WELDS

After the welds have been x-rayed and accepted (the x-ray machine is kept in the weld hut and one of the divers is a trained x-ray technician), the bare pipe within the weld hut must be coated with epoxy, or Splash Zone.

After the weld hut and the line-up frame have been lifted off of the pipeline and returned to the surface, the areas of pipe that were covered by the weld hut seals, and any areas of concrete damaged by the line-up frame clamps, must be coated with epoxy.

RETURN TO THE P.T.C.

When all of this has been accomplished, and accepted by the inspecting divers, you and your diving partner return to the P.T.C. and it is lifted to the surface and mated to the deck chamber. Most saturation complexes have the shower located in the P.T.C. After every dive, strip out of your gear and shower before you leave the P.T.C. After every dive, pass all of your garbage, used towels, leftover sandwiches, etc., out of the P.T.C., to be sent out of the service lock. Always leave the P.T.C. clean and shipshape for the next crew.

DECOMPRESSION

When the job is completed, the long, monotonous decompression must be endured. During the prosecution of the job, you will be very tired each time you return to the D.D.C. and most of your time within it will be spent sleeping. During decompression you will also sleep a lot, especially for the first day or so, but after that the boredom becomes oppressive.

The usual rate of ascent from a saturation dive is four feet an hour with periodic lengthy stops at each of the last four atmospheres. For instance, the ascent might be interrupted for a one-hour stop when you reach 132 feet, a two-hour stop at 99 feet, a three-hour stop at

66 feet, and a four-hour stop at 33 feet. These are arbitrary times, but they have resulted in 99 percent successful decompressions from a saturated condition.

Using this decompression procedure, the ascent time could be as long at 110 hours from 400 feet, or better than four and a half days. This is a long time when you have nothing to do, and confined within the close limits of the standard saturation complex with five or seven disparate companions. Good books and a facility for introspective amusement will be your only salvation during this period.

It is important to refrain from annoying or disturbing your chamber mates with excessive noise or chatter, especially if some of them are trying to sleep. Although food is generally tasteless and insipid within the chamber because of the rapid heat dissipation in the helium atmosphere, mealtimes are a welcome time-consuming break.

If you are unfortunate enough to be cooped up with a loudmouthed, insufferable ass (as often happens), bear your cross in silence. The interior of the deck chamber is no place for a fistfight, especially when you may have to endure two or three days before final escape. Stay cool. When the outside hatch finally opens, it will seem a deliverance. The fetid odors you have been living with (and to which you have become accustomed) will be sharply counterpointed by the first draughts of fresh air. The realization that you *can* smoke at *last*, will probably quickly cancel out any resolves you may have made to quit smoking during your long confinement. The first few puffs on your weed will affect you with the impact of a double whisky, but that won't keep you from having two of them going simultaneously.

After decompression from a saturation dive, you should remain close to the chamber for a period of at least 12 hours. Oddly enough, after having slept for a good part of the last four days, you will become very sleepy soon after emerging into the fresh air, so most of this post-dive mandatory period aboard the barge will be spent in an air-conditioned bunk. You will have plenty of time to spend your money tomorrow, regaling the girls with heroic tales of high adventure under the sea.

You are now a deep-water saturation diver, the apogeic culmination of the skills and technologies of your trade.

Chapter 15

DIVER'S PAY; UNIONS; SUMMARY OF SAFETY PRECAUTIONS

PAY RATES

Diving can be a lucrative profession. There are variations in pay scale as well as in the amount of available work, from place to place around the country. However, if you are capable and willing to hustle and move around a little, you can earn $20,000 a year easily.

The Gulf Coast of the United States is probably the center of more concentrated year-round diving activity than any other single place in the world. A recent count disclosed more than 30 diving service companies operating in southern Louisiana and Texas.

There is no cohesive or effective union organization in this area, and most companies operate on the open-shop principle. Because of this, pay rates vary considerably among companies and as a result of severe competition for work in recent years, these rates have been declining—especially for depth pay.

Most of the work in the Gulf is done far offshore, and the crews live aboard the work barges. The pay is usually based on a 12-hour day, with 8 hours at straight time and 4 hours at time and one-half. The average hourly rate paid to divers by the large companies is $6.00. This amounts to $84.00 for a 12-hour day. Some companies sweeten this up by paying the diver $10 or $15 a day for equipment rental. This brings the average pay for a Gulf-diver working for one of the larger companies to $100 a day, give or take a few dollars.

The hourly rate for tenders is between $2 and $3. The payment of depth bonus is subject to wide variation among Gulf Coast companies. I know of only one company that still pays the diver the full depth bonus; that is, $1.00 per foot from 50–100 feet; $1.50 a foot from 100–150 feet; and $2.00 a foot from 150–200 feet. Some companies pay no depth bonus whatsoever, and I sense an ominous trend in this direction. However, most companies presently operating in the Gulf will pay a depth bonus ranging from 40 to 70 percent of the above figures.

Two or three of the largest companies in the area have established a ceiling of about $450 a day for all saturation diving, regardless of the depth. This seems like a lot of money until you consider that it is equivalent to a day's pay plus depth bonus for an oxy-helium dive to about 212 feet. Until a year ago, saturation diving to 300 feet was worth about $1,000 a day.

Overseas diving pay is subject to even wider variations, but is usually less than domestic pay. Depth bonus is rarely paid overseas and, rather than a daily pay rate, the diver is likely to be hired on an 18-month contract, with a fixed monthly salary and a bonus for completing the 18 months. Salaries range from $1,000 to $3,600 a month, and it is impossible to establish an average.

UNIONS

In those parts of the country under strong trade-union control, the daily rates and depth bonus are usually much higher than in the Gulf but, unfortunately, most of these places do not have the volume of work or the depth of water to be found in the Gulf.

Diving unions in the United States are affiliated with the United Brotherhood of Carpenters and Joiners of America. This at first seems unusual, until you consider that the evolution of commercial diving in this country coincided with the development of our numerous ports and harbors. Years ago most divers were also pile drivers, and bridge, wharf, and dock builders.

In New York Harbor, the hourly rate for union divers is $8.40, and for tenders, $6.15. There is no depth pay to 60 feet, and from that depth to 74 feet it is only $ .22 a foot. From 75–125 feet, a bonus of $ .70 a foot is paid. Beyond 125 feet, the depth bonus is negotiated between the contractor and the diver. This is perhaps the lowest depth bonus in the country, but the majority of diving work in New York Harbor is in water less than 60 feet deep, so depth bonus is not an important factor.

In Detroit, a diver is paid $115.50 plus $34.50 for equipment rental, or a total of $150.00 per 8-hour day. Depth bonus is at a standard rate of $1.00 per foot for depths over 30 feet.

The diving unions on the West Coast, Los Angeles, San Francisco, Seattle, and in Alaska, are unquestionably the strongest in the country and exert a greater beneficial effect for their membership, in terms of pay, depth bonus and enforced safety regulations, than any other diving organizations anywhere.

In Los Angeles, a diver is paid $100.00 a day for an 8-hour day or fractional part. If he works over 8 hours, he is paid for another full day at double-time rates. If he supplies any of his own diving equipment, he receives an additional $30.00 a day. If a diver supplies his own gear, tender and insurance, the rate is $230.00 per 8-hour day or fractional part.

A depth bonus is paid for compressed-air dives beyond 50 feet, according to the following table:

Depth in Feet	Bonus ($ per ft.)	Depth in Feet	Bonus ($ per ft.)
50—100	$ 1.00	200—225	$ 3.00
100—150	1.50	225—250	4.00
150—200	2.00	250—275	8.00

If a diver happens to be one of those truly rare supermen, capable of making an effective dive beyond 275 feet on compressed air, his depth bonus is negotiated but according to union rules, it cannot be less than $10.00 per foot beyond 275 feet.

For mixed-gas diving, the compressed air rates apply down to 200 feet. From 200—225 feet, the premium is $4.00 per foot: from 225—250 feet, $5.00; from 250—275 feet, $6.00 per foot; beyond 275 feet, the depth bonus is negotiated, but cannot be less than $6.00 per foot.

Union rules specify that for all jobs 100 feet and over, there must be a recompression chamber at the site and a standby diver, dressed in and ready to go. The union also clearly specifies optimum bottom time limits for helium-oxygen diving: 50 minutes at 225 feet; 40 minutes at 250 feet; 30 minutes at 300 feet; 25 minutes at 350 feet, and 20 minutes at 400 feet. In water depths over 100 feet, only one optimum dive is allowed within any one 24-hour period. In water depths under 100 feet, only one optimum dive, as specified in the Navy diving tables, will be allowed within any one 12-hour period. This is very significant when compared with the common practice in the Gulf of making repetitive dives around the clock for days, if the exigencies of the job require it.

Most unions also specify a bonus payment per foot for lateral entrances into tunnels and wrecks, and require another diver, tending hose, at such entrances.

There is a much greater difference between union and non-union pay rates for tenders than there is for divers. A good rate for tenders in the Gulf is $3.00 an hour, where on a union job, a tender is always paid the equivalent of a foreman carpenter's wages. Nowhere in the country is this less than $6.00 an hour. Some locals even specify a footage bonus for tenders, based upon the length of hose he has to handle. A union tender is responsible for only one diver during an 8-hour shift. If he tends more than one man during an 8-hour shift, he receives an additional day's pay for each man tended.

If you wish to join a union in your area, seek out the local chapter of the United Brotherhood of Carpenters and Joiners of America, or the Pile Drivers, Carpenters, Bridge, Wharf and Dock Builders.

An association of professional divers has recently been chartered in New Orleans, with the declared purpose of establishing standard safety procedures for divers throughout the country. If you intend to become a professional diver, I recommend that you join this association. (See Appendix for address.)

SUMMARY OF SAFETY PRECAUTIONS

Diving is a hazardous profession. The secret of survival in any hazardous profession does not lie in the mystique of macho, or daredeviltry, but rather in a careful appraisal of all the risks involved, and their determined and systematic elimination. The fewer risks you take, the less you will have to worry about, with the consequent freeing of your mind for concentration upon the mechanical problems of your job. The prudent diver, the careful diver, the safe diver, is usually the most productive diver, because he has covered all bets and he is secure and confident in his medium.

I would like to end this book with what will be, to a large extent, a recapitulation of previously stated diving safety precautions, plus a few random hints that I may have missed in the text.

To begin with, take the time and spend the money to assemble a decent kit. Every truly professional tradesman is proud of the quality and condition of his tools.

Buy a custom-tailored wet suit of the thickness required to keep you warm in the water you intend to work in. Carry patching material, glue, silicone grease for zippers and talcum powder for suit storage in your kit. Wear coveralls to protect your suit. Carry boots or large-size tennis shoes for working on the bottom, and a pair of flippers for when swimming might be important. Make or buy yourself a good substantial weight belt and carry a variety of extra weights to cover all possible variations of wet-suit buoyancy and working conditions. Have at least two good-sized rings on your weight belt for securing tools. Have a stout safety belt or harness to fasten your hose assembly to. If your weight belt or harness is leather, carry neat's-foot oil in your kit. Carry your own safety quick-release hook. By splicing a 6-inch grommet made of ¼-inch polypropylene line into it, you can attach it to any hose assembly you might be required to use by taking several round turns and passing it through the end loop.

If you are more comfortable and work better with a specific type of mask or helmet, buy it and carry it with you. Carry spare phone speakers, a nonreturn valve, and any other obvious spares for this unit. If you provide your own telephone, carry a spare battery and fuses. It would be well to always carry your own hand light and extra bulbs and batteries. Always have two or three good diving knives and a file and sharpening stone. You should always carry a few rudimentary tools with you: adjustable wrenches, pliers, screwdrivers, a 4-pound hammer, and a 4-inch stiff-blade putty knife or scraper. Gloves are a constant necessity and six pairs are not too many. Always have with you legible tables, a watch or stopwatch, a notebook and several ball-point pens. A wrist depth gauge is also a wise investment. Round out your kit with a roll of tape, teflon tape, hose and pipe fittings. Buy or build a suitable box to conveniently carry this basic kit.

When you get a job call, try to find out as much about the job as possible—depth of water, type of equipment, breathing media, type of work to be performed, etc.—so that you can properly plan and anticipate your tool and equipment requirements.

When you arrive on the job, check out the equipment immediately, starting with the compressor. You may be tired of hearing this, but the air intake must be in a location free from all possible contamination from the drive engine or from any other engine or source. Other things to check on the compressor are: adequate volume and pressure for the intended dive; if diving over 100 feet, adequate pressure and volume to also supply a standby diver and to run the recompression chamber; full fuel; lube oil up to the marks on both the compressor and drive unit. Detergent lubrication oil should not be used on the compressor unit. The compressor sump should never be filled with oil over the dipstick full mark and oil of a viscosity less than SAE 30 should not be used because of the possibility of pumping oil into the air supply. Check all piping, hoses and valves for leaks. Check the pressure-release valve, the check valve, and the automatic unloader. Before diving, be sure the wing nut on the unloader is in a compressing position, the clutch is engaged and all proper valves are open. Drain all moisture from the volume tank and filters and be sure the petcocks are properly closed. Drive belts are a frequent cause of compressor problems. Be sure they are properly adjusted and not slipping when the compressor is under full load. There must be an accurate pressure gauge on the air receiver.

Check out the supply pipes and hoses from the compressor to the diving stations and to the chamber. Be sure the hoses are run in protected locations, free of the possibility of machinery being dropped on or run over them. Eliminate all sharp kinks and beware of hoses running across hot areas, such as exhaust manifolds and steam lines. Keep hoses out of oil spills. If hoses run close to any such areas, tie them off so that they cannot be accidentally moved.

Each diving station should have a volume tank and a gauge indicaing supply pressure. Barge air or some other emergency air supply must be piped into each diving station with valves and separate pressure gauge. There should be an accurate gauge with a large face for the pneumofathometer, and this should be hooked up to the emergency air supply. In this way the diver will carry two air systems with him on the bottom, and in case of a primary air-supply failure, he can insert the end of the pneumohose into his mask or helmet, and make a safe ascent. There must be adequate room at the diving station to coil the diver's hose and the standby diver's hose. Keep the diving station policed and free of welding leads, mooring lines and tool hoses.

The recompression chamber should be in a shaded place away from the vibrations of deck machinery. The chamber must be checked for easily operated hatches, proper hatch seals, valves not leaking, gauges working and calibrated, oxygen masks in good condition and a supply of oxygen on hand. Never rap on the chamber ports with tools, the telephone, or any other object with the purpose of attracting attention, either inside or outside. The chamber should be hooked up to an emergency air supply. Chambers must be kept clean, free of oil and grease and all other combustibles. Chamber lights should never be larger than 25 watts. The diver's clothes must be checked before they are put into the chamber to be sure there are no cigarette lighters or matches in them. The chamber occupant must keep himself and his clothes clear of the intake and exhaust openings. The chamber operator must keep the occupant under close visual observation at all times, especially during oxygen decompression. The chamber must be ventilated frequently.

Before you dive, check out all equipment. Don't make decompression dives without a recompression chamber at the site. Use a harness or a safety belt. If the dive is over 100 feet, carry a bailout bottle. Be sure your tender has tables and a watch on his person. Check the air to your helmet or mask before you put it on. Know all there is to know about your expected job before you hit the water. Have whatever tools or special materials you may need readily available. If there is to be a standby diver, be sure he and his gear are ready to go. Never dive without a good sharp knife.

Check your communications before you enter the water, and again immediately after entering the water. Avoid long, tiring swims on the surface before a dive. If necessary, have a tender swim a line over to whatever you are going to descend on; never swim if you can pull yourself along a line. Don't jump into the water if you can climb. Never enter the water by any means until you are sure that there is a ladder for you to get back out. Before you descend, thoroughly check out all your equipment in the water. Pay strict attention during your descent to avoid spiraling around the down line. On the bottom, take a minute or two to get oriented before you start working.

Don't overexert yourself. If you feel odd in any way, try to analyze why. If an unusual physical feeling persists or worsens, surface immediately. Stay out from under suspended loads and always cross over the top of pipelines, cables or other obstacles. Keep your hose free and clear at all times. If your hose becomes fouled, clear it immediately before continuing with your work. At all times, keep your hose slack to a minimum to avoid fouling. If communications fail at any time, terminate the dive.

Don't crowd the tables; leave the bottom in plenty of time. Let your tender pull you up. You can help him by climbing, of course, but be sure he is in control of your slack and that you are not ascending too fast. Try to have something to sit on during your water-decompression stops so that your muscles will be relaxed and your blood circulation unimpeded. Don't shorten or skip water-decompression stops. On deck, get into the chamber as fast as possible. If your suit has any oil or grease on it, remove your suit before you enter the chamber. Don't fall asleep while breathing oxygen. Avoid repetitive dives whenever possible, without a full 12-hour rest between optimum dives.

Don't live-boat at night, or when your air bubbles will be obscured by whitecaps or sea-foam. Don't live-boat without a bailout bottle and emergency flotation vest. Don't live-boat with a green tender or an inexperienced boat operator. Don't live-boat from the stern or from a large, difficult-to-maneuver boat. Don't ever handle or use electric blasting caps under water. Use Primacord for setting and hooking up all underwater charges.

Always wear rubber gloves when burning or welding under water. Don't change rods until the power is off. Use a welder's lens when burning in clear water. Always beware of stress and the possibility of sudden severe movement of the object you are severing. Beware of gas pockets collecting above or behind the steel you are burning.

Don't use scuba gear unless you have been trained with it. Every diver should know how to use scuba gear, and the easiest way to learn is by taking a YMCA, NAUI, or other accredited instruction course. Don't use scuba gear at night, in rough seas, or in strong currents. Don't make working dives or decompression dives with scuba gear.

The next statement might seem like a put-on, but it is not. If you don't know how, *learn to swim.* Take a Red Cross instruction course in first aid, especially for the treatment of drowning victims. Drowning is perhaps the greatest single cause of death in and around diving operations. Learn to protect yourself and others from this often complacently regarded hazard.

Finally, diving under the influence of alcohol or drugs is as patently suicidal as driving in the same condition. There are more than enough inherent narcosis and potential psychedelia in a compressed-air dive to 170 feet to cope with when you are perfectly straight.

As a beginning professional diver, it is incumbent upon you to read and learn everything possible about your trade. Every public library is a source of months of reading on diving and associated subjects.

Magazines and periodicals are an excellent way to keep abreast of new developments in your trade. A list of some of the most interesting and pertinent ones is included in the Bibliography.

MAKE A MILLION!

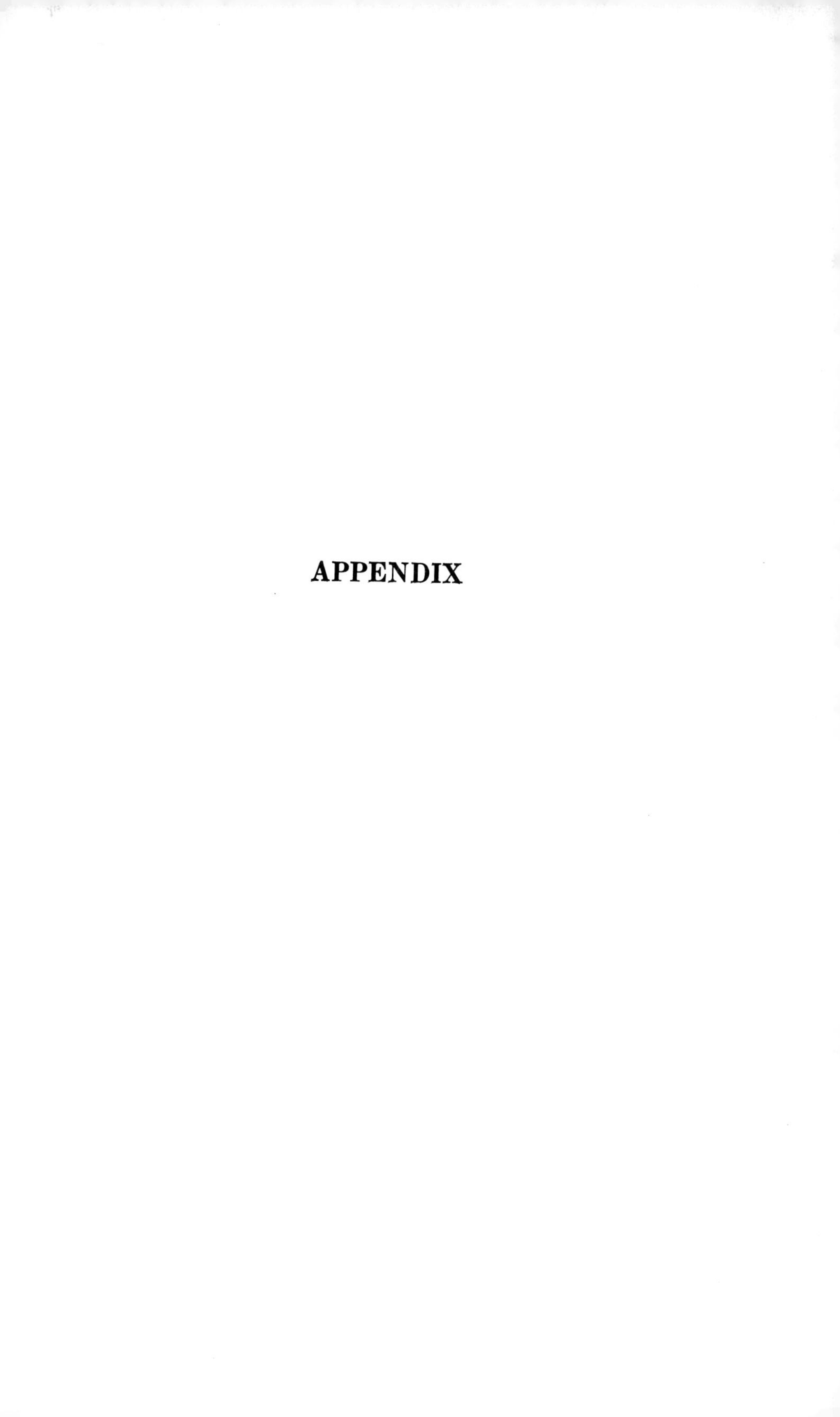

APPENDIX

BIBLIOGRAPHY

UNDERWATER WORK, SALVAGE, CUTTING and WELDING

Brady, E. M., Marine Salvage Operations. Cambridge, Md., Cornell Maritime Press, Inc., 1960. (Chapters on cutting and welding plus other fine material on underwater operations.)

Cayford, J. E., Underwater Work., Cambridge, Md., Cornell Maritime Press, Inc., Second Edition 1967.

Equipment for the Working Diver, Washington, D. C., Marine Technology Society.

Roberts, Fred, Basic Scuba, New York, N. Y., Van Nostrand Reinhold Co., 1966.

Thompson, F. E., Diving, Cutting and Welding in Underwater Salvage Operations, Cambridge, Md., Cornell Maritime Press, Inc., 1944.

Underwater Welding, Cutting and Hand Tools, Washington, D. C. Marine Technology Society.

U. S. Navy, Diving Manual, Washington, D. C., Superintendent of Documents, United States Government Printing Office. (A comprehensive coverage for the professional diver.)

U. S. Navy, Underwater Cutting and Welding Manual, Washington, D. C., Supt. of Documents, U. S. G. P. O. (Out of print and not available from publisher.)

EXPLOSIVES

E. I. DuPont de Nemours and Co., Inc., The Blaster's Handbook, Wilmington, Delaware, Fifteenth Edition 1969. (An excellent work on the general use of explosives.)

U. S. Navy, Use of Explosives in Underwater Salvage, Washington, D. C., Supt. of Documents, U. S. G. P. O. (Out of print and not available from publisher.)

ROPES, KNOTS and RIGGING

Cornell, F. M., and Hoffman, A., American Merchant Seaman's Manual, Cambridge, Md., Cornell Maritime Press, Inc., Fifth Edition 1967.

Graumont, R., and Hensel, J., Encyclopedia of Knots and Fancy Rope Work, Cambridge, Maryland, Cornell Maritime Press, Inc., Fourth Edition 1952.

Graumont, R., and Hensel, J., Splicing Wire and Fiber Rope, Cambridge, Md., Cornell Maritime Press, Inc., 1945.

Graumont, R., Handbook of Knots, Cambridge, Md., Cornell Maritime Press, Inc., 1945. (Knots for all purposes are included and are indexed by name and use.)

Leach, Robert, Riggers Bible, Robert Leach, Springfield, Mo.
Rossnagel, W. E., Handbook of Rigging: For Construction and Industrial Operations. New York, N. Y., McGraw-Hill Book Co., Inc., 1964.

PHYSICS, PHYSIOLOGY and MEDICINE

A Bibliographical Source Book of Compressed Air, Diving and Medicine, Washington, D. C., Supt. of Documents, U. S. G. P. O.
Bascom, W., A Hole in the Bottom of the Sea. New York, N. Y., Doubleday & Co., Inc. (An excellent work on the beginnings of deep-water drilling.)
Bennett, P. B., and Elliot, D. H., Editors, Physiology and Medicine of Diving and Compressed Air Work. Baltimore, Md., Williams and Wilkins Co., 1969.
Halstead, Bruce W., M. D., Dangerous Marine Animals: That Bite—Sting—Shock—are Non-Edible, Cambridge, Md., Cornell Maritime Press, Inc., 1959. (illustrates these organisms, their noxious effects, treatment and how to avoid them.)
Lambertson, C. J., Editor, Underwater Physiology, Proceedings of the Third Symposium (NAS - NRC, ONR). Baltimore, Md., Williams and Wilkins Co.
Livingston, R. B., and Brandner, H., Physiology and Physics of Diving. University of California, Institute of Marine Resources. (Pamphlet) (An excellent and simplified treatment.)

WATER SAFETY

Lanone, F., Drownproofing: A new technique for water safety. Englewood Cliffs, N. J., Prentice-Hall, Inc., 1963. (An outstanding work on survival.)

DIVING PERIODICALS

Sport and recreational aspects of diving. Invaluable as a marketplace for new equipment and for items of general diving interest.

Dive Magazine
P. O. Box 7765
Long Beach, Cal. 90807

East Coast Diver
P. O. Box 2109
Grand Central Station
New York, N. Y. 10017

Faceplate Magazine
528 East Dixie
West Carrolton, Ohio 45449

NAUI News (Members)
22809 Barton Road
Grand Terrace (Colton), Cal. 92324

Skin Diver Magazine
8490 Sunset Blvd.
Los Angeles, Cal. 90069

The Triton
40 Grays Inn Road
London WCI, England

The following publications covering the entire range of ocean activity, with frequent articles about the application of diving to such diverse endeavors as offshore mining, harvesting of shellfish, harvesting of seaweed, mariculture, hyperbaric medicine, oceanography, etc. Send for at least one copy of each of these to help you make a choice concerning a subscription.

Aerospace Medicine
Aerospace Medical Association
Washington National Airport
Washington, D. C. 20001

American Oceanography
854 Main Building
Houston, Tex. 77002

The Armed Forces Journal
1710 Connecticut Avenue NW
Washington, D. C. 20009

Commercial Fisheries Review
1801 N. Moore Street
Room 200
Arlington, Va. 22209

G. L. A. C. D. News
P. O. Box 2787
Hollywood, Cal. 90028

Gulf Review
Gulf University Research Corporation
1611 Tremont Street
Galveston, Tex. 77550

Innerspace
P. O. Box 700
Coronado, Cal. 92118

International Journal of Oceanology and Limnology
P. O. Box 395
Haddonfield, N. J. 08033

Journal of Hydronautics
1290 Avenue of the Americas
New York, N. Y. 10019

Journal of Physical Oceanography
45 Beacon Street
Boston, Mass. 02108

Limnos
Great Lakes Foundation
2200 North Campus Blvd.
Ann Arbor, Mich. 48105

Marine Biology Digest
P. O. Box 404
Mt. Arlington, N. J. 07856

Marine Technology Society Journal (members)
1730 M Street NW
Washington, D. C. 20036

MTS Memo
1730 M. Street NW
Washington, D. C. 20036

National Fisherman
Camden, Maine 04843

Navy
Navy League of the U. S.
818 18th St. NW
Washington, D. C. 20006

Ocean Oil Weekly Report
P. O. Box 1941
Houston, Tex. 77001

Ocean Science News
National Press Bldg.
Washington, D. C. 20004

Oceanic Citation Journal and Ocean Index
Box 2369
La Jolla, Cal. 92037

Oceanography News Letter
P. O. Box 191
La Jolla, Cal. 92037

Oceanology
1156 15th St., NW
Washington, D. C. 20005

Oceanology International
Industrial Research, Inc.
Beverly Shores, Ind. 46301

Pacific Search
200 2nd Avenue N.
Seattle, Wash. 98109

Science
1515 Massachusetts Ave., NW
Washington, D. C. 20005

Science and Technology
205 E. 42nd Street
New York, N. Y. 10017

Sea Frontiers
10 Rickenbacker Causeway
Miami, Fla. 33149

Sea Grant 70's
Sea Grant Program Office
Texas A & M University
College Station, Texas 77893

Seas
2351 Research Blvd.
Rockville, Md. 20850

PROFESSIONAL PERIODICALS AND JOURNALS

Devoted to commercial diving. Published in the center of Gulf Coast diving, by a professional oil-field diver, it is an essential guide to current diving activities:

Undercurrents
P.O. Box 2383
New Orleans, La. 70116

Focus on military diving. They cover government trends in new equipment and techniques.

Faceplate
Supervisor of Diving
Naval Ship Systems Command
Washington, D.C. 20360

R. N. Diving Magazine
H. M. S. Vernon
Portsmouth, England

Brass-tacks trade journals for the offshore oil and gas industries. They cover the hard facts of drilling and pipelining throughout the world:

Ocean Engineering
Pergamon Press
122 E. 55th Street
New York, N. Y. 10022

Ocean Industry
Box 2608
Allen Parkway
Houston, Tex. 77001

Offshore
211 South Cheyenne
Box 1260
Tulsa, Okla. 74101

The Oil and Gas Journal
P.O. Box 1260
Tulsa, Okla. 74101

Petroleum Today
433 W. 21st Street
New York, N.Y. 10011

Pipeline Industry
P.O. Box 2608
Houston, Tex. 77001

Undersea Technology
Compass Publications, Inc.
Suite 1000
1117 N. 19th Street
Arlington, Va. 22209

World Dredging and Marine Construction
P.O. Box 20810
840 Van Camp Street
Long Beach, Cal. 90802

World Oil
3301 Allen Parkway
Houston, Tex. 77001

DIVING EQUIPMENT MANUFACTURERS AND DISTRIBUTORS

Advanced Diving Equipment & Mfg., Inc.
1800 Newton Street
Gretna, La. 70053

Diving helmets, air and mixed gas, diving masks, telephones, bailout rigs, belts, burning torches, rods, compressors, recompression chambers, diving hoses, Bailey suits.

A&P Auto Supply
435 Sala Avenue
Westwego, La. 70094

Desco masks and lightweight helmets, diving hoses, small compressors, telephones, complete scuba equipment, wet suits.

General Aquadyne
333 E. Haley Street
Santa Barbara, Cal. 92626

Air and mixed-gas helmets, masks.

Bolstad Sales and Service Corp.
401 South Center Street
San Pedro, Cal. 90731

Diving compressors, all sizes; recompression chambers.

Craftsweld Equipment Corp.
2424 Jackson Avenue
Long Island City, N.Y. 11101

Complete diving equipment, including hard-hat gear and burning and welding equipment.

C-Vu
P.O. Box 125
Costa Mesa, Cal. 92626

Diving masks.

Diving Equipment and Supply Co. (DESCO)
240 North Milwaukee Street
Milwaukee, Wisc. 53202

Complete diving equipment and accessories, including hard-hat gear.

Divequip
P.O. Box 339
Melbourne, Fla. 32901

Complete diving equipment, specializing in Aquadyne products.

David Clarke Co., Inc.
360 Franklin Street
Worcester, Mass. 01604

Air and mixed-gas diving helmets.

Equitable Equipment Co.
410 Camp Street
Dept. R
New Orleans, La. 70130

Large air compressors and recompression chambers.

H & F Supply Co.
441 Peters Road
Harvey, La. 70058

Diving hoses and fittings; diving bell umbilicals.

J & J Marine Diving Co., Inc.
P.O. Box 4117
Pasadena, Tex. 77502

Japanese hard hats and dresses, hydraulic tools and power units.

O'Nil Landry
1031 Bonnabel Boulevard
Metairie, La. 70005

The Landry diver's phone.

Miller Diving Mfg. Co.
R.R. #3
Box 1005
Morgan City, La. 70380

Diving helmets; weight belts.

Morse Diving Equipment Co.
Sleeper Street
Boston, Mass. 02210

Complete hard-hat equipment.

Perry Submarine Builders & Undersea Eng., Inc.
P.O. Box 10448
Riviera Beach, Fla. 33404

Submarines, diving bells, and recompression chambers.

Savoie Research and Development Co.
P.O. Box 98
Boutte, La. 70039

Air and mixed-gas diving helmets.

Scott Aviation
225 Erie Street
Lancaster, N.Y. 14086

Diving masks, demand masks for chamber O_2 breathing, mixed-gas diving equipment, overboard-gas dump systems for O_2 in chambers and other breathing gases in underwater habitats.

Scubapro
3105 E. Harcourt
Compton, Cal. 90221

Small chambers, complete scuba equipment and accessories.

Skin Diving Unlimited
8716 La Mesa Boulevard
La Mesa, Cal. 92041

Wet suits, hot-water suits, complete hot-water units.

Underseas Industries, Inc.
17000 S. Broadway
Gardena, Cal. 90247

U.S. representative for Galeazzi chambers and other diving equipment.

U.S. Divers, Commercial Diving Division
3323 W. Warner Avenue
Santa Ana, Cal. 92702

Complete diving equipment, specializing in the Kirby-Morgan line.

Widolf
26096 Getty Drive
Laguna Niguel, Cal. 92677

Diving masks.

FOREIGN DIVING EQUIPMENT MANUFACTURERS

Drager
Dragerwerk-D-24 Lubeck 1
Moislinger Allee 53/55
West Germany

Complete diving equipment, including hard-hat gear, masks, etc., recompression chambers, saturation complexes and underwater habitats.

Siebe Gorman & Company, Ltd.
Neptune Works, Davis Road
Chessington Surrey, England

Complete diving equipment, including hard-hat gear, masks, etc., recompression chambers and saturation complexes; diving bells and submersible decompression chambers.

Roberto Galeazzi, Ltd.
Via Oldoini, 75
La Spezia, Italy

Complete diving equipment, including hard-hat gear, masks, etc., recompression chambers, and saturation complexes; diving bells and submersible decompression chambers.

The Yokohama Diving Apparatus Co., Ltd.
Yokohama, Japan

Hard-hat equipment and dresses.

DIVING SCHOOLS

Coastal School of Deep Sea Diving
320 29th Ave.
Oakland, Cal. 94601

Commercial Diving Center
School of Deep Sea Diving
201 West Water St.
Wilmington, Cal. 90744

Divers' Institute of Technology, Inc.
P.O. Box 5102
Pier 106 Ballard Marine Center
Seattle, Wash. 98107

Divers' Training Academy
1915 NE 15th Ave.
Ft. Lauderdale, Fla. 33305

Santa Barbara City College
Marine Diving Technician Program
721 Cliff Drive
Santa Barbara, Cal. 93105

Underseas Technician Program
Highline Community College
Midway, Wash. 98031

ASSOCIATION OF PROFESSIONAL DIVERS

International Association of Professional Divers, Inc.
P.O. Box 23641
New Orleans, La. 70123

MANUFACTURERS OF SUB SEA DRILLING AND COMPLETION EQUIPMENT

Cameron Iron Works
P.O. Box 1212
Houston, Tex. 77001

Vetco Offshore Industries, Inc.
250 W. Stanley Ave.
Ventura, Calif. 93001

Shaffer Tool Works
P.O. Box 4008
Beaumont, Tex. 77704

National Supply Division
Armco Steel Corp.
Chamber of Commerce Bldg.
Houston, Tex. 77002

UNDERWATER BURNING AND WELDING SUPPLIERS

Advanced Diving Equipment & Mfg. Co.
Newton Street
Gretna, La. 70053

Arcair
P.O. Box 406
Lancaster, Ohio 43130

Craftsweld Equipment
Jackson Avenue
Long Island City, N. Y. 11101

Desco
212 N. Broadway
Milwaukee, Wisc. 53202

PIPELINE REPAIR SUPPLIES (SLEEVES AND CLAMPS)

Pipeline Development Co. (Plidco)
1831 Columbus Road
Cleveland, Ohio 44113

INDEX

"Abalone" diving rig, 328-329
Acetylene, 151
Adjusting mask and hood, 20-21
Advanced Diving Equipment, 64, 67, 71, 74, 76, 303
Aeroquip, 74
A-frame derrick, 243
Air breathing, 51-53
Air-control valve, 20, 22, 24, 61
Air embolism, 115
Air pump, 57
AIRCO diving gas control panel, 318-319
Air silencer, 67
Air tuggers 219, 221, 224, 265, 294-295
Ammerman, Herb, 61
Anchor winches, 267
Anodes, installing, 262-264
Anoxia, 60
Aquadyne, 56, 303
Aqua-Lung, 65, 77-78
Aqua-Salvors, Inc., 263
Arcair torch, 171
Ascent, 33, 50, 84, 310-311, 314
  rate, 315-316
Ashley Book of Knots, 135
Associated Divers, 306

Bailout bottle, 17, 18, 32, 79, 216, 251, 257, 274, 354
Baralyme (National Cylinder Gas Co.), 313
Barton, Otis, 320
Bascom, Willard, 65
Beaumont clay, 245
Beckman Instruments, Inc., 64, 70, 319
Becksted, John, 66, 160
Beebe, William, 320
Bell guide, 267
Bells, 296, 320-348
  "Cachalot," 325, 330-332
  Cal-Dive, 335
  Dick Evans, 328
  Mark DCL, 327
  P.T.C. (personnel transfer chamber), 323-333
  "Purisima," 321-322, 325
  Seatask, 326
  S.D.C. (submersible decompression chamber), 321-323, 333
Bends, 11, 37, 113, 311
  symptoms, 39
Berg, Victor, 63
Bert, Dr. Paul, 82
Bethlehem Steel Corp., 202
Bight, 132
Bio-Marine, 319
Black Magic, 178, 181
Blaster's Handbook, The, 180
Blasting
  clay or cemented sand, 191-192
  machine, 177, 179
  preparations, 178-182
  rock, 189
Blowing hole at riser, 248-249
Blowout preventer stack, 289-292
Blowup, 60-61
Blowup (involuntary), 315
Blueprints, 6, 211, 219-220, 289
Bock, Carl, 68
Bolstad-Lister Co., 75, 77
B.O.P., 290-294, 297, 299
Bottom time, 25, 45, 49, 85, 321
Bouey-Fenzy, 17, 32, 216
Boyle's Law, 9-11, 31, 81
Breathing oxygen, 51-53
Brown and Root, Inc., 198, 201, 214, 243, 244, 267
Brown, Jack, 63
Buckled pipeline
  cutting, 342-343
  lining up, 343-344
  positioning weld hut, 344
  repairs, 334
Bungi cord, 192
Buoy, 265, 297
  placing, 278-279
Buoyancy, 60
Burning Bar, 166
Burning
  cable, 232
  electrodes, 155-156
  guide, 343
  pipeline, 234
  preparations, 153-155
    topside, 159
    underwater, 150-160

Caisson disease, 81, 113
Cal Divers, 70, 315
California Divers, Inc., 298, 335
Camera
  Calypso, 79
  Nikonos, 79
  television, 277
  underwater, 79
Cameron Iron Works, Inc., 288, 297
Carbon dioxide, 11-12, 37, 302, 333
Carbon dioxide poisoning, 314
Carbon monoxide, 11, 27, 28
Cathodic protection system. *See Anodes*
Chacon, Earl, 66
Changing rams, 292-293
Checking the ditch, 251
  pipe, 252
  precautions, 251
Checking sled and hoses, 252

"Christmas tree," 297, 298, 299
Church, Ron, 298
Clamps
  bolting, 223-224
  lowering into position, 221-223
Clark, David, 70
Clark helmet, 70-71
Claw, 241, 244, 253, 255
Closed-circuit diving rigs, 319
Clucas thermal-arc equipment, 166, 169
Coiling rope, 136-138
CO_2, 70, 207, 300, 303, 313, 314
CO_2 absorbent canisters, 313, 331, 338
CO_2 scrubber, 308, 309, 317
Collins, "Frenchy," 188
Come-alongs, 149-150, 220, 230, 250
Comex, 331
Commercial Diving Division, 64, 70
Commercial Diving Division, U.S. Divers, 307
Communications, 338
Compressed air, 77, 316
Compressed-air diving, 300, 305
Compressor, 17, 19, 75, 205, 215, 274, 295
Conductor, 267
Conductor housing, 290-291
Conductor pile, 265
Conger, 303
Con Shelf series, 330
Contaminants, 27
Control tower, 203, 242
Cousteau, Jacques, 330
Craftsweld Equipment Co., 63
Cranes, 147, 198, 213-214, 225-227, 231-233, 235
Crew boat, 276
Crew changeover, 339
Crossovers, 254-256
Cutting
  buckled pipeline, 164
  jacket-leg piles, 164-165
  nonferrous metals, 165
  pipelines, 160-163

David Clark Co., 309
Davis, Sir Robert H., 302-303, 320-322
Davis Submersible Decompression Chamber, 320
Davits, 204, 227, 231-232, 233, 236, 239, 247
Deck-decompression chamber, 321
Decompression, 13, 38-40, 83, 347-348, 354
  sickness, 38, 84, 113-114, 316, 317
  stops, 34, 50, 84
  treatment tables, 81-91
  using
    compressed-air, 311-312
    mixed gas, 309-311
Deep-diving systems, 296
Deep-sea gear, 55, 57, 300, 312, 314
  dressing out in, 57-60
Deep Submergence Systems, Inc., 321
Deep-water wellheads, 321
Denayrouse, 55
Derrick, 264, 285
Derrick barge, 265-267
Descending, 21-22
Desco mask, 303, 307, 309
Detasheet, 186, 192
D.D.C. (main deck chamber), 323
  one-atmosphere observation bell, 327
"Dick" Evans, Inc., 324
Divcon, Inc., 170, 239, 325, 326, 332
Diver's
  kit, 79-80, 352
  pay, 349-351
Diving
  career, 1-6
  bell, 316-317, 320-327
    Purisima, 321-322
  from drilling rig, 281-286
  gear
    "Abalone," 329
    Aquadyne, 329
    Clark, 329
    Kirby-Morgan Band, 329
    Swindell, 329
  knife, 17, 19, 137, 178, 261, 352
    case rigger's, 137
    jackknife, 14, 17
  panel control operator, 313
  school, 41, 366
    stage, 296, 298
Diver's tools, 4-5, 80, 158, 244, 263, 295, 352
Diurine suit, 33, 78. *See* also hot water suit
Dope coating, 258, 277
Dope station, 202-203
Draeger, 55, 67
Drager mixed-gas diving outfit, 332
Dressing the diver, 47
Drilling
  ocean-bottom well, 287, 289
  platform, 266-267
  precautions during, 294-299
  rigs, 296
  subsea system, 286-287
Drill ship, 285-286, 295
Dry atmosphere welding under water, 334
Dry welding habitat, 346
  working in the hut, 344-346
Dynamite, 173-174, 176, 178
  cutting steel with, 185-187
  firing, 184
  misfire, handling, 184-185
  placing, 182-183
  rules for handling, 194-197

Ear discomfort, 20, 33-34
Edel, Peter, 307
E.I. DuPont de Nemours, 179, 192
Electric blasting caps, 175-176
  testing of, 180
Embolism, 29-30
Emergency
  ascent, 20, 30-32
  devices, 17
End, Dr. Edgar, 304

Ensign-Bickford Co., 176
Epoxy coating. *See* Splash Zone
Equipping the chamber, 336-337
Equitable Equipment Co., 75
Exiting the bell, 341
Explosives
  handling rules, 194-197
  use, 173-197

Faceplate fogging, 63
Field joints, 203
Field-joint tins, 275
Fire danger, 334
Flanges, 234-236, 273
Flux-coated rods, 345
Free descent, 274
Free-flow control valve, 314

Galeazzi, 244
Galliazi, 55, 77
Galvanometer, 177, 179
Gardner-Denver Co., 285
Gardner-Denver swivel, 285
Gas cutting 151-153
Gates Rubber Co., 74
Gelatin dynamite, 175
General Aquadyne, Inc., 65, 67, 72, 157, 308
General Electric, 319
Grantz, Bob, 15
Guide base, 288-289
Guide cables, 289, 295
Guide (or down) line, 341
Guideposts, 288, 289

Haldane, J. S., 82
Hand-jetting, 60, 234, 254
Hand light, 352
  Birns, 336
  Dacor, 77, 260
  Ikelight, 77, 260
  Sawyer, 336
Handling nuts and bolts, 221
Hand signals, 45, 49, 148
Hang-off bar, 84, 313
  line, 274
"Hard Hat," 55, 62-63, 68
Hard hat gear, 63, 307
Harness, 296, 315, 336, 339, 352, 354
Heinke, 55
Heliox, 318. *See also* Helium-oxygen diving equipment
Helium, 67, 301-303
  atmosphere, 331, 337
  dives, 67
Helium-oxygen
  decompression tables, 312
  diving equipment, 318-319
    gear, 312-315
  mixtures, 303-305
  safe diving depth, 317
Helmets, 68-72
  Advanced, 70
  Aquadyne, 71, 72
  Clark, 70
  Desco, 71, 158
  Kirby-Morgan, 71
Helmets (*cont.*)
  Miller, 71
  Ski Hat, 72
  Swindell, 71
Henry's Law, 11-12, 37
Herman, Art, 133, 135, 138, 176
Hitch, 132-133, 191, 207-208
Hooks and chain slings, 143-145, 214
Hose, diving, 19, 32, 47, 52, 74-75, 208, 231, 261, 274, 340, 353
Hot-water suit, 78, 313, 336, 338, 341-342
Hughes, Mike, 306
Hydraulic clamps and rams, 344
Hydraulic control pods, 290
Hydrogen, 151, 300
Hydro-pak, Scott, 65

Impact wrench, 238-240, 246
Inserting bolts in flanges, 237-238
Inspecting
  crossovers, 279
  jetting operation, 279-280
  jobs, 216-218
  welds, 347
Inspection dives, 273, 276
Intrepid, 282

Jacket leg, 215, 226-227, 267-268, 276
  clamps, 219-220, 222
  template, 264
Jack-up rig, 282
Janssen, P.J.C., 301
"Jap" hat, 61, 63
Jet barge, 241, 243, 273, 279
Jet hose, 250
Jet nozzle, 241, 244, 249
Jet sleds, 242-243, 244, 246, 257
J. Houston-McRoot, 205
J. & J. Diving Equipment Co., 63
J. Ray McDermott Co., Inc., 199, 209-210, 225, 228, 266, 324, 328, 346

Kell, Larry, 56
Keller, Hannes, 331
"Kelly," 285-287
Ketchman, Norman, 306, 307
Kirby-Morgan Gas Hat, 306-307
Kirby, Richard, 306
Knife, 17, 19, 137, 178, 261, 336, 352
Knife-switch, 155
Knots, 133-136, 227
Knudsen, Norman, 306-307
Krasberg, Alan, 330-331

Landry phone, 73
Lattice type derrick, 286
L. B. Meaders, ii, 198
Lenz, 74
Lexan, 165
Lifting and pulling devices, 149, 213
Lifting slings, 264
Lindbergh, Jon, 321
Linear shaped charge cutter, 186
Lining up flanges, 236-237
Link, Ed, 330
Live-boat, 212, 217, 354-355
Live-boating, 273-275

Locating leaks, 268-269
Lockyer, Sir Norman, 301
Long, Dick, 78
Louisiana, 284. *See* Semi-submersible rig

MacLeish, Kenneth, 331
Magnetometer, 254
Making mold impressions, 277
Marine Contractors, Inc., 330, 331
Marine riser, 293-294
Mark VI diving gear, 298, 327
Mask, 13, 63-67
  Aquadyne, 65, 67, 157
  Currin, 67
  C-Vu, 62, 67
  Desco, 13, 14, 20, 47, 65, 67
  Kirby-Morgan, 18, 64, 67
  Scott, 14, 67
  Swindell, 64, 67
  Widolf, 66, 67
McCann submarine rescue bell, 320
Miller, Ben, 68, 74
Miller-Dunn, 55
Miller helmet, 68
Mixed-gas diving, 300-319
Mixed-gas diving gear
  Advanced, 307
  Aquadyne, 308
  Swindell, 307
Mobile drilling rigs, 282
  bottom supported, 282, 283
  floaters, 282, 285
Molly Hogan (or Flemish eye), 141-142
Morse Diving Equipment Co., 55, 63
Mousing a hook, 138-139
Munroe effect, 186

Neck dam, 72
Neck dam seal, 314
Nemrod, 17, 32, 216
"Nico-Press," 262
Nitrogen, 8, 38, 82, 300, 302, 334
Nitrogen narcosis, 11, 26-27, 300, 302, 305, 311
Nohl, Max Gene, 303, 304
Normal-air mask, 67
Notes on recompression, 121-125
Nuts and bolts, 236

O_2 decompression, 334
Ocean Systems, 67, 334
O'Neill, Jerry, xi, 33, 329, 330
Open-circuit demand mask
  Advanced, 309
  Aquadyne, 309
  Kirby-Morgan, 309
Oral-nasal mask, 309
O-ring charge cutters, 188-189
O-ring, inserting the, 236, 237, 238
Oxy-arc cutting 152, 163
  torches
    Arcair, 156
    Craftsweld, 156-157
    Desco, 156
    Swindell, 156, 170

Oxygen, 8, 13, 15, 30, 35, 51, 53, 115, 205, 301, 303
Oxygen breathing, 310-311, 313
Oxygen decompression, 316
Oxygen partial pressure sensor, 331
Oxygen poisoning, 300

"Pack-off shoe," 287
Pad eye, 220, 222-223, 243, 246, 291
Partial pressure, 318, 337
Perry Submarine Builders recompression chamber, 76-77
Phips, Sir William, 320
Pigs, 344
Pingers, 234, 297
Pipe-lay barge, 198-240
Pipeline
  assembly, 200
  dredge barge, 241-243
  inspection of, 235
  measuring, 226-228
    accuracy needed, 228-230
  walking, 275
  welding, 200-201
Pipeline Development Co. (Plidco), 268-269, 270-271, 276
Placing buoys, 278-279
Plastic pipe tape
  Pycoflex, 259
  True-X, 259
Pneumatic impact wrench, 224
Pneumofathometer, 19, 43, 74, 315
Pneumogauge, 222, 340
Pneumohose, 50, 74-75, 206, 211, 222
Pneumoreadings, 256
Poisoning
  Carbon-dioxide, 11, 29, 60
  Carbon-monoxide, 11, 27-29
  Oxygen, 11, 35-37
Pontoon, 204, 207, 246
  getting rid of, 225-226
  flooding, 211-213
  pushing pipeline into, 213-215
  testing, 211
  types, 208-211
Pontoon pneumogauge, 217
Primacord, 176-179, 181-185, 192-197, 335
P.T.C., 334, 338, 339, 344, 347
Pup joint, 345
"Purisima," 321-322

Quick eye. *See* Molly Hogan
Quincy compressor, 75

Ratcliffe, Bob, 69
Ratcliffe helmet, 66, 69, 298
Reading and Bates Seashell, 296, 334
Recirculating diving rig
  Clark, 316
  Swindell, 316
Recirculating systems, 302-303
Recompression chamber, 13, 35, 40, 43, 52-53, 75-77, 205, 276, 317, 353
Recompression helium-oxygen dive, 316-317
Repairing pipeline, 234-236

Repair sleeves, 268
Repetitive dive tables, 85, 92, 104
Reports, 218
Rigging, 132-150
Riser clamps, 218-219, 280
  bolting, 223-224
  installing, 221-223, 230-232
Riser clamp and cap setting, 226-228
Riser-clamp faceplate, 230
Rope
  Manila, 227, 261
  nylon, 261
  polypropylene, 227, 261
  strengths, 141
  use in rigging, 359, 360

Safety precautions, 351-355
Saturation diving, 321, 329-348
  commercial application, 330-331
  descent begins, 337-338
  equipping the chamber, 336-337
  open-sea test, 330
  physical plant, 331-333
Saturation Diving Complex, 324
Savoie helmet, 52, 68-69
Savoie, Joe, 69
Schrader, 55, 169
Scott Aviation Co., 77
Scott mask, 307
Scuba equipment, 77-78
"Seachore" (Divcon), 296
Sea Lab tests, 330
Searching, 261-262
Seibe, Augustus, 57
Seibe-Gorman, 55
Seizing, 132, 139-140
Semiclosed-circuit recirculating mixed gas diving systems
  Advanced Diving, 309
  Aquadyne, 309
  Clark, 309
  Mark VI, 309
Semi-submersible rig, 284-286, 295
Service boat, 282
Setting a structure, 264-268
Setting the sled, 245-246, 248, 254
  in heavy seas, 256-257
  on buried pipeline, 256
Setting up diving equipment, 205-207
Shackles, 144-146, 278
Shackling the sling, 215
Shaffer Tool Works, 292, 293
Shaped charges, 186-189, 190
Signals
  crane, 149
  hand, 45, 49, 148
Sizing pig, 234-235
Skin Diver Magazine, 15, 66, 73, 330
Skin Divers Unlimited, 78
Slings, 225-226, 231-232
Soda-Sorb (Ohio Chemical Co.), 313
Spanish windlass, 138-139
Splash Zone, 192, 258-259, 277, 347
Spelter socket, 288, 289
Splicing, 138, 140-141
  eye, 138, 140
  short, 138, 140
Split sleeves, 269
Snatch blocks, 220-221, 230
Squalus, 303, 320
Square knot (reef), 133
Squeeze, 313
  ear, 23
  face, 23-24
Stafford, Paul, 209-210, 225, 229
Stage (diving), 315
Standby diver, 315, 322, 327
Stationary drilling platform, 281, 284
Stenuit, Robert, 330
Stinger, 242. *See* Wishbone
Submarex, 281, 285
Submarine wellhead. *See* Christmas Tree
Subsea drilling system, 286-287
Sub-Sea International, 305
Suction strainers, 261
Surfacing, 34-35
Swindell, George, 70, 170

Taylor Diving & Salvage Co., Inc., 169, 188, 307, 309, 324, 327, 334
Teamwork, 240
Telephone, diver's, 18, 19, 43, 47, 65, 73-74, 147, 322
Teller, Albert, 192
Template, 160, 165
Tenders, 41-42, 52, 231, 320
  duties, 42-45, 49-54
  kit, 45
Tending, 274
Tensiometer, 242
Tensioning shoes, 204
Thermal-arc cutting equipment, 167-169
Thermal lance, 166
Thomas, Jim, 282
Thompson, Dr. Elihu, 301-302
Tie-in valves, 253
Todd, Gerry, 306
Tools, 5, 80, 339
Torch
  care of, 163-164
  Craftsweld, 80
  Swindell, 80
Treatment of unconscious diver, 116-127
Trimix, 312
Tugger cables, 295
Turk's head, 132
T.V., 293
Tzimoulis, Paul J., 15, 66, 330

Underseas Industries, Inc., 77
Underwater Cutting and Welding, 152
Underwater welding, 151-172
Unions, 350-351
U.S. Divers Co., 65
U.S. Navy Tables, 36-39
  Exceptional (Extreme) Exposures, 92, 100-104
  Repetitive Dive, 85, 92, 94-95

U.S. Navy Tables (*cont.*)
  Standard Air Decompression, 83-84, 86-91
  Surface Decompression Using Air, 93, 106-109
  Surface Decompression Using Oxygen, 93, 105, 110-112
  Surface Interval Credit Table for Air Decompression Dives, 96-97

Venturi educator principle, 302
Vertigo, 33-34
Vest, inflatable, 17, 32, 79, 216, 251, 257. *See also* Bouey-Fenzy and Nemrod
VETCO Offshore Industries, Inc., 288, 290
Violette, John, 15, 202

Walking pipeline, 275
Water blaster, 342, 343
Water-decompression stop, 311, 313, 315-316
Webb, Dennis, 188
Weight belt, 16-17, 19, 27, 61, 74, 137, 336, 340, 352
Weight coating, 200, 228, 258, 275, 277, 342
Weld+, 169
Weld+ ends, 217-271, 276
Weld hut, 169, 334
Welding lens, 158, 355
Welding, 169-172
  dip-transfer, 347
  MIG (metallic inert gas), 347
  short-arc, 347
  TIG (tungsten-inert gas), 347
  underwater, 151-172
  wire, 346
Wet suit, 13, 15-16, 78, 260
  Bailey, 78
  gear, 313
  Imperial, 78
Westinghouse, Inc., 174, 190, 298, 305, 325, 328, 330, 332, 334
Wheel jobs, 260-261
Whipping, lineman's, 132, 137-138
Widolf, Phil, 65, 309
Wilson, Dan, 66, 305-306, 321-322
Wishbone (stinger), 242-243
Wrist compass, 262
Wrist depth gauge, 352
World Wide Divers, 306

X-rayed, 345, 347
X-ray station, 201

Yokahama Diving Equipment Mfg., 55, 63
Young, George R., 335

Zapata Norness, Inc., 282, 284
Zetterstrom, Arne, 301
Zinkowski, Sue, 148, 187